Next Generation Compliance

Next Generation Compliance

Environmental Regulation for the Modern Era

CYNTHIA GILES

Oxford University Press is a department of the University of Oxford. It furthers the University's objective of excellence in research, scholarship, and education by publishing worldwide. Oxford is a registered trade mark of Oxford University Press in the UK and certain other countries.

Published in the United States of America by Oxford University Press
198 Madison Avenue, New York, NY 10016, United States of America.

Library of Congress Cataloging-in-Publication Data
Names: Giles, Cynthia J., author.
Title: Next generation compliance : environmental regulation for the modern era / Cynthia Giles.
Description: New York, NY : Oxford University Press, [2022] | Includes index.
Identifiers: LCCN 2022010818 (print) | LCCN 2022010819 (ebook) | ISBN 9780197656747 (hardback) | ISBN 9780197656761 (epub) | ISBN 9780197656754 (updf) | ISBN 9780197656778 (online)
Subjects: LCSH: Environmental law—United States.
Classification: LCC KF3817 .G548 2022 (print) | LCC KF3817 (ebook) | DDC 344.7304/6—dc23/eng/20220613
LC record available at https://lccn.loc.gov/2022010818
LC ebook record available at https://lccn.loc.gov/2022010819

DOI: 10.1093/oso/9780197656747.001.0001

1 3 5 7 9 8 6 4 2

Printed by Sheridan Books, Inc., United States of America

To Carl, for everything

Contents

Acknowledgments

Many people supported and challenged (just as important!) me on the road from the first ideas of Next Gen to the publication of this book.

David Hindin was by my side throughout the growing pains of Next Gen at EPA. He was a champion, a creative collaborator, and a practical implementer who put his shoulder to the wheel and helped turn Next Gen from idea to as-good-as-we-could-do-it implementation. Next Gen would not have made the progress it did, and I wouldn't know what more was required to move that mountain, without his tireless and imaginative effort. My hat is off to this devoted public servant.

It isn't possible to have a more supportive, thoughtful, and willing to work outside the box co-conspirator than Joe Goffman. He patiently tolerated my hectoring him during our years together at the Obama EPA, and then inspired me to translate Next Gen into a fully developed argument that resulted in this book. We also served together on the Biden-Harris EPA transition team, where I had the chance to experience 24/7 Joe's vision, excellent judgment, and good humor under pressure. I feel privileged to have him as a colleague and a friend.

Multiple extremely generous people read portions of the manuscript and offered valuable insights that improved the book and expanded my horizons, including that most rare of collegial virtues, telling me when I was wrong. I am greatly in their debt: Alex Barron, Rob Glicksman, Joe Goffman, David Hindin, Dan Ho, Mike Levin, Janet McCabe, Gina McCarthy, Dave Markell, Joel Mintz, Bill Reilly, Ricky Revesz, Lisa Robinson, and Mike Shapiro.

Numerous experts took the time to patiently explain things to me and answer my detailed questions. Some of these public-spirited citizens also let me twist their arm to review portions of the manuscript. I hasten to add that any errors of thought or fact (or math!) despite their educated input are mine alone. And that the positions I take in this book do not necessarily reflect the views or positions of EPA or any other organizations at which these capable people are employed. Many thanks to: Julius Banks, Mike Barrette, David Hawkins, Palmer Hough, Jim Jones, Seema Kakade, Jim Kenney, Heather Klemick, Tyler Lark, Mike Levin, Maria Lopez-Carbo, Al McGartland, Cindy Mack, Edward Moriarty, Matt Morrison, Elizabeth Ragnauth, Tim Searchinger, Courtney Tuxbury, and Will Wheeler.

The Harvard Environmental and Energy Law Program has given me a home and a platform for getting my ideas into the world, for which I am forever grateful.

Special thanks to Jody Freeman and Rich Lazarus for their inspiring example of engaged scholarship, and to Joe Goffman and Robin Just, who guided my first iteration of some chapters to publication. The initial version of the introduction, and what turned into chapters 1, 2, and 6 through 9 were originally posted on the website of Harvard Law School's Environmental and Energy Law Program.

The country benefits every day from the dedication, determination, and boundless expertise of the career staff of EPA. Every American is healthier as a result of the behind-the-scenes commitment that these folks consistently bring to their jobs, no matter how trying the circumstances. My heartfelt thanks to these steadfast experts, both for what they taught me, and for how reassuring it is to know that the country's environmental health is in their hands.

And finally, I thank my husband, Carl Bogus, who has supported my career in every possible way and beyond, and who helped me work out Next Gen ideas over the dinner table for untold hundreds of hours and thoughtfully edited many iterations of drafts. He is the best writer I know, and I aspire to someday approach his level of skill in turning complex ideas into clear and compelling communication. He is my biggest fan, and I his.

Introduction

What Is Next Gen and Why Does It Matter?

Senior environmental officials at both the state and federal level often give the public the same reassurance about environmental compliance. *Almost all companies comply*, they say. *The large companies comply; it is mainly the small ones that have compliance issues*. Does the evidence agree? In a word: no.

The data reveal that for most rules the rate of serious noncompliance—violations that pose the biggest risks to public health and the environment—is 25 percent or more. For many rules with big health consequences the serious noncompliance rates for large facilities are 50 percent to 70 percent or even higher. And those are just the ones we know about; for many rules, the US Environmental Protection Agency (EPA) has no idea what the rate of noncompliance is.

Rampant violations have real consequences: areas of the country that are not achieving air pollution standards, impaired water quality for half of the nation's rivers and streams, contaminated drinking water, public exposure to dangerous chemicals, and avoidable environmental catastrophes with health, ecological, and economic damages.

When the public expresses outrage about serious environmental violations, all eyes turn to enforcement. We need more enforcement. Smarter enforcement. Tougher enforcement. It is taken as given that compliance is the job of enforcement, so if there are violations, then enforcers need to up their game. This perspective—held by most environmental policy practitioners, including government regulators, regulated companies, legislators, academics, and advocates—assumes compliance is about what happens after environmental rules are written. If only we had the right combination of inspections, assistance, and enforcement, we would achieve the goal of widespread compliance.

This is the wrong way around. By far the most important driver of compliance results is the structure of the rule itself. A well-designed rule that makes the most of creative strategies to set compliance as the default can produce excellent compliance rates with very little enforcement involvement. Poorly designed rules that create opportunities to evade, obfuscate, or ignore will have dismal performance records that no amount of enforcement will ever fix. Compliance isn't

Next Generation Compliance. Cynthia Giles, Oxford University Press. © Cynthia Giles 2022.
DOI: 10.1093/oso/9780197656747.003.0001

consistently achieved by force-fitting it on the back end; it results from careful design up front.

This book proposes a solution for rampant environmental violations. *Next Generation Compliance* turns conventional wisdom on its head by insisting that we can achieve much better compliance only if we acknowledge that implementation outcomes are controlled by rule design. If we wait until after a rule is written to try to demand compliance, as the traditional paradigm has us do, we will continue to suffer from pervasive violations. Robust enforcement is an essential part of any compliance program, but it can't close the giant gap created by a poorly designed rule.

The ideas of Next Gen are based on my decades of experience in environmental protection, including eight years as the Senate-confirmed head of the Office of Enforcement and Compliance Assurance at EPA from 2009 to 2017 under President Barack Obama. In this role I was responsible for enforcement of all federal environmental laws. Earlier in my career I prosecuted civil violations of federal environmental laws as an Assistant United States Attorney and led the enforcement office for EPA Region 3, serving Delaware, the District of Columbia, Maryland, Pennsylvania, Virginia, and West Virginia. I also ran the water protection office for the Commonwealth of Massachusetts and just prior to my nomination by President Obama I was an environmental advocate seeking changes in state laws and regulations to address climate change. I have been in the trenches, and I have been at the highest levels of policy. I have served at the local, state, and federal levels of government and in the private sector. My experience is broad and deep.

In every one of these positions, I found a giant chasm between the commonly assumed levels of compliance and the extensive violations that we routinely found in the field. I was determined to understand the persistent mismatch between what most people believed and what I saw on the ground. I discovered that comprehensive evidence supports what I had anecdotally observed over decades of experience: significant violations are far more prevalent than everyone thinks. Strong national compliance records are the exception, not the rule. The violations are serious with real consequences for people's health. And that's just for the programs where EPA has data. For far too many programs, EPA doesn't know what the national compliance picture is, but there is ample reason to think it is probably bad.

I learned that many policy makers do not see the evidence in front of them—about the widespread violations or the reasons for them—because those facts don't align with ideas that have governed environmental strategy for decades. Long-held assumptions, some of which are enshrined in policy, are blinding us to the facts. The belief that most companies comply, and that noncompliance should be left to enforcers, has been stubbornly persistent, despite the reality that serious violations are common, and the design of the environmental rules themselves is the main driver of good, or bad, performance.

Next Gen isn't about making regulations enforceable. Of course they should be enforceable because otherwise why bother writing a rule. But Next Gen goes way beyond that. It's about designing a rule so that compliance is the default. Where implementation will be strong regardless of enforcement attention.

EPA under President Obama was starting the shift toward Next Gen. The Trump EPA's single-minded focus on deregulation pushed all that aside. Under President Joe Biden there are hopeful signs of a renewed commitment to Next Gen ideas. That's essential if we are going to reduce the wide gap between regulatory ambition and on-the-ground reality.

Why Compliance Matters

Although federal environmental laws differ in approach, they have something in common: all force change through compliance obligations. Congress sets broad goals, and EPA's regulations translate them into actions required by facilities. Through regulations the general becomes specific: How much pollution from the individual regulated facilities is allowed? What actions are the regulated facilities required to take or avoid taking?

Compliance is where the rubber meets the road. We only get public health benefits from our laws and regulations when the regulated companies do what the rules require. When they take steps to control pollution, or conduct the required monitoring, or implement process controls to reduce the risk of catastrophic releases, the standards in the rules translate to real protection on the ground. Rules are just words on a page unless they cause action. This is compliance: Do rules result in the necessary action in the real world?

Environmental practitioners know that compliance is not an all-or-nothing proposition. Some firms completely ignore or deliberately evade the rules, but it is also very common to find companies that partially comply or take steps to comply that don't work. They take unreasonable risks, or they install pollution controls but then don't operate them properly. These failures can lead to pollution that is many times the allowed levels. Neighboring communities, and sometimes people hundreds of miles downwind or downstream, bear the burden.

Some rules don't just limit how much pollution can be emitted; they aim to prevent releases altogether. The regulations concerning hazardous waste, for example, define how that waste is to be stored, handled, transported, and disposed, all with a goal of preventing it from ever being released into the environment. If a company violates one of those work practice standards—by storing the waste in unstable conditions or using a trucking company not licensed to haul that waste—it makes a future release more likely. Sometimes we get lucky, and even though the rules were violated, no one is harmed. Other times things can go disastrously awry. An unlicensed pesticide applicator uses a pesticide too close to a home and injures or kills people. Or a pipe explodes, and oil or chemicals contaminate the neighboring community. When these incidents occur, the investigation nearly always reveals that the company failed to take the preventive measures required by law, that is, there was a violation. No one can predict which violation will combine with bad circumstances to cause real harm, so it is important to protect people from that risk by insisting on compliance with prevention requirements.[1]

It's not just pollution violations that matter. The rules also impose obligations to monitor and to report. When companies don't check their pollution or their compliance, or don't report as required, regulators don't know about potentially serious problems. If a company seeking approval of a chemical doesn't reveal a negative health study, a dangerous chemical might be mistakenly approved. If your drinking water provider isn't checking to see if the water you drink is contaminated before sending it to your home, you probably would not dismiss that as a minor problem.

For all these reasons, the success of laws that protect the health of people across the country depends on companies doing what they are supposed to do. If every facility is meeting its obligations and following the rules, we have a good chance of achieving clean air and water and reducing our risk of exposure. If they aren't, we don't.

The impact of compliance failures isn't equal; it falls disproportionately on people already overburdened by pollution. Low-income and minority neighborhoods not only suffer from more contamination, they also face lower quality housing and a disparity in access to healthcare that makes protecting their families more difficult. Violations are troubling no matter where they happen, but they are an even bigger problem for people in communities with environmental justice concerns who already face greater health risks from pollution.

One of the reasons we get bogged down and people resist even starting this discussion is the morality frame that many adopt in discussing compliance. Yes, there are definitely bad guys who knowingly commit crimes and endanger people; our obligation in rules is to be open-eyed about that and not make criminal acts easy to do or get away with. But often there is no one actively deciding

[1] Here are a few examples of preventive obligation violations that led to disaster: a huge chemical spill into Charleston, West Virginia's primary source of drinking water, shutting it down for days, after Freedom Industries failed to inspect and repair corroding tanks and containment walls; a family of four poisoned after a chemical banned for indoor applications was used by Terminix to fumigate residential units; one person killed and multiple people injured after Tyson Foods failed to implement required risk-management procedures, resulting in releases of the dangerous chemical anhydrous ammonia into the air.

to violate, just a series of ill-advised choices, or failures to choose, that result in a serious violation. People who try to characterize this as either "companies want to do the right thing" or "companies put profits above people" create barriers to solutions by putting a moral frame around what is actually just a practical issue. When government decides to make things happen to protect the public, it is on government to design the rule to make it more likely we achieve the goal. That's it. Some rules do that. Most don't.

There is no question that the laws and regulations to date have made a huge difference. Our air and water are much cleaner, and our environment better protected than when EPA was created 50 years ago. Gone are the days when rivers caught on fire and air pollution was so thick you couldn't see to the end of the block. Federal laws, like the Clean Air and Clean Water Acts, and the regulations that followed, have dramatically reduced pollution.

Public outrage has propelled progress so far, and the public continues to insist on strong pollution controls. Polls consistently show that over 70 percent of Americans, including a majority of Republicans, favor tougher pollution standards for business and industry and more stringent enforcement of environmental laws.[2]

The drop in pollution and the strong public demand for environmental protection are impressive national achievements. Yet for all the improvement, progress on the ground is falling far short of where it needs to be. Widespread violations, including many with serious and proven impacts on health, occur across pollution-control programs. Some of the violations have proven tenaciously resistant to change and are interfering with our ability to achieve the health protections in our environmental laws. As the nation confronts the urgent problems before us right now—like a quickly changing climate and the necessity of addressing environmental injustice—we cannot afford to repeat past regulatory mistakes. Next Gen shows us what needs to change to get far more reliable results.

What's in This Book?

This book presents the evidence for the necessary paradigm shift, discusses some of the biggest barriers, and outlines strategies to get there from here.

[2] A 2018 Gallup poll found that 74% of Americans favor setting higher emissions and pollution standards for business and industry (61% Republicans, 84% Democrats) and 73% of Americans favor more strongly enforcing federal environmental regulations (51% Republicans, 89% Democrats). Frank Newport, "Americans Want Government to Do More on Environment," *Gallup*, March 29, 2018, https://news.gallup.com/poll/232007/americans-want-government-more-environment.aspx.

Chapter 1: Rules with Compliance Built In

The thesis of Next Gen is that there is a way to significantly improve compliance, but it requires us to jettison a central assumption that has persisted for decades: that regulations set standards and it is the job of enforcers to come in afterward and make it happen. The idea that compliance is the job of enforcers and not rule writers is so deeply ingrained that most people don't even recognize they are adopting that paradigm. They take it as a given; it's just the way things are.

In reality, the most important determinant of compliance is the structure of the regulation and the extent to which it adopts—or ignores—strategies to make compliance the default. The structure of the rule makes all the difference. A robust enforcement program will always be needed. Some problems can only be solved in that way. But enforcement alone will never get us there.

Because rules are the foundation of compliance outcomes, and case studies are the most powerful illustration of why that's true, the first chapter is an in-depth exploration of the compliance design of eight programs: four successful and four compliance disasters. The examples include regulations about air pollution, water pollution, drinking water, pesticides, and even one compliance catastrophe from South Korea. Two air examples concern the same sector—coal-fired power plants, one of the most polluting sources in America—but have dramatically different outcomes: the brilliantly designed Acid Rain Program (compliance over 99 percent) and the perfect storm of bad compliance design in New Source Review (over 70 percent of the largest coal-fired companies in serious and continuing violation). This stark contrast in results for the identical facilities vividly makes that point that success on the ground doesn't depend on the compliance culture of the regulated companies: it's the design of the rule that matters.

Chapter 2: Noncompliance with Environmental Rules Is Worse Than You Think

People may be persuaded by chapter 1 that rule design had a big impact for those case studies, but they may wonder if those examples are outliers, and if compliance overall is still pretty good. Chapter 2 provides the answer: serious violations are widespread, across all programs and industry types. It presents the most complete accounting anywhere of compliance with the nation's environmental laws. Here are just a few examples:

- The vast majority of US petroleum refining companies were sued by EPA for dangerous air pollution violations, resulting in 37 enforcement agreements covering 112 refineries, responsible for over 95 percent of the nation's petroleum refining capacity.
- All of the top five, and nine of the top ten US cement manufacturers—responsible for 82 percent of the total US cement production—were sued by EPA for significant air pollution violations.
- Nearly every large city consistently violated the Clean Water Act and was eventually sued by EPA to cut discharges of raw sewage to the nation's rivers and streams.

The dismal records cited here are for programs that have received persistent regulatory attention. For many programs, the data is far less complete, but the available evidence points to compliance problems that are just as bad. There are also some programs for which EPA purports to have national compliance data, but repeated audits prove that the actual rate of significant violations is far worse than claimed. For far too many regulations, EPA does not have enough information to make an educated guess about serious noncompliance.

I have asked many people—including some who have spent their entire careers working on environmental protection—to guess what the rate of serious noncompliance is with environmental rules. The most frequent response: 5 percent to 10 percent. You wish. It is nowhere near that low for most major environmental programs for which EPA has national data.

Chapter 3: Rules about Rules

Because this book is about changes that are needed in the design of environmental rules, it requires a very brief introduction to the laws and practice that govern rule-writing: the rules about rules. Nuance and complexity aren't included in this chapter. It is just a very basic outline of how it happens that agencies come to write rules, and what standards govern how they do it. Some of those are set by Congress in laws, some are directives from the president or from courts, and some are practices that the agencies themselves adopt. The purpose of this chapter's short overview is to provide context about the constraints agencies face in writing regulations. The good news is that nothing in the rules about rules blocks Next Gen: every rule can include Next Gen strategies, although some will have more latitude than others.

Chapter 4: Getting in Our Own Way: How EPA Guidance Reinforces Faulty Compliance Assumptions

The biggest hurdle to getting Next Gen ideas into regulations isn't Congress or the courts, as chapter 3 briefly explains. It's how regulators get in their own way by solidifying incorrect compliance assumptions into agency practice. The beliefs that compliance is good, and that it is up to enforcement to ensure it is, are part of the zeitgeist. No one has to tell rule writers that, it's something most of them already believe. But EPA makes the problem worse by baking those inaccurate compliance assumptions into written policy and agency practice. Those choices happen largely behind the scenes and are a formidable barrier to including better compliance design in rules.

It starts with the EPA guidance about rule-writing process. The way that process is implemented makes it easy for rule writers to ignore compliance pitfalls until it is far too late to address them, and to box out anyone from the compliance office that might try to insert a different perspective. The written process guidance doesn't mandate that result—in fact, as written, it even suggests things should be different—but in reality that's how it works. The other relevant EPA guidance governs how the benefits and costs of the rule should be tallied up. That all-important comparison—benefits have to be higher than costs for most rules to have a prayer—explicitly blesses a dismissive attitude about compliance, encouraging rule writers to assume they will achieve 100 percent compliance. Assuming the result you want won't make it happen in real life of course, as the mountain of evidence about widespread violations shows. But it is a heavy lift to persuade overloaded rule writers to embark on the hard work of careful compliance design when they are invited to bypass that effort by assuming compliance happens by magic.

These seemingly arcane agency policies and practices are almost never discussed in the fierce policy debates about EPA rules. The policies didn't create the erroneous compliance assumptions. But by accepting them as gospel, these policies have a powerful, and sometimes determinative, impact on the ultimate effectiveness of EPA rules. Chapter 4 explains how that happens, and what needs to change to turn it around.

Chapter 5: Next Gen Strategies: A Playbook

Once we understand that violations are serious and widespread, and that rule design is the key to better results on the ground, what should we do? Although the best regulatory designs carefully match solutions to their specific problems, there are some Next Gen ideas that perform well in many contexts. These are the workhorses of Next Gen, the go-to ideas that should always be considered as part of the suite of regulatory strategies to drive strong implementation.

New technologies allow real-time monitoring, electronic reporting, and sophisticated analytics that will put pressure on companies for better performance at the same time that they make it harder to hide. These solutions are likely to be a game changer for many intractable compliance challenges of the past.

All of the Next Gen basics emphasize simplicity, the underappreciated powerhouse of strong compliance. Elegant design—whether that's achieved by deployment of innovative high tech or the lowest-of-low-tech like posting signs or mandating an unfavorable assumption when reports are missed—can simplify even very complex issues, as we saw in the Acid Rain Program. Or a solution can just be simple, such as making the dangerous action impossible. Chapter 5 describes the fundamentals that should be part of the playbook for every rule.

Chapter 6: The Ideologues: Performance Standards and Market Strategies

Next Gen is practical. It asks what will work and eschews ideology, so in Chapter 6, I examine some of the current ideology-heavy approaches—such as performance standards and market methods—to see if they will produce better compliance. These ideas show promise in some situations, but they require even more mandatory safeguards than other approaches do. And all regulations, including market and performance rules, require use of what is often dismissively referred to as "command and control."

Chapter 7: Ensuring Zero-Carbon Electricity

This chapter applies the lessons of Next Gen to a top priority for cutting climate-forcing emissions: electric generation. The extremely good news is that we know how to achieve zero-carbon electricity with very high rates of compliance. The compliance challenge comes from proposals to connect clean electricity to investment in energy efficiency. Ramping up energy efficiency is indispensable for climate action, but for a host of reasons its results are far less reliable. That's why linking efficiency to clean power inserts high degrees of uncertainty into the must-have clean electricity goal. Robust investment in energy efficiency is essential, but not by putting the otherwise sure thing of zero-carbon electricity in jeopardy.

Chapter 8: Don't Double Down on Past Mistakes with Low-Carbon Fuels

The promise of renewable fuels is compelling: portable energy-dense fuels with no carbon impacts. The actuality hasn't been as good. This chapter explains the extensive carbon impacts that current strategies overlook, and why enforcement is useless as a solution. Instead, we should revise existing programs to ensure that they are closer to the mark and restrict use to the few instances where there is no other option. It closes with a reminder of the ever-present risk of fraud, which can pull the rug out from under carbon-reduction goals, and how robust rule design can make fraud much less likely.

Chapter 9: Innovative Strategies Are the Only Way to Cut Methane from Oil and Gas

Oil and gas companies know how to slash methane from oil and gas production. They just aren't doing it. Tough standards requiring use of currently available and affordable controls are the first step. Achieving widespread compliance with those standards at over a million diverse and far-flung sources is harder, especially because we currently lack reliable real-time monitoring. Methane control is a poster child for the obvious impossibility of enforcement as the solution. The only way to reliable implementation for this challenging sector is through creative use of unconventional regulatory approaches. And taking heed of the lesson learned from already abandoned wells: rules have to make it close to impossible for companies to dump their compliance responsibilities on the taxpayer.

Chapter 10: Updating Federalism

Another significant but often ignored challenge for compliance is the relationship between federal and state governments. Federal laws adopted by Congress protect people across the nation and assure a level playing field. States implement the federal laws in a way that fits local circumstances. There has always been a dynamic tension between the imperative to protect all Americans equally and the different circumstances of each state. This tension has fueled many environmental advances. But the dynamic has gotten out of balance, to the detriment of compliance.

States have a chokehold on environmental compliance information. States' failure to share that information with the feds—even when sharing is required by law—obscures violations, reduces protection, and prevents us from understanding how well our national laws are working. At the same time, the federal government resists the states' important role as innovators and laboratories for new approaches.

Next Gen provides a way out of this stalemate by recognizing that the model of federalism that made sense in the 1970s when most environmental laws were created is not working for us now. We need more reliable information about environmental compliance and increased autonomy for states to try new approaches. Both problems have a common solution: use today's monitoring and information technologies to build a federalism model for the modern era that strengthens protection and innovation at the same time. Chapter 10 outlines what that updated federalism could look like.

Chapter 11: Environmental Enforcement in the Next Gen Era

Tough enforcement—both civil and criminal—will always be essential to environmental protection. Bad guys are infinitely creative in trying to get around environmental rules, and regulators have to be just as insistent in ferreting out bad behavior and holding companies accountable. We know deterrence works, and it is an essential strategy in any robust compliance program.

Next Gen's focus on building compliance into rules doesn't replace enforcement. Next Gen will strengthen enforcement. Enforcers have long been given the impossible task of trying to get millions of companies to play by the rules. Once regulations force stronger compliance out of the gate, enforcers will be better able to do what they do best: focus on the worst violators. With robust Next Gen provisions in rules, regulators will know who those violators are; they will stand out because serious violators will be less common. Spending less time chasing after routine violations—because there will be far fewer of them—gives enforcers more time to tackle the most egregious problems.

The same paradigm shift needed for rules is also important for enforcement. Enforcement remains stuck on the tired and boring debate between tough enforcement and compliance assistance (or just ignoring violations, which is more an abdication of responsibility than a strategy). This has been the dynamic for decades. It is time to get over it. While the world has changed around us, the enforcement and compliance debates are mired in last-century thinking. Chapter 11 describes revitalized enforcement for the Next Gen era.

Conclusion: What's the Bottom Line?

When we admit that violations are common, and that rule design is the most important factor in achieving better compliance, we see the work of environmental and public health protection in a new light. That's the power of a paradigm shift; opportunities appear where previously we saw only problems. Next Gen is fundamentally an optimistic idea. Yes, it can feel discouraging to face up to the reality that left to their own devices many companies violate the rules. But that's not the end of the story, it's the beginning. We have the tools to change the outcome. We just have to decide to use them.

1

Rules with Compliance Built In

Some environmental regulations achieve widespread and consistent compliance. Most don't. As I explained in the introduction, the principal explanation for the difference isn't enforcement after the fact. Noncompliance can be common in industries that have received significant enforcement attention and rare even when there have been almost no cases brought against violators. It isn't because some companies have a stronger compliance culture; companies that comply with one rule sometimes ignore another.

The biggest reason for widespread compliance is the structure of the rule. Does the rule make compliance the path of least resistance? Is it designed to make compliance the default? If yes, compliance can be the norm, even with little enforcement effort. If no, violations will be rampant, no matter what enforcement may do.

Enforcement serves an essential role in holding violators accountable. A compliance program can't succeed without it. But a handful of enforcers will never be able to ensure general compliance at millions of facilities. We will only be able to protect the public from serious harms if we write environmental rules with compliance built in.

What rule structures succeed in making compliance the default? What design flaws lead to pervasive violations? The best way to understand what strategies succeed is to learn from existing rules. Some have worked remarkably well. Most have not. Serious violations are widespread and too often we have no idea what the compliance picture is. This chapter explores four programs that functioned well because they were built to work in the real world, with all its complexity and messiness. I also examine four programs that were compliance failures; they contain design mistakes that resulted in all-too-predictable serious violations.

Every rule is different. There is no one answer for every compliance design challenge. The rules that succeed build in strategies that take the world as it is. The rules that fail rely on hoping for the best. A close examination of these examples shows how rule design makes all the difference.

Next Generation Compliance. Cynthia Giles, Oxford University Press. © Cynthia Giles 2022.
DOI: 10.1093/oso/9780197656747.003.0002

Programs with Strong Compliance Outcomes: Four Examples

Some rules achieve amazingly good levels of compliance. Why? The regulations that work blocked the exits and smoothed the path toward compliance. Sometimes the solution is simple and elegant. But most successful rules build a unified whole using an array of structural provisions. This section describes four rules that achieved impressive compliance outcomes.

1. *Air Pollution: Acid Rain Program*

EPA set up the Acid Rain Program in 1995 in response to Congress's direction to do something about the acidic rain that was devastating forests and fresh water in many parts of the country. The principal culprit was sulfur dioxide (SO_2) emissions from coal-fired electric utilities. Those emissions traveled long distances through the air and caused serious damage when they ultimately landed on forests, rivers, and lakes. How best to cut those emissions?

As I described in the introduction, the Acid Rain Program was a textbook example of thoughtful and effective program design. The basic elements of that design were established in the legislation creating the program, showing that Congress knows how to build compliance into laws when it wants to. The Acid Rain Program set a cap on the amount of SO_2 that could be emitted by all of the coal-fired utilities collectively and issued allowances: one allowance per ton of allowable emissions. Utilities could buy and sell allowances. Hence the name: cap and trade. The compliance determination was straightforward: Does the utility have an allowance for every ton of emissions? The pollution-reduction goals were achieved on time, and at lower cost than many had predicted. And compliance was reported at over 99 percent with very little enforcement required. How did they do it?

The beauty of the Acid Rain Program design wasn't any one thing by itself, it was how the pieces worked together. Omit any one of these features, and EPA might have had a very different result. The excellent summary by John Schackenbach describes the elements of the program.[1]

[1] John Schackenbach et al., "Fundamentals of Successful Monitoring, Reporting, and Verification under a Cap-and-Trade Program," *Journal of the Air & Waste Management Association*, Vol. 56 (2006): 1576. See also Lesley McAllister, "Enforcing Cap-and-Trade: A Tale of Two Programs," *San Diego Journal of Climate & Energy Law* Vol. 2 (2010): 1 (providing an analysis of the Acid Rain Program and the Regional Clean Air Incentives Market (RECLAIM)). For an over-the-top testimony, see "The Invisible Green Hand," *The Economist*, July 4, 2002.

Continuous emissions monitoring systems (CEMS) for SO_2 were the central feature of the program. The CEMS continuously measured the amount of SO_2 being emitted. Companies were required to monitor, and report, SO_2 emissions using the CEMS data. With well-functioning CEMS the utilities and the government would know exactly how much SO_2 was emitted from each utility every quarter, allowing companies and government to track progress and plan ahead. Reliable and accurate data were also the foundation of the market for allowances: everyone knew that one allowance from any company actually equaled a ton of emissions, and everyone had the same data. Companies could trade with confidence and government could know that its pollution-reduction goals would be achieved.

How could EPA ensure that the CEMS were functioning well? Here's where an inspired but underappreciated detail was key: when the monitoring equipment was not working properly, the utility was required to report emissions using very conservative assumptions. If CEMS weren't operating reliably, the company had to assume emissions that were most likely much higher than its actual emissions.[2] Assumed higher emissions increased the cost of CEMS errors and downtime because they required more money to be spent buying allowances. These substitute data requirements provided a powerful incentive for utilities to assure that their CEMS were operating and operating properly.

EPA set up a centralized electronic reporting system to receive quarterly reports. Standardized electronic reporting streamlined recordkeeping and allowed EPA to track performance before the end-of-year reconciliation. The standardized e-reporting included a data-checking system that flagged inconsistencies and inadvertent omissions, spotting problems before they turned into violations. Much like you can't submit an online order if you leave out your address or credit card information, automated checking for obvious problems reduces errors and improves accuracy. And EPA performed its own electronic audits on the data, comparing companies to each other and checking company data against external information to verify accuracy. Having the data submitted electronically allowed EPA to conduct these audits efficiently from Washington without the necessity of field visits.

At the end of the year, utilities had to "true up," by proving they had purchased enough allowances to equal their emissions. How did EPA avoid the situation we see too often today: sources waiting to be caught and only then doing what's required? Two key provisions helped: (1) Simplicity: did the number in Column A (verified emissions) match the number in Column B (allowances)? Yes or

[2] The more data that was missing or that failed quality assurance tests, the more the substitute data provisions required overestimating actual emissions. McAllister, "Enforcing Cap-and-Trade," 6. See also Missing Data Substitution Procedures, 40 C.F.R. Part 75, Subpart D, §§ 75.30–75.37.

no? Violations were hard to miss. (2) Automatic penalties for companies that didn't have enough allowances to cover their emissions and a reduction in emissions cap the following year. These penalties were deliberately higher than the cost of buying an allowance. Why wait and pay more? It was cheaper to comply. Automatic penalties had the additional advantage of reducing the time to bring and resolve enforcement actions.

These features combined to create one of the most effective and efficient pollution-reduction programs in EPA's history. All the design elements were combined into one elegant program that achieved its goals early, at lower cost, and with compliance rates than most programs can only dream of. It is especially worth noting that while the overall compliance structure embraced simplicity (1 allowance = 1 ton), the underlying technology is complex, which is reflected in the hundreds of pages of EPA guidance on monitoring and reporting.[3] Simplicity of rule design can be entirely consistent with technical complexity.

There are a few caveats that should make us wary of using the Acid Rain Program as an all-purpose model and remind us that each program has to be designed to the circumstances of the problem it is addressing. Although the key success of the Acid Rain Program is that it achieved its pollution-reduction goals, it is often cited as evidence that markets can get reductions at a much lower cost. The first caveat is that while the market strategy for acid rain had lower costs than projected, much of the reduced cost was likely the result of reduced prices for low sulfur coal as a result of rail deregulation.[4] While markets in the right circumstances do hold promise for improved efficiency, that wasn't the only, or maybe even the main, reason for reduced costs in this case.

Second, the sources being regulated were sophisticated and homogenous. That made use of CEMS and standardized electronic reporting much easier. When oil- and gas-fired units were added to the program, alternative methods for calculating emissions were necessary.[5]

Third, California had a worse experience with a very similar cap-and-trade program, the Regional Clean Air Incentives Market (RECLAIM).[6] Like the

[3] McAllister, "Enforcing Cap-and-Trade," 5. See also EPA Clean Air Markets Division, "Clean Air Markets—ECMPS Reporting Instructions," https://www.epa.gov/airmarkets/clean-air-markets-ecmps-reporting-instructions.

[4] See Richard Schmalensee and Robert Stavins, "The SO_2 Allowance Trading System: The Ironic History of a Grand Policy Experiment," *Journal of Economic Perspectives*, Vol. 27, No.1 (2013): 103, 110–12 (noting that rail deregulation was a significant factor in lower costs). See also Bradley Karkkainen, "Information as Environmental Regulation: TRI and Performance Benchmarking, Precursor to a New Paradigm?," *Georgetown Law Journal*, Vol. 89 (2001): 257, 276 (noting that unanticipated advances in scrubber and fuel blending technologies also helped reduce costs).

[5] See McAllister, "Enforcing Cap-and-Trade," 5–6 (noting that over 95% of the emissions were at sources using CEMS).

[6] This discussion draws from McAllister, "Enforcing Cap-and-Trade."

Acid Rain Program, RECLAIM made use of CEMS and electronic reporting as well as tough substitute data provisions to inspire accurate reporting. However, RECLAIM experienced technical malfunctions with monitoring equipment and electronic reporting early on. In addition, RECLAIM struggled with the fact that its regulated sources were not homogenous, which prevented the state from establishing a uniform emissions calculation tool. These difficulties resulted in a state decision to verify emissions through a time-consuming and costly field inspection and audit of each facility each year. Adding to the administrative burden, the state did not have automatic penalties like those in the Acid Rain Program. Therefore, not only did the state do detailed facility specific audits, it had to pursue time-intensive enforcement actions to address the violations found.[7] These delays not only put a large administrative burden on government, they led to uncertainty in the market as audits and adjudications stretched well into the next compliance period. The California electricity crisis in 2000 and 2001 exacerbated these problems.

The experience of Congress and EPA with the Acid Rain Program shows that it is possible to hit a home run in rule design by thoughtfully combining elements tailored to the specific circumstances of the sector(s) being regulated. But the comparison to California RECLAIM confirms that solutions that work in one program are not necessarily completely transferable: variation in the types of sources being regulated can make the job much harder. The entire program has to be structured to be self-executing: leaving out just one element that results in significant need for continuous government intervention has the potential to derail program effectiveness. It also helps when unexpected events break in your favor, rather than the other way around.

It is worth noting that markets don't spring from the earth fully formed. They are created by tough, prescriptive regulations that dictate outcomes and direct how and when they must be achieved. Like the Acid Rain Program. The market for SO_2 allowances was intended to reduce the costs of the rule. That's a worthy goal, but it isn't what cut emissions. Credit for that remarkable achievement goes to the mandatory use of continuous emissions monitoring, tough substitute data requirements, obligatory electronic reporting, automatic penalties, and the other directives in the Acid Rain Program rules. These are all classic regulatory mandates. Markets are created by and built on the foundation of regulatory command and control, as the Acid Rain Program so powerfully demonstrates.[8]

[7] The most common violations were late or missing emissions reports, followed by emissions exceeding allowances. McAllister, "Enforcing Cap-and-Trade," 19–20.

[8] See the discussion of the Acid Rain Program's dependence on command and control in chapter 6.

Market strategies require the same hard work and careful Next Gen structure that all rules need.[9]

2. *Water Pollution: Secondary Treatment for Sewage Treatment Plants*

In the years before the Federal Water Pollution Control Act of 1972,[10] Congress had called upon the states to confront the problem of water pollution. Excessive loading of organic matter, nutrients, sediment, and pathogens into the nation's rivers and streams was leading to widespread low dissolved oxygen, fish kills, and bacterial contamination.[11] One of the chief culprits was the large and growing discharge of sewage from municipally owned sewage treatment plants, commonly called publicly owned treatment plants, or POTWs.[12] In 1968, many sewage treatment plants had only primary treatment, in which some of the solids were removed.[13] Pollutant loads were large and increasing. Congress expected states to set water quality standards and to go after the facilities impairing water quality.

It didn't work. States proved unable or unwilling to step up to the plate. Acknowledging that the states-first approach "has been inadequate in every vital aspect,"[14] in October 1972 Congress opted instead for something much more directive and centrally controlled: every sewage treatment plant would be required to install secondary treatment,[15] and more stringent limits would be imposed on sewage plants where necessary to achieve local water quality standards.

Every POTW was required to have a permit, and this mandate was underscored by making any discharge without a permit unlawful. Every permit would specify

[9] Getting markets right is why EPA has benefited from having a division that specializes in markets: the Clean Air Markets Division in the Office of Air and Radiation. They are responsible for the Acid Rain Program and the Cross State Air Pollution Rule (CSAPR) and two other prior air market rules: The NOx Budget Trading Program and the Clean Air Interstate Rule (CAIR), "Eight Things to Know: Program Highlights," EPA, https://www.epa.gov/airmarkets/eight-things-know-program-highlights.

[10] Amendments to the federal water law were later called the Clean Water Act, and that is the term used elsewhere in this book to reference the federal law governing water pollution.

[11] US EPA Office of Water, "Progress in Water Quality: An Evaluation of the Environmental and Economic Benefits of the 1972 Clean Water Act," Report No. EPA-832-R-00-008, June 2000, ES-1. See also William L. Andreen, "Success and Backlash: The Remarkable (Continuing) Story of the Clean Water Act," *George Washington Journal of Energy & Environmental Law*, Vol. 4 (Winter 2013): 25.

[12] Andreen, "Success and Backlash," 25.

[13] EPA, "Progress in Water Quality," ES-1.

[14] S. Rep. No. 92-414, at 7 (1972), reprinted in 1972 *United States Code Congressional and Administrative News*, 3668, 3674.

[15] Section 301(b)(1)(B) of the 1972 law, 33 U.S.C. §1311(b)(1)(B). The pollution limits for secondary treatment were to be defined in regulation by EPA. See Andreen, "Success and Backlash," 25–26.

the pollution limits applicable to that particular plant.[16] And every POTW had to regularly monitor its own wastewater and submit that data to the government. Congress also set aside funding to help POTWs achieve the standards.

So how did it go? With impressive alacrity, EPA finalized secondary treatment standards on August 17, 1973, a short 10 months after the law was enacted.[17] Who says agencies can't move fast? The nation now had a federally mandated and federally enforceable one-two punch: every sewage plant had to meet the minimum technology standards plus more stringent limits where necessary to address local water quality problems.

A little over 20 years later, 99 percent of the nation's 16,024 POTWs met the requirement for secondary treatment.[18] Discharges of organic pollution were cut by 45 percent, even though the volume of sewage treated increased by 35 percent.[19] And water quality improved as a result of this big reduction in loading: 69 percent of the river reaches saw improvements in dissolved oxygen.[20] What were the keys to success?

The first was clear, uniform, technology-based performance standards. The rule set unambiguous standards that applied to every plant, making it crystal clear what the rules of the road were. Years of attempts to start with water quality and work back to individual limits on dischargers were a failure.[21] Starting with ambient conditions seems logical, and economists praise its economic efficiency. But it didn't work. Why not? Determining the impact of individual sources on water quality is technically complex, subject to endless site-by-site debate and litigation. And it requires both professional expertise and political backbone at all levels of government. The reality is that pressure on publicly owned sewage plants to upgrade performance translates directly into increased rates for the local community. That puts huge political pressure on the state and local decision makers. Experience shows that many local and state governments can't find a way to get around politicians who strenuously object. Uniform national standards bypassed much of that debate.[22]

[16] The name given these permits reflected the ambitions of the Clean Water Act: they are called National Pollutant Discharge Elimination System or "NPDES" permits.

[17] *Federal Register*, Vol. 38 (August 17, 1973): 22298.

[18] EPA, "Progress in Water Quality," 2–24.

[19] EPA, "Progress in Water Quality," ES-5, 2–43 (regarding loading of carbonaceous biochemical oxygen demand (CBOD5)). During the same period industrial BOD pollution, also subject to federally enforceable technology-based standards, fell 40%. Andreen, "Success and Backlash," 28.

[20] EPA, "Progress in Water Quality," ES-10; Andreen, "Success and Backlash," 26–29.

[21] Andreen, "Success and Backlash," 25.

[22] Not surprisingly, there were amendments, exceptions and modifications eventually built into the statute and the rules, in response to the problems that emerged after the law was initially passed. See EPA, "Progress in Water Quality," 2-20 to 2-25. These changes did insert greater complexity into the rules, but the fundamental structure was not changed: POTWs had individual permits with definite limits and facilities had to monitor and report hard numbers that unambiguously showed whether they were in violation.

The marriage of uniform performance standards as the floor with more stringent standards when necessary to protect local water quality allowed the best of both worlds. Strong state programs could make a big difference in how well their waters were protected. But whether the state had a strong program or not, its waters and people would be defended by the minimum secondary treatment standards.

Federally enforceable limits gave the local and state governments a reason to insist on compliance. All the National Pollution Discharge Elimination System (NPDES) permit limits were federally enforceable, including any more stringent state water quality limits.[23] EPA is usually less subject to the "small p" political pressures that loom large at the local level, so was more likely to follow through with suits to require action. Federal enforceability is essential to the national effectiveness of water pollution controls. The value of federal enforceability isn't just the cases that it allows EPA to bring, it is the knowledge by states and POTWs that EPA could bring such cases. Defiance at the state or local level is therefore unlikely to succeed. Federal enforceability also strengthens the state's hand with the permitted facilities because there is no way around having to meet the standards. We are so used to federal enforceability as a fundamental component of federal environmental laws that we can overlook its structural value.

Compliance also benefited from clear permits, self-monitoring, and uniform reporting. The law said that any discharge without a permit was a violation. Therefore, every POTW had to apply for a permit. That permit set out in very clear terms what performance limits applied to that individual facility. Under EPA's regulations, every POTW had to sample its own discharge and report on that to government under penalty of perjury. Determining who was violating was a very simple matter of comparing the reported discharge to the permitted amount. If the facility was over the limit, it was in violation. This structure established under the Clean Water Act is often cited as a model of pollution regulation because it establishes both the limit and the monitoring sufficient to determine if the limit is exceeded and requires the facility to self-identify its discharges as in violation.[24] This clarity and definitiveness make violations much easier to identify and harder to evade, putting increased pressure on sources to comply.

[23] The permits are concurrently enforceable by EPA and by the state that issued the NPDES permit. State enforcement is always important, and even more so when the federal government's enforcement efforts falter. Andreen, "Success and Backlash," 26.

[24] Noting the success of the Clean Water Act's permitting, monitoring, and reporting structure, Congress attempted to establish a similar structure for the Clean Air Act in the Title V permits required by the CAA amendments of 1990. That effort has been hemmed in by the courts and by administrations that don't support tighter monitoring so has not achieved the goals that were originally envisioned. See Adam Babich, "The Unfulfilled Promise of Effective Air Quality and Emissions Monitoring," *Georgetown Environmental Law Review*, Vol. 30, No. 4 (Fall 2018): 569, 590–96.

In addition, it helped that Congress provided funding to support the upgrade of POTWs. Between 1970 and 1995, about $61 billion in federal funds were distributed to facilities or to state funding programs to help POTWs install the pollution controls necessary to meet the standards.[25] Achieving pollution reductions from publicly owned facilities can be challenging. Public entities are often boxed in by local laws and approving bodies that make it hard to obtain approval for necessary upgrades. They have less freedom to make investment decisions than do private firms, and they have fewer pathways for obtaining funding and recovering the costs of pollution control modernization. Making public money available therefore smoothed the way for sewage plant upgrades, although the time needed to set up programs and infrastructure for distributing funds meant that the expenditures didn't happen quickly.

Sewage treatment plants upgraded, pollution was significantly reduced, and water quality improved as a result of Congress's vision back in 1972 and the strong implementing regulations EPA adopted. The program was intended to cut pollution discharged directly from sewage treatment plants, and it did that very effectively.[26]

The experience with secondary treatment shows the value of directive one-size-fits-all approaches in some circumstances. The nuanced, flexible, local control strategy beloved by economists didn't work here. Congress recognized that a more forceful response was needed to address this urgent public health problem. So Congress said to sewage treatment plants: you will achieve at least this minimum level of control. Period, full stop. That unambiguous directive was what was needed to overcome the huge political and practical barriers to better water quality. Inefficient? Yes. Effective? Absolutely.

3. *Chemicals: Paraquat*

Sometimes the simple answer is the best. Paraquat dichloride—commonly called paraquat—is one of the most widely used herbicides in the United States for the

[25] Andreen, "Success and Backlash," 28. State and local governments invested in capital improvements of approximately the same magnitude. William Andreen, "Water Quality Today—Has the Clean Water Act Been a Success?," *Alabama Law Review*, Vol. 55, No. 3 (2004): 552.

[26] Andreen, "Success and Backlash," 29; Andreen, "Water Quality Today," 546. What it didn't do was address the pollution that never got to the treatment plant. Raw sewage and contaminated stormwater that are discharged before reaching the treatment plant continue to be a significant challenge. The regulatory structure that worked well for end-of-pipe treatment plant discharges wasn't designed for this more dispersed stormwater-related pollution. Andreen, "Success and Backlash," 30. That's why EPA was forced to sue nearly every major city to clean up these health-threatening discharges. See "National Compliance Initiative: Keeping Raw Sewage and Contaminated Stormwater Out of Our Nation's Waters," EPA, https://www.epa.gov/enforcement/national-compliance-initiative-keeping-raw-sewage-and-contaminated-stormwater-out-our.

control of weeds and as a defoliant in many agricultural settings.[27] It is also extremely dangerous to people. For this reason, all paraquat products registered for use in the United States are Restricted Use Pesticides (RUPs), which can only be used by certified applicators.[28]

Since 2000, there have been 17 deaths—three involving children—caused by accidental ingestion of paraquat. These deaths resulted from the pesticide being illegally transferred to beverage containers and later mistaken for a drink and consumed. A single sip can be fatal.[29]

In one tragic example, an eight-year-old boy drank paraquat that had been put in a Dr. Pepper bottle, which he found on a windowsill in the garage. He died in the hospital 16 days later. His older brother had used the product on weeds around the house and put it in the bottle. The older brother obtained the product from a family friend who was a certified RUP applicator.[30]

EPA's solution was quite simple: require a redesign of the packaging, so that these tragic mistakes could no longer happen. The 2016 EPA decision required new closed-system packaging that would prevent transfer or removal of the pesticide except directly into proper application equipment. No more pouring it into beverage containers because that literally would be impossible.

And the regulatory requirement isn't complicated or long. Here's the whole provision on packaging:

EPA is requiring that all paraquat non-bulk (less than 120 gallon) end use product containers comply with EPA-approved closed system standards. The closed system packaging for paraquat products must be engineered so that paraquat can only be removed from the container using closed system technology meeting the following EPA-approved standards:

- The closed system must connect to the container in a way that the closed system is the only feasible way to remove paraquat from the container without destroying the container; therefore, a screw cap for the pourable closure on a typical pesticide container is not sufficient; and
- The closed system must remove the paraquat from its original container and transfer the paraquat to the application equipment through connecting hoses, pipes and couplings that are sufficiently tight to prevent exposure of the mixer or loader to the paraquat (except for the negligible escape associated with normal operation of the system).

[27] "EPA Takes Action to Prevent Poisonings from Herbicide" EPA, https://www.epa.gov/pesticides/epa-takes-action-prevent-poisonings-herbicide.

[28] "Paraquat Dichloride," EPA, https://www.epa.gov/ingredients-used-pesticide-products/paraquat-dichloride.

[29] EPA, "EPA Takes Action."

[30] "Paraquat Dichloride: One Sip Can Kill," EPA, https://www.epa.gov/pesticide-worker-safety/paraquat-dichloride-one-sip-can-kill.

- All paraquat closed system packaging must be approved by EPA.[31]

That's it. Short. To the point. Don't rely on good judgment or attention to warnings to solve a persistent human health threat. EPA already knew that wasn't working. Just make it impossible.

4. *Reporting: Greenhouse Gas Reporting Program*

The Greenhouse Gas Reporting Program (GHGRP) collects annual greenhouse gas information from the top emitting sectors of the US economy, such as power plants, oil and gas facilities, refineries, chemical manufacturers, and others—about 8,000 facilities in total.[32] This is a reporting rule only; it requires companies to report their emissions but doesn't set any emission limits.[33]

To achieve its impressive 98 percent compliance rate, the GHGRP uses many Next Gen strategies. EPA provides a handy online tool for sources to figure out if they are required to report.[34] Reporting must be done electronically using a common template, making reporting fast and the information immediately available. EPA makes the most of e-reporting technology to screen electronic reports before they are submitted; reports that are incomplete or contain obvious errors are not accepted.[35] Companies must certify their submission as true, accurate, and complete when submitted. EPA then checks the filed report against other data and notes possible inaccuracies for discussion with the company.[36]

[31] EPA, "Paraquat Dichloride Human Health Mitigation Decision," December 14, 2016, at 8, available at regulations.gov, docket EPA-HQ-OPP-2011-0855. Another useful feature of the decision is a requirement that paraquat products intended for handheld and backpack equipment (which also have to meet the closed system packaging requirements) should contain an indicator dye to aid in early detection of paraquat leaks and spills.

[32] "GHGRP Reported Data," EPA, https://www.epa.gov/ghgreporting/ghgrp-reported-data (reporting year 2020).

[33] Although the GHGRP doesn't require emissions reductions, it appears to have spurred some companies to cut greenhouse gas (GHG) emissions, although that effect may be offset by increases in emissions by nonreporting facilities. Lavender Yang, Nicholas Z. Muller, and Pierre Jinghong Liang, "The Real Effects of Mandatory CSR Disclosure on Emissions: Evidence from the Greenhouse Gas Reporting Program," *National Bureau of Economic Research*, July 2021, https://doi.org/10.3386/w28984, discussed in chapter 5.

[34] See "Applicability Tool," EPA, https://www3.epa.gov/ghgreporting/help/tool2014/index.html.

[35] Note that mandatory e-reporting can spur development of private sector reporting tools, which are sometimes more nimble and responsive to customer needs than government developed reporting tools. See, e.g., HIS Markit, "IHS Introduces Software that Streamlines Compliance with EPA Mandatory Reporting Rule for GHG," Press Release, September 14, 2011, https://news.ihsmarkit.com/press-release/ehs-sustainability/ihs-introduces-software-streamlines-compliance-epa-mandatory-report.

[36] EPA, "Greenhouse Gas Reporting Program Report Verification," https://www.epa.gov/sites/production/files/2015-07/documents/ghgrp_verification_factsheet.pdf.

That's not all. EPA puts considerable effort into finding all of the facilities that are required to submit reports. It doesn't just wait for the facilities to self-identify. EPA contacts facilities that seem like they should be reporting but aren't. Just in case any facility thought it might avoid detection or let an error slide, EPA announces to the world which facilities—by name and address—aren't complying: either they didn't meet the verification requirements (orange flag) or stopped reporting without a valid reason (red flag). All of that information is public, very easily searchable, and available on EPA's website.[37] Anyone can check to find out if companies near them have orange or red flags. That's the beauty of transparency strategies; the pain of violations being listed for the world to see inspires many companies to conclude that it is less trouble just to comply.

This combination of Next Gen strategies sets the gold standard for reporting programs. Over the last eight years (the years for which violations data is available on the web), this program averaged a noncompliance rate of less than 2 percent.[38]

Programs with Pervasive Violations: Four Examples

Rules that have widespread violations provide an opportunity to learn what we did wrong. Seeing how the rules failed to achieve broad compliance can be illuminating, especially when it reveals the flaws of strategies that are still widely used.

This section presents an in-depth look at four rules that didn't get it right, leading to avoidable deaths, significant health issues, and a failure to know what the health impacts really are. The examples of clean air and drinking water rules are still in effect in the United States today. The first example is from an accident that happened in South Korea. While this doesn't involve an environmental issue and isn't in the United States, it helps to illustrate how rules can create the opportunity for criminal violations that cause serious harm. Analyzing a situation not in our backyard also makes it easier to see the regulatory flaws that our emotions might obscure in an example closer to home.

[37] EPA, "Facility Level Information on GreenHouse gases Tool (FLIGHT)," https://ghgdata.epa.gov/ghgp/main.do.

[38] See EPA, "Facility Level Information on GreenHouse Gases Tool (FLIGHT)." Percentage noncompliance is calculated by comparing the number of violators (red or orange flags) with the total number of reporting facilities for each year.

1. *Crimes: Sinking of the Sewol Ferry*

In 2014, the South Korean Sewol Ferry sank, and more than three hundred people died. The investigations that followed revealed that the ferry was carrying too much cargo on the day of its demise and that an overly sharp turn caused the top-heavy vessel to list. The overloaded cargo then shifted, sinking the ship. Investigators also discovered bribery, falsified documents, lax inspectors, and insufficient emergency response capability. The disaster led to a huge public outcry. There were criminal convictions, and ultimately, it contributed to the fall of the national government.

The Sewol Ferry routinely ignored the limits on the amount of cargo it was permitted to carry. On the day it sank, it was carrying over twice the allowed cargo weight. While the public uproar focused on demands for tougher enforcement, less attention was given to the regulatory structure that had failed to prevent this horrific accident.

A key fact, only mentioned in a small number of stories, was that the company running the ferry could not make money if it stayed within the cargo limit.[39] The load that was safe was economically impossible to sustain. News reports and official investigations highlighted the fact that the company made almost $3 million from illegal cargo, citing this as evidence of the company putting profits over safety. Far less noticed was that the Sewol Ferry was operating at a loss of about $750,000 in the year preceding the accident.[40] Coastal ferries in South Korea are small and the profit margins thin.[41] Statements from other ferry owners suggest that it was commonplace for ferries to exceed cargo weight limits.[42] In fact, two other overloaded ferries previously sank in South Korea, resulting in the deaths of over 600 people.[43]

Certainly, people and companies are responsible for their own criminal behavior, and it is appropriate to bring criminal charges when criminal conduct results in entirely preventable loss of life. Punishment is important, and it also deters others from violating. But government should design programs that make

[39] Jung-yoon Choi, "South Korea Ferry Was Routinely Overloaded," *USA Today*, May 4, 2014, https://www.usatoday.com/story/news/world/2014/05/04/south-korea-ferry-was-routinely-overloaded/8686733/.

[40] In-Soo Nam, "South Korea Ferry Probe: Cargo Was Three Times Recommended Maximum," *Wall Street Journal*, April 23, 2014, https://www.wsj.com/articles/south-korea-expands-probe-over-sunken-ferry-1398243668.

[41] "The Sewol Tragedy: Part II—Causes and Contributing Factors," *Ask a Korean!*, May 2, 2014, http://askakorean.blogspot.com/2014/05/the-sewol-tragedy-part-ii-causes-and.html.

[42] Jeyup S. Kwaak, "In South Korea, Lessons from Ferry Disaster Slow to Take Hold," *Wall Street Journal*, April 12, 2015, https://www.wsj.com/articles/in-south-korea-lessons-from-ferry-disaster-slow-to-take-hold-1428874202.

[43] David A. Tyler, "Sewol Disaster Demonstrates the Danger of Ignoring Cargo Load Limits," *Professional Mariner*, December 5, 2014, http://www.professionalmariner.com/December-January-2015/Sewol-disaster-demonstrates-the-danger-of-ignoring-cargo-load-limits/.

such criminal behavior far less likely and also easier to spot before it harms people. It is predictable, even certain, that some companies will violate safety standards if complying makes it impossible for them to break even. This is a no-brainer. When you position a government rule in opposition to a company's survival, you create the circumstances for the unscrupulous to violate. This is not to say that government shouldn't have rules to protect the public that interfere with profit-making. That's a central role of government: advancing the public interest over private gain. But when government decides to take that kind of action, it has to acknowledge that violations will be rampant, and so government must create regulatory structures to prevent violations and catch violators. On this score, the regulatory design for ferry safety in South Korea was a monumental failure. The entities policing compliance had built in conflicts of interest, the limits on cargo weight were known only by the standard setting arm of government and not the compliance arm, and the compliance system was both weak and laughably easy to evade.[44] Here are just a few of the key flaws in the regulatory structure:

- In an effort to allow more passengers, the vessel made modifications that increased the weight and raised the center of gravity of the ship. In response, the government entity responsible for the safety review cut the approved cargo by half and increased ballast requirements by the same amount.[45] Experts said later that the ship never should have been cleared to operate under these conditions because it could not make money with the drastically reduced cargo load limits.[46]
- Having approved the operation of the revamped Sewol, albeit with tighter operating restrictions, the regulatory structure then completely collapsed by not requiring the operating restrictions to be communicated to the entity responsible for policing the limits. It was therefore easy for the Sewol owners to lie with impunity about its cargo limits and impossible for inspectors to discover there was a violation.
- The organization responsible for assuring compliance was an industry group funded by the marine companies—a built-in conflict of interest.[47] Any structure that puts enforcement in the hands of the regulated should expect widespread violations.
- The incentives created by the method used to check compliance made the safety problem worse. The changes made to the Sewol vessel that resulted in

[44] See Chico Harlan, "Soul-searching in South Korea after a Disaster Waiting to Happen," *The Guardian*, May 2, 2014, https://www.theguardian.com/world/2014/may/02/korea-ferry-disaster-economic-growth-safety.

[45] "Sinking of MV Sewol," *Revolvy* [article not dated], https://www.revolvy.com/page/Sinking-of-MV-Sewol.

[46] Choi, "South Korea Ferry Was Routinely Overloaded."

[47] Choi, "South Korea Ferry Was Routinely Overloaded."

> a reduction in allowable cargo weight also required additional ballast to increase the stability of the ship. But the inspectors "checking" for overloading only looked to see if the vessel was riding low in the water. Cheaters who added illegal cargo could evade detection by decreasing ballast, an outcome that was entirely predictable and exactly what happened in the case of the Sewol.[48] In this way, feeble safety compliance checks made the safety problem worse.

When catastrophes like the Sewol Ferry sinking occur, and it turns out that multiple violations contributed to the disaster, it is common to hear calls for tougher enforcement. The same thing happens when environmental disasters occur in the United States. That's fine as far as it goes. But what we should learn from this and other calamities is that a failure in regulatory design not only created the opportunity for violations, but virtually ensured they would happen.

It is not enough to set a regulatory standard, expect compliance, and prosecute criminal violations. In high-stakes settings where lives are on the line, government cannot ignore the pressures that firms subject to regulations will face. When violations are likely, all it takes is a confluence of unfortunate circumstances to result in catastrophe. We see the same thing in criminal violations of US environmental laws. Government's obligation is to design stronger countervailing pressures so that the public interest will prevail. The more pressure on the regulated, the stronger and more robust the regulatory design has to be.

It isn't unusual to hear people say, after one of these disasters involving criminal conduct, well, they were bad guys, what can you do? No. We can do a lot. We know there are unscrupulous companies that will break the rules. It's on government to design rules that block the bad guys so that the worst outcomes don't happen.

2. *Drinking Water: Pathogens*

Americans care about access to clean drinking water.[49] Public opinion polling finds that over 87 percent of the public thinks clean drinking water is very important to their daily life, ranking even higher than clean air.[50] Pollution of drinking water is regularly at the top of peoples' concerns.[51]

[48] Choe Sang-Hun, "Legacy of a South Korean Ferry Sinking," *New York Times*, April 11, 2015, https://www.nytimes.com/2015/04/12/world/asia/legacy-of-south-korea-sewol-ferry-sinking.html.

[49] James Conca, "Super Majority of Americans Worry about Clean Drinking Water," *Forbes*, June 29, 2017, https://www.forbes.com/sites/jamesconca/2017/06/29/super-majority-of-americans-worry-about-clean-drinking-water/#685a5c5c41e8.

[50] See Nestle Waters, "Perspectives on American's Water," June, 2017, https://www.nestle-watersna.com/content/documents/pdfs/perspectives_on_americas_water-june2017.pdf.

[51] See Gallup polling, https://news.gallup.com/poll/1615/environment.aspx?version=print.

Although drinking water in the United States is among the safest in the world, it is not as clean as government pronouncements would have the public believe. The rules designed to keep our drinking water safe have serious compliance design flaws, which have resulted in many more violations than are officially claimed. EPA regularly asserts that fewer than 10 percent of public water suppliers violated one or more drinking water health-based standards each year.[52] Ten percent with such serious violations is too many. Unfortunately, it's also incorrect; this section explains that the actual number of health-based violators is substantially higher, although flaws in the regulations make it impossible to know the real number.

One vivid illustration of the problem is the rule to protect us from pathogens: the bacteria or viruses that can cause disease and illness. Pathogens can and do contaminate drinking water. Bacteria can be in the source water—the surface or ground water that the water system uses as its water supply—or can be introduced in the pipes that convey the drinking water to the consumer. Millions of people in the United States are sickened every year from pathogens in their drinking water.

A number of rules adopted under the authority of the Safe Drinking Water Act require drinking water systems to control pathogens. The rules require both treatment of the water before it leaves the drinking water facility and monitoring throughout the distribution system to ensure that the treatment is working to keep the water safe. While those rules have helped, pathogens in drinking water still contribute significantly to illness in the United States: a 2006 EPA report estimated that pathogens in drinking water from community water systems in the United States cause 16.4 million cases of acute gastrointestinal illness a year.[53]

The principal regulation controlling bacteria in drinking water is the Total Coliform Rule (TCR), finalized in 1989. That rule set standards for total coliform, an indicator that more dangerous kinds of bacteria might be present.[54] And it required sampling of water throughout the distribution system to make sure the water met the standard.

[52] See, e.g., EPA, "Providing Safe Drinking Water in America: National Public Water Systems Compliance Report" (2014–2016 national snapshots), https://www.epa.gov/compliance/providing-safe-drinking-water-america-national-public-water-systems-compliance-report. Buried in the fine print are caveats revealing that EPA does not stand behind the accuracy of these numbers. For good reason, as the discussion in this section explains. "Health based" describes violations that are directly about contamination of drinking water and does not include monitoring and reporting violations.

[53] Michael Messner et al., "An Approach for Developing a National Estimate of Waterborne Disease Due to Drinking Water and a National Estimate Model Application," *Journal of Water and Health*, Vol. 4, Suppl. 2 (2006): 201–40.

[54] Total coliform isn't itself proof that dangerous bacteria are in the water; it only suggests that a problem may exist. Under TCR, if a sample tested positive for total coliform, it had to be retested for evidence of fecal coliform or E. coli. If those more dangerous bacteria were found, that was an "acute" TCR violation. The TCR violation that is discussed in this section is the monthly average total coliform limit, the so-called "chronic" TCR violation.

Total coliform has been by far the single biggest cause of reported safe drinking water violations by community systems.[55] More than 300 million Americans—roughly 94 percent of the US population—got at least some of their drinking water from a community water system in 2017.[56] In 2007, the first year of EPA's Report on the Environment, EPA said that 10.6 million people were served by community systems that self-reported a violation of the TCR's health based standards.[57] Between 1993 and 2003, there was an annual average of almost 10,000 TCR self-reported health-based violations a year.[58] Those numbers are high, but the actual levels of noncompliance were far worse. The evidence described in this section shows that violations were significantly underreported. Why?

The first reason is that the structure of the monitoring requirements allowed drinking water systems to avoid violations. For larger systems, violations of TCR were based on the percentage of samples exceeding the threshold.[59] So if a system were in danger of exceeding the percentage threshold, one option was to take more than the required number of samples and thus bring the percentage exceeding the standard to below violation levels. This strategy is called "sampling out."[60] A study of drinking water systems in one state found very strong evidence that systems were sampling out to avoid triggering a TCR violation.[61]

[55] Maura Allaire et al., "National Trends in Drinking Water Quality Violations," *Proceedings of the National Academy of Sciences*, Vol. 115, No. 9 (2018): 2078–83. Of all the reported violations of health-based standards by community water systems for the period 1997 to 2003, 37% were violations of TCR. Allaire, 2079.

[56] EPA, "2018 Report on the Environment, Drinking Water," https://www.epa.gov/report-environment/drinking-water#roe-indicators. A word about nomenclature. There are over 150,000 regulated public water systems in the United States. These are systems that are required to follow the rules adopted by EPA for safe drinking water. Within that total, there are about 50,000 "community" public water systems, which are public water systems that supply drinking water to the same populations year around. The remainder of the public water systems supply water to facilities like schools or offices ("non-transient non-community" public systems) or to locations used infrequently, like gas stations or campgrounds ("transient non-community" public systems). "Information about Public Water Systems," EPA, https://www.epa.gov/dwreginfo/information-about-public-water-systems.

[57] EPA, "2008 Report on the Environment," 3–55, https://cfpub.epa.gov/roe/documents/EPAROE_FINAL_2008.PDF.

[58] EPA, "Analysis of Compliance and Characterization of Violations of the Total Coliform Rule, 2007," at 17, http://citeseerx.ist.psu.edu/viewdoc/download?doi=10.1.1.174.7626&rep=rep1&type=pdf.

[59] Systems that were required to take 40 or more samples a month would be in violation if more than 5% of those samples tested positive for total coliform. See Lori Bennear et al., "Sampling Out: Regulatory Avoidance and the Total Coliform Rule," *Environmental Science & Technology*, Vol. 43, No. 14 (2009): 5176, 5177. Note that states are permitted to have more stringent rules, and some do.

[60] Under the rules, additional samples are supposed to be approved by the state regulator and should be "representative" of the system, but it does not appear that these supposed constraints interfered with systems' ability to oversample.

[61] Bennear, "Sampling Out." This study was done in Massachusetts because of the relatively complete data it had on drinking water system compliance. Full disclosure: between 2001 and 2005, I was the Assistant Commissioner for the office in the Massachusetts Department of Environmental Protection that had responsibility for oversight of drinking water systems. My admittedly biased perspective is that Massachusetts had a very robust drinking water program; I think it is unlikely that Massachusetts had more sampling out than other states experienced.

The researcher estimated that, as a result, almost one-third of what otherwise would have been TCR violations in the state went undetected. Although she recommends caution in extrapolating these results to the national data, she estimates that sampling out may have masked an additional 3,000 to 4,000 TCR violations per year.[62]

The second source of underreporting of pathogen health-based violations is systems that didn't report at all. Self-reporting a TCR violation had significant consequences for drinking water systems: among other things they had to notify the public of the violation within 14 days. A system with a reporting violation doesn't have to disclose that until the summary end-of-year notice to consumers. Therefore, not reporting at all had fewer serious consequences for the water supplier than reporting a health-based violation. These disproportionate incentives caused some systems to take the path of least resistance by not monitoring or not reporting in some months, rather than disclosing a health-based violation.[63] According to data supplied to EPA by the states, for the period between 1997 and 2003 there were over 31,000 TCR monitoring or reporting violations a year.[64] It is unknown how many TCR health-based violations might have occurred in systems that didn't report, however, a Government Accountability Office (GAO) investigation found that a monitoring violation was a strong and statistically significant predictor of health-based violations.[65] If you are thinking that 31,000 TCR monitoring and reporting violations revealed to EPA by states each year puts a ceiling on the total possible actual TCR health-based violations, read on.

The third reason for underreporting of TCR violations is the state not telling EPA about them.[66] The regulations require states to put all violation information into the national data base EPA uses to assess and report on program performance.[67] Nevertheless, state reporting of violations to EPA is notoriously incomplete. One assessment found that about 17 percent of the TCR health-based violations that the states knew about were not reported to EPA.[68] And states failed to tell EPA about a stunning 71 percent of the monitoring and reporting

[62] Bennear, "Sampling Out," 5181.

[63] See EPA, "Economic Analysis for the Final Revised Total Coliform Rule," September 2012, at 4–5: "Low compliance with monitoring and reporting may occur if systems would rather incur a Monitoring/Reporting violation rather [*sic*] than risk an MCL violation by sampling." https://nepis.epa.gov/ (search in search bar for "Economic Analysis for the Final Revised Total Coliform Rules, 815R12004").

[64] EPA, "Analysis of Compliance," 17.

[65] GAO, "Drinking Water: Unreliable State Data Limit EPA's Ability to Target Enforcement Priorities and Communicate Water Systems' Performance," GAO 11-381 (2011), 16.

[66] This is distinguished from violations that the states themselves don't know about due to sampling out or systems not reporting violations to the state. The statistics in this paragraph are only about violations that appear in the states' files.

[67] 40 C.F.R. § 142.15(a)(1).

[68] EPA, "2006 Drinking Water Data Reliability Analysis and Action Plan," EPA 816-R-07-010, March 2008, at 19.

violations.[69] Here's the math: states told EPA about 31,000 TCR monitoring and reporting violations per year. If that's only 29 percent of the violations states knew about, then there may have been over 100,000 monitoring and reporting violations a year.

All of this evidence shows that the actual number of people consuming water from drinking water systems with violations of the TCR rule was many multiples of the 10 million EPA reported.

The TCR regulation made a number of structural choices that contributed to the obviously gross underestimate of TCR violations. (1) It defined a violation as a percentage threshold and allowed systems to include in their averaging all samples taken in a month.[70] That created an easy pathway for the strategic behavior of sampling out, and it appears that a substantial number of systems made use of that pathway. (2) It structured the consequences so that incurring a monitoring or reporting violation was comparatively better than conceding a health-based violation. Requiring any organization to self-disclose violations creates an uncomfortable dynamic. There must be a strong counterweight, or some organizations will take the easy way out by admitting a monitoring or reporting violation rather than confessing to a pollution standards violation. (3) The third structural choice applies to drinking water rules across the board, not only TCR. Revealing violations to EPA creates hassle and intrusion and aggravation for the state, so many states would rather not. Plus, it takes time and effort to put data into the national database. Beyond the legal requirement, what's the state's motivation to spend time doing that? Investigations have repeatedly shown that many states don't divulge information about violations to EPA. There are virtually no consequences to states for not reporting, whereas admitting to violations is likely to bring unwanted attention.

The net effect of these structural problems is that EPA really does not know how widespread TCR violations have been, except that they are many times higher than what EPA was saying in public reports. The drinking water program may be helped by the fact that many drinking water system operators know that they are engaged in a public trust and see a direct line between their choices and their own and their neighbors' health.[71] That's an advantage, but the evidence of

[69] EPA, "2006 Drinking Water Data," 18.

[70] As noted earlier, states are permitted to have more stringent drinking water rules, and some states prohibited using additional samples as part of the compliance determination. Bennear, "Sampling Out," 5181.

[71] However, that public spirited attitude is far from universal. For example, an employee of a drinking water system in North Carolina took samples from one location, falsely claiming that they were from multiple locations within the distribution system; he pled guilty to criminal charges in 2016. See DOJ, U.S. Attorney's Office Eastern District of North Carolina, "Former Town of Cary Employee Pleads Guilty to Falsifying Drinking Water Sampling Results," Press Release, September 26, 2016, https://www.justice.gov/usao-ednc/pr/former-town-cary-employeepleads-guilty-falsifying-drinking-water-sampling-results. Leaders in one Chicago suburb secretly used contaminated well water in the drinking water system and lied about it to authorities and the public for

the TCR rule shows that it is obviously not sufficient. Rules put pressure on the regulated to do things that take time and money. Failure to do them can have negative consequences on the world of course, but also on the regulated entity and the people who run it. If the rule structure gives them an out, many will take it.

EPA recently revised the TCR rule. The new rule makes it much harder to know if systems are experiencing pathogen contamination. Having 5 percent of samples exceed the total coliform threshold is no longer a violation. Instead, exceeding the 5 percent threshold triggers an obligation to conduct a self-assessment.[72] There is only a violation if a system fails to do a self-assessment or undertake the corrective measures it selected in its self-assessment. How does the state know if the system triggered the obligation to do a self-assessment and then did one? Systems are supposed to self-disclose violations, but if they don't, it is nearly impossible for the state to discover violations on its own. Piled on top of this already feeble compliance structure is an additional incentive not to report: systems with good compliance records can reduce the amount of sampling they must do. Any system that might be inclined to disclose a violation will think twice. Not surprisingly, almost no violations of total coliform requirements were reported by states in FY17.[73]

3. *Drinking Water: Lead*

As a result of the catastrophe in Flint, Michigan,[74] almost everyone is aware of the hazards of lead in drinking water. Ingesting high levels of lead can cause liver and kidney damage as well as brain dysfunction and behavioral disorders. Young children are particularly vulnerable.[75] Flint—and more recently, Newark, New Jersey—have refocused national attention on the important problem of

over 20 years. See Press Release, DOJ, U.S. Attorney's Office Northern District of Illinois, "Former Crestwood Water Officials Sentenced for Concealing Village's Use of Well in Drinking Water Supply," Press Release, November 21, 2013, https://www.justice.gov/usao-ndil/pr/former-crestwood-waterofficials-sentenced-concealing-village-s-use-well-drinking-water.

[72] EPA, "Revised Total Coliform Rule: A Quick Reference Guide," September 2013, https://nepis.epa.gov/Exe/ZyPDF.cgi?Dockey=P100K9MP.txt (rule promulgated in 2013, effective in 2016).

[73] EPA, "Report on the Environment, 2018," Exhibit 3: "U.S. Population Served by Community Water Systems with Reported Violations of EPA Health-based Standards, By Type of Violation, Fiscal Year 2017," https://cfpub.epa.gov/roe/indicator.cfm?i=45#3.

[74] Merrit Kennedy, "Lead-Laced Water in Flint: A Step-by-Step Look at the Makings of a Crisis," *NPR*, April 20, 2016, https://www.npr.org/sections/thetwo-way/2016/04/20/465545378/lead-laced-water-in-flint-a-step-by-step-look-at-the-makings-of-a-crisis.

[75] "Basic Information about Lead in Drinking Water," EPA, https://www.epa.gov/ground-water-and-drinking-water/basic-information-about-lead-drinking-water#health.

childhood lead exposure and the reality that communities with environmental justice concerns are the most affected.[76]

The Lead and Copper Rule (LCR) that governs federal standards for lead in drinking water includes multiple places where compliance can go off the rails. And it does. EPA's data say that about 10 percent of water systems were in violation of the LCR as of the end of 2016.[77] Is it that bad? As this section will show, it's actually much worse. The rule makes it easy for drinking water systems to miss elevated levels of lead. Most of the violations that systems do admit are never reported to EPA. The incentives set up by the rule encourage unreliable monitoring and failure to report. All these dropped balls result in significant undercounting of lead rule violations. Below I describe how the rule's design creates these problems.

Lead usually isn't in the source water; it leaches into drinking water from lead in underground pipes or fixtures in the home. The traditional way to prevent that is by treating the water so that it won't corrode the inside of the pipes. This treatment—referred to as corrosion control—is the main line of defense against lead contamination in drinking water.[78]

Systems are directed to find out if their corrosion control approach is working by checking for elevated lead in homes. They are supposed to check at locations of highest risk: in places with lead pipes.[79] If more than 10 percent of the samples exceed the action level, additional requirements to address lead contamination kick in.[80]

Rules that direct regulated entities to select sampling sites that are most likely to reveal a violation and give them considerable discretion in selecting sampling locations invite bad sampling practice. Many systems don't know where the lead pipes are, despite prior direction to identify them, so aren't able to follow this directive. Of course, knowing where the lead pipes are also allows the unscrupulous to avoid sampling in areas likely to produce a high reading.[81]

[76] EPA, "Lead and Copper Rule Revisions White Paper," October 2016, at 4, https://www.epa.gov/sites/production/files/2016-10/documents/508_lcr_revisions_white_paper_final_10.26.16.pdf.

[77] GAO, "Additional Data and Statistical Analysis May Enhance EPA's Oversight of the Lead and Copper Rule," September 2017, at 19. See also 40 C.F.R. §§ 141.80–141.91 (Lead and Copper Rule).

[78] EPA Office of Water, "Lead and Copper Rule Monitoring and Reporting Guidance for Public Water Systems," EPA 816-R-10-004, March 2010, at 5, https://nepis.epa.gov/Exe/ZyPDF.cgi?Dockey=P100DP2P.txt.

[79] EPA, "Lead and Copper Rule Monitoring," 15.

[80] The action level is more than 10% of samples exceeding 15 parts per billion (ppb) of lead. The 15-ppb number is not a safe level of exposure to lead: the only safe level of lead is zero. The action level is intended as a system wide assessment of the effectiveness of corrosion control and is not a measure of the safety of water in an individual residence. EPA, "Lead and Copper Rule Revisions White Paper," 11.

[81] See, e.g., Brenda Goodman, Andy Miller, Erica Hensley, and Elizabeth Fite, "Lax Oversight Weakens Lead Testing of Water," a joint investigation by WebMD and Georgia Health News. The study found that about half the drinking water systems in the Georgia study had falsely claimed to test at higher risk sites, noting that the state "has relied on an honor system, trusting utilities to test

These unreliable monitoring requirements are compounded by setting the action level as a percentage of samples over the threshold. A system can take additional cleaner samples, or the state can disqualify the highest readings, to bring the system below the 10 percent threshold.[82] Whether systems deliberately obfuscate, take advantage of ambiguity in the rules, or just make sampling mistakes, the flexibility in monitoring makes it hard to know if a system has a lead contamination problem.[83]

Having created pathways to avoid discovering high lead levels, the rule then adds powerful incentives to use them. Once a system exceeds the lead action level, the rule requires an escalating series of measures to fix the problem.[84] Failing to take those steps is a violation.[85] Each of the required actions costs money and ratchets up public scrutiny. If the problem continues, the drinking water provider must excavate and replace the lead pipes, which can be expensive. Once a system steps on the conveyor belt that starts with exceeding the action level, it might never get off. Fear of what might happen next creates significant pressure to avoid that position. And exceeding the action level doesn't just bring negative consequences, it also puts some benefits out of reach: systems that report being below the action level for a year can reduce monitoring and thereby save money.[86] Strong incentives to avoid finding a problem coupled with lots of ways to accomplish that are a dangerous combination.

homes that qualify under federal rules." https://www.georgiahealthnews.com/2017/06/lax-oversight-dilutes-impact-water-testing-lead/.

[82] The Association of State Drinking Water Administrators (ASDWA)—the body that represents state drinking water program managers—describes the provision that allows systems to take additional samples to get below the 10% threshold as a "loophole." ASDWA, "Comment Letter on Long-term Revisions to the Lead and Copper Rule," March 8, 2018, at 6, https://www.asdwa.org/2018/03/08/asdwa-submits-detailed-comments-on-lead-and-copper-rule/. Many of these strategies are alleged to have occurred in Flint. See, e.g., Ron Fonger, "Documents Show Flint Filed False Reports about Testing for Lead in Water," *MLive*, November 12, 2015 (Flint incorrectly claimed in reports to the state that it only tested tap water from homes with lead pipes); Mark Brush, "Expert Says Michigan Officials Changed a Flint Lead Report to Avoid Federal Action," *Michigan Radio NPR*, November 5, 2015 (the state disqualified two samples submitted by Flint, pulling the city below the action level for lead).

[83] GAO, "Drinking Water: Unreliable State Data," 12–13: "In addition, numerous stakeholders have criticized the current rule as providing too much discretion in sampling approaches and providing opportunities for systems to implement their sampling procedures to avoid exceeding the action level, even in circumstances where corrosion control has not been optimized."

[84] Systems exceeding the action level are required to take action: additional monitoring, corrosion control (start if not already doing it, otherwise optimize the existing system), source water treatment if needed, public notice, including a press release to television, print and radio, public education about reducing exposure to lead, and lead line replacement if treatment does not bring the lead to below action levels. EPA, "Lead and Copper Rule Monitoring and Reporting Guidance," 4, 11, 13.

[85] An action level exceedance is not a violation. However, a system is in violation if after exceeding the action level it fails to do the required follow-up steps, including public notification and commencing corrosion control. EPA, "Lead and Copper Rule Monitoring and Reporting Guidance," 4, 11, 13.

[86] GAO, "Additional Data," 11, n.79; EPA, "Lead and Copper Rule Monitoring and Reporting Guidance," 17–18.

Despite all these off-ramps, many systems report that they are in violation. There are "health-based" violations: a system discovers lead above the action level and fails to take the mandatory steps to fix the problem. There are monitoring and reporting violations: a system doesn't sample or samples incorrectly or fails to file the required reports. Health-based violations are obviously concerning but monitoring and reporting violations can be just as serious. If a system violates the law by not sampling, or not telling the state what's going on, serious health issues can be occurring that no one knows about. A GAO study found that monitoring violations were a strong and statistically significant predictor of health-based violations.[87]

What does the official record show regarding violations of the lead rule? EPA's data say that at least 6,567—about 10 percent—of public water systems had nearly 13,000 violations of the LCR as of December 2016.[88] A 2016 study by the Natural Resources Defense Council (NRDC) looking at EPA's data on community water systems found LCR violations at more than 5,300 community systems serving over 18 million people.[89] That's not good, but is it the outer boundary of the LCR compliance problem? Nowhere near. An EPA review discovered that states only told EPA about 8 percent of the LCR health-based violations.[90] *Eight percent.* EPA's investigation also found that states failed to tell EPA about 71 percent of the monitoring and reporting violations.[91] A GAO analysis uncovered even bigger problems, finding that states didn't disclose 84 percent of the LCR monitoring and reporting violations.[92] This is a reporting system in full failure mode.

States are required to tell EPA about violations.[93] But obviously they are seldom doing that. Whatever the reasons, the fact is that the national data about violations of lead standards are grossly understated. EPA continues to issue national reports relying on what it knows is deeply flawed information, because that is the only information it has.[94]

The actual number of people potentially affected by lead rule violations is unknown, but it is many times the 18 million people suggested in EPA's official

[87] GAO, "Drinking Water: Unreliable State Data," 16.

[88] GAO, "Additional Data," 19, 23.

[89] Erik D. Olson and Kristi Pullen Fedinick, "What's in Your Water? Flint and Beyond," Natural Resources Defense Council, June 2016, at 5, https://www.nrdc.org/sites/default/files/whats-in-your-water-flint-beyond-report.pdf.

[90] EPA, "2006 Drinking Water, Data Reliability Analysis and Action Plan for State Reported Public Water System Data in the EPA Safe Drinking Water Information System/Federal Version (SDWIS/FED)," 2008, at i, 19.

[91] EPA, "2006 Drinking Water Data Reliability Analysis," 18.

[92] GAO, "Drinking Water: Unreliable State Data," 16.

[93] GAO, "Additional Data," 15.

[94] EPA OIG, "EPA Claims to Meet Drinking Water Goals Despite Persistent Data Quality Shortcomings," Report No. 2004-P-0008 (2004) (by reporting performance using a database that omits a large number of violations, EPA portrayed an incorrect picture of the percentage of people drinking water that met all health-based standards).

database. Just as with TCR, rule design is the reason: many ways to prevent violations from being discovered, incentives that motivate systems to steer around rule requirements, state failure to report violations to EPA, and almost no way to discover what's really going on.[95]

4. *Air Pollution: New Source Review for Coal-fired Power Plants*

The 1970 Clean Air Act announced a new day for environmental protection. Congress stated firmly and clearly that it expected to cut the air pollution choking the nation. Congress envisioned a two-part strategy for major stationary sources: EPA would set technology-based standards for all *new* plants in listed categories (called "New Source Protection Standards," or NSPS), and states would impose controls on all *existing* plants within their borders as necessary to achieve ambient air quality standards established by EPA.[96]

Allowing existing plants to have less stringent standards than new plants is commonly referred to as "grandfathering."[97] But the Clean Air Act also set up a transition: as existing sources were replaced or modified, they too would be subject to the federal NSPS standard. In this way Congress envisioned that the existing stock of polluting sources would gradually be cleaned up as they were modernized or replaced.[98]

[95] EPA revised the LCR in January 2021. EPA, "National Primary Drinking Water Regulations: Lead and Copper Rule Revisions," *Federal Register*, Vol. 86 (January 15, 2021): 4198. This revision attempts to close some of the LCR loopholes and includes some interesting Next Gen strategies, for example, treating all service lines as containing lead unless shown otherwise. However, the revisions ratchet up the pressure on systems to avoid going over the action level (or the new "trigger level") while obscuring violations behind newly introduced complexity. Depending on states to report lead violations to EPA has not worked; this gigantic hole in the foundation will only get worse as additional burdens are piled on underfunded states and the incentives to avoid reporting increase. The Biden EPA has stated it intends to develop a new rule to strengthen key elements. EPA, "Stronger Protections from Lead in Drinking Water: Next Steps for the Lead and Copper Rule," December 2021, https://www.epa.gov/system/files/documents/2021-12/lcrr-review-fact-sheet_0.pdf.

[96] See Thomas O. McGarity, "When Strong Enforcement Works Better Than Weak Regulations: The EPA/DOJ New Source Review Enforcement Initiative," *Maryland Law Review*, Vol. 72 (2013): 1204, 1208; Jonathan Remy Nash and Richard L. Revesz, "Grandfathering and Environmental Regulation: The Law and Economics of New Source Review," *Northwestern University Law Review*, Vol. 101 (2007): 1677, 1681.

[97] McGarity, "When Strong Enforcement Works Better," 1209. This unfortunate term has its origin in voting rights laws after the Civil War: people whose ancestors were allowed to vote were exempt from the new literacy requirements. If your grandfather were allowed to vote, so could you. This term came to describe any case where new requirements didn't apply to existing situations. Alan Greenblatt, "The Racial History of the 'Grandfather Clause,'" *NPR*, October 22, 2013, https://www.npr.org/sections/codeswitch/2013/10/21/239081586/the-racial-history-of-the-grandfather-clause.

[98] See Nash and Revesz, "Grandfathering and Environmental Regulation," 1681–82, nn.18 and 19; McGarity, "When Strong Enforcement Works Better," 1209.

Harmful emissions from the nation's coal-fired power plants were very much on Congress's mind when legislating for clean air.[99] Coal-fired power plants were among the largest sources of SO_2, particulate matter (PM) and oxides of nitrogen (NO_x), major contributors to a wide variety of serious diseases.[100]

In 1977, Congress established permit requirements for major new and modified existing sources.[101] Before a major new source could be built, or an existing source modified, the source of pollution had to obtain a permit that would impose tough pollution limits. These permits were known as New Source Review, or NSR, permits.[102] New or modified coal-fired power plants were one of the categories of facilities that needed such preconstruction permits.

Whether a source was being "modified" and therefore had to go through NSR and install modern controls was a case-by-case determination that was fiercely contested. Sources claimed that their changes should be classified as "routine maintenance, repair and replacement" and thus exempt from NSR. Facilities argued that the emissions resulting from plant changes didn't trigger NSR and fought over the way to calculate emissions and whether government can challenge the accuracy of the calculations. Industry attacked the regulations in both the political arena and in the courts. Different EPA administrations changed the regulations, courts struck them down, and EPA changed them again.[103]

What didn't change was the case-by-case determination that was very complicated and deeply fact-intensive.[104] The rules also stipulated that, in the first instance, companies themselves decided whether they had "modified" their

[99] David Spence, "Regulation of Coal-Fired Electric Power under U.S. Law," *American Bar Association's Section of Environment, Energy, and Resources*, January 16, 2014, at 9, https://www.americanbar.org/content/dam/aba/administrative/environment_energy_resources/resources/spence_coal_electric.pdf. See also David Spence, "Coal-fired Power in a Restructured Electricity Market," *Duke Environmental Law and Policy Forum*, Vol. 15 (2005): 187, 189. The discussion of NSR in the introduction and in this chapter as an example of poor compliance design resulting in a compliance break down focuses on coal-fired power plants because they were such an important instance of NSR violators and clearly illustrate the point about rule design. NSR applied to many other sectors also, and the design challenges that contributed to violations by coal-fired power plants occurred in other sectors as well. See McGarity, "When Strong Enforcement Works Better," 1267–68.

[100] See McGarity, "When Strong Enforcement Works Better," 1209; Bruce Barcott, "Changing All the Rules," *New York Times*, April 4, 2004. See also GAO, "Wider Use of Advanced Technologies Can Improve Emissions Monitoring," June 2001, at 19 (stating that coal-fired utilities produced 52% of criteria pollutants emitted by large stationary sources in 1998).

[101] See McGarity, "When Strong Enforcement Works Better," 1213. "Major" was defined as sources that had the potential to emit over a defined threshold of pollution; the threshold varied by type of pollutant.

[102] NSR includes both Prevention of Significant Deterioration (PSD) for areas in attainment and NSR permits for non-attainment areas. See Nash and Revesz, "Grandfathering and Environmental Regulation," 1683.

[103] For a history of the long regulatory battle over NSR, see McGarity, "When Strong Enforcement Works Better"; Nash and Revesz, "Grandfathering and Environmental Regulation"; and Barcott, "Changing All the Rules."

[104] McGarity, "When Strong Enforcement Works Better," 1217–28; GAO, "EPA Needs Better Information on New Source Review Permits," June 2012, at 12–16.

facilities, and they didn't have to inform government of their decision.[105] If a company thought it shouldn't have to or didn't want to go through NSR, it just did the renovation project and didn't apply for NSR approval.

So far this might seem like normal regulatory tussles. Congress passes a law with a clear directive and leaves it to the agency to figure out the details. Regulated parties try to get the best definition of details that they can. And everyone wrestles over the rules as government and the regulated industries gain experience with applying the rules to specific instances.

Here's where coal-fired power was different: the costs of compliance. The technologies to control SO_2, PM, and NO_x were established and known. For example, scrubbers to remove SO_2 cut pollution by 95 percent.[106] The benefits of modern controls were huge. But the controls were also very expensive. Compliance costs of hundreds of millions to over a billion dollars were not unusual.[107]

It is easy to predict what happened next. Determining whether the rule applied was extremely complicated and subject to a highly technical debate. The complexity of the rules and the flexibility inherent in the case-by-case decision-making created an opening for utilities to argue—speciously in many cases—that they weren't sure whether the modifications they undertook were subject to NSR.[108] And company decisions about NSR were invisible to government; only an extensive investigation could reveal a violation.[109] EPA might never catch them, but if companies were caught, they could begin time-consuming litigation,[110] after which they probably would have to install the controls. But meanwhile they would save tens, if not hundreds, of millions of dollars by dragging out their compliance obligation. Penalties, which are intended to prevent exactly this kind of thinking by recovering the economic benefit of violating, weren't going to work this time. EPA had never imposed hundreds of millions of dollars

[105] See McGarity, "When Strong Enforcement Works Better," 1226. Not only did the rules allow the company to make the NSR applicability decision, they didn't even require the company to keep records of the changes to the plant or the resulting emissions. GAO, "EPA Needs Better Information on NSR," 12–16.

[106] See Barcott, "Changing All the Rules."

[107] To give an idea of the scale of costs we are talking about, here are the amounts that some coal-fired utilities spent to come into compliance with NSR after litigation with EPA, in non-inflation adjusted numbers: $1.2 billion (Virginia Electric and Power Company, 2003), $600 million (Wisconsin Electric Power Company, 2003), $400 million (South Carolina Public Service Authority, 2004), $500 million (Illinois Power Company and Dynegy Midwest Generation, 2006), $650 million (East Kentucky Power Cooperative, 2007), $4.6 billion (American Electric Power Service Corporation, 2007), $1.1 billion (Ohio Edison Company, 2009), $500 million (Weststar Energy, 2010), $3–5 billion (Tennessee Valley Authority, 2011), $1 billion (Wisconsin Power and Light, 2013), $1 billion (Consumer's Energy, 2014). For a partial listing of settlements with links to information about each case, see "Coal-fired Power Plant Enforcement," EPA, https://www.epa.gov/enforcement/coal-fired-power-plant-enforcement.

[108] See McGarity, "When Strong Enforcement Works Better," 1279 and 1286 (citing evidence that many plants knew their projects should have triggered NSR).

[109] See GAO, "EPA Needs Better Information on NSR," 16–17.

[110] See McGarity, "When Strong Enforcement Works Better," 1243.

in penalties against individual stationary source violators, and probably wasn't going to get a federal court to do that now.

The list of companies EPA eventually sued tells you how common NSR violations were for coal-fired power plants. It's the "who's who" of coal-fired electric utilities,[111] and includes over 70 percent of the top 25 coal-fired companies.[112] The evidence that emerged in investigations showed that what happened is exactly what should have been predicted: many of the nation's largest power companies had engaged in significant renovations without undergoing NSR.[113] Companies hid behind what they claimed were ambiguities in the regulations to avoid complying. They learned that if you don't think you will like the answer, just don't ask. They were advised to dress up big plant overhauls as routine maintenance.[114] They knew that it would take EPA years and reams of documents and many experts to catch the violators. And the list of violators was long, so that would also slow down EPA. Meanwhile, pollution controls for the nation's largest sources of air pollution would either be years delayed or avoided altogether.

Seeing that the violations were causing huge health impacts and that neither new rules nor general deterrence were going to ride to the rescue, EPA enforcers made a decision. They would go after the violators one at a time. They would do the investigations, and when they found violators, they would ask federal judges to order the plants to install the controls, as the rules required. The effort would be enormous, and EPA wouldn't win every case, but the health impacts were just too overwhelming to ignore. And thus did armies of lawyers end up fighting over modified: yes or no? The resulting enforcement dominated the docket at EPA and also the Department of Justice (DOJ) for the next two decades.[115] But it generated correspondingly huge benefits for public health.[116] The battle isn't

[111] See GAO, "EPA Needs Better Information on NSR," 20 (stating that EPA has alleged violations at over half of the coal-fired units EPA has investigated). One senior EPA official described the mound of evidence of wrongdoing uncovered in EPA NSR coal-fired power plant litigation as the environmental equivalent of the tobacco litigation. Barcott, "Changing All the Rules"; McGarity, "When Strong Enforcement Works Better," 1230–31.

[112] See "Ownership of Existing U.S. Coal-fired Generating Stations," Center for Media and Democracy, https://www.gem.wiki/Existing_U.S._Coal_Plants (the table titled "Ownership of Existing U. S. Coal-fired Generating Stations" lists the top 25 coal-fired utilities in 2005). Sixteen were sued by EPA for violating the Clean Air Act. Two others were sued by the Sierra Club for the same kind of violation. For a partial list of EPA cases, see "Coal-fired Power Plant Enforcement," EPA, https://www.epa.gov/enforcement/coal-fired-power-plant-enforcement. Sierra Club sued MidAmerican Energy and Entergy. There are many other coal-fired power settlements with companies not on the top 25 list.

[113] McGarity, "When Strong Enforcement Works Better," 1230–31.

[114] McGarity, "When Strong Enforcement Works Better," 1224, 1236, 1268, 1279.

[115] McGarity, "When Strong Enforcement Works Better," 1257–58 (describing the notable hiatus on new cases during the George W. Bush administration).

[116] "Coal-fired Power Plants," DOJ, last updated May 14, 2015, https://www.justice.gov/enrd/coal-fired-power-plants. See also McGarity, "When Strong Enforcement Works Better," 1290.

over, but many more coal-fired power plants have modern pollution controls today.[117]

There is an expression that the exception proves the rule. That's true here. It is exceedingly rare that states or EPA have the time to individually sue just about every source subject to a regulation. This brute force method of obtaining compliance doesn't make sense except in the very unusual circumstance where the benefits to be gained are enormous and the number of identified sources is small (enough). Ultimately, EPA got what Congress wanted. One at a time was the right decision. But it took two decades (and counting) and an incredible amount of government resources. That is not practical, or even possible, for the vast majority of environmental compliance problems. And while in limited instances EPA can get there eventually, it guarantees that the health benefits of the rules will be delayed as EPA slogs it out in court.

One other thing makes NSR an unusual case for thinking about regulatory structure. EPA was beset from within during much of the period after the Clean Air Act was passed. Changes in administrations brought in EPA leadership that was hostile to the idea of controls on coal-fired power. Again and again EPA political leadership attempted to change the rules to eviscerate NSR and give utilities a safe harbor (they actually called it that!) from Congress's directive to clean the air.[118] The same thing happened during the Trump administration.[119] Not only did EPA enforcers confront the problem of violators trying to get by them, they were being tackled from behind by their own team.

Most of the regulations discussed in this book are included to illustrate how we might learn to prevent widespread noncompliance. I recognize that regulations *intended* to give the regulated a way out are in a different category. For them, noncompliance is a feature, not a bug. But we can still understand from those examples what kinds of regulations make violations more likely.

[117] For a map of the pollution control status of the coal-fired power plants in the United States in 2020, see EPA, "2020 Coal Controls for SO_2 and NO_X, https://www.epa.gov/sites/default/files/2021-02/documents/coalcontrols_needsv6_feb2021.pdf, slide 5.

[118] See McGarity, "When Strong Enforcement Works Better," 1244–56; Barcott, "Changing All the Rules." The safe harbor regulation has been characterized as giving utilities "perpetual immunity" from NSR rules. See Nash and Revesz, "Grandfathering and Environmental Regulation," 1703. The courts eventually invalided the safe harbor rule as inconsistent with the Clean Air Act. Nash and Revesz, 1704–05. For a history of the twists and turns of the regulatory proposals, see McGarity, "When Strong Enforcement Works Better," and Nash and Revesz, "Grandfathering and Environmental Regulation."

[119] See Joseph Goffman, Janet McCabe, and William Niebling, "EPA's Attack on New Source Review and Other Air Quality Protection Tools," Harvard Law School Environmental & Energy Law Program, November 2019, http://eelp.law.harvard.edu/wp-content/uploads/NSR-paper-EELP.pdf; "New Source Review Preconstruction Permitting Requirements: Enforceability and Use of the Actual-to-Projected-Actual Applicability Test in Determining Major Modification Applicability," Memorandum from EPA Administrator Scott Pruitt, December 7, 2017; "EPA's National Compliance Initiatives—Say Goodbye to NSR Enforcement," *Foley Hoag, Law & the Environment Blog*, February 19, 2019, https://www.jdsupra.com/legalnews/epa-s-national-compliance-initiatives-31961/.

What can we learn from the coal-fired utility NSR experience?

Next Gen thinking needs to apply to grandfathering in legislation too. There are compelling arguments against protecting existing sources from pollution rules.[120] Congress didn't protect existing sources in 1972 when it passed the Clean Water Act, showing that it isn't always a political necessity.[121] If Congress decides to include grandfathering, it likely will be because regulated sources have pushed back hard. That means that existing sources have demonstrated their strong interest in looking for ways to avoid or delay meeting the new requirements. This is exactly when a strong countervailing pressure is needed, as countless examples in this book demonstrate. One option is for laws to include a specific expiration date for all grandfather provisions.[122] There may be other strategies too. But grandfathering when substantial amounts of money are at stake, then not adopting regulations that cut off firms' ability to resist pollution control upgrades, is likely to result in major violations and delay turnover from old polluting plants to modernized and cleaner plants.[123]

As described in the introduction, coal-fired NSR was the perfect storm of regulatory structure problems. Compliance was expensive so the pressure to evade was strong. And opportunities to evade were everywhere. The rules were extremely complicated, with many exceptions and exemptions and complex calculations and applications of judgment. All of this was applied case by case, so each decision was unique with its own factual complexity. In addition, companies made the choice to comply or not in near complete privacy. Violations would only be found by aggressive and lengthy investigations involving a wide variety of specialists. Everyone knew that enforcers would have a tough time finding violators, and even when caught most companies would pay less in fines than they saved by violating. Looked at in this light, the ensuing extensive violations seem not just likely but inevitable.

Once again states had the authority and the public health imperative to solve this problem, but they didn't. Some states took strong action—especially against plants located in other states—but overall state regulators didn't take

[120] See Nash and Revesz, "Grandfathering and Environmental Regulation."

[121] See Environmental Law Institute Dialogue, "Grandfathering Coal: Power Plant Regulation Under the Clean Air Act," *Environmental Law Reporter*, Vol. 46, No. 7 (July 2016): 10541, 10544.

[122] Richard Revesz recommends this approach. See ELI, "Grandfathering Coal," 10550.

[123] The experience of coal-fired power plants is a great illustration. Congress expected in 1970 that the life of a coal-fired power plant was 30 to 40 years. Nash and Revesz, "Grandfathering and Environmental Regulation," 1682, n.19. However, in 1985, the Congressional Research Service reported that the retirement age for power plants had increased from 30 years to as long as 60 years. McGarity, "When Strong Enforcement Works Better," 1220 (citing Larry B. Parker et al., Cong. Research Serv., 85-50 ENR, "The Clean Air Act and Proposed Acid Rain Legislation: Can We Get There from Here?" at 46 (1985)). Coal-fired power plants built in the 1940s and 1950s were still operating as of 2014. Steve Mufson, "Vintage U.S. Coal-fired Power Plants Now an 'Aging Fleet of Clunkers,'" *Washington Post*, June 13, 2014. See also GAO, "EPA Needs Better Information on NSR," 2–3.

on the biggest polluters.[124] In fact, they allowed sources causing local pollution problems to build high stacks and ship the pollution to downwind communities.[125] It is helpful to remember these examples when legislators or regulators get too romantic about the idea that states can be counted on to solve tough pollution problems on their own.[126]

Many companies violate. Wake up and smell the coffee. All things equal, would companies prefer to comply than to violate? Certainly. But all things are never equal. When government, for solid public policy reasons, requires companies to spend significant money to achieve a public health objective, some, if not most companies will look for a way out. Regulations that have little countervailing pressure and enough of a gray zone around compliance will result in a lot of violations. When responsible companies see that violations are rampant, they will understandably question why they should suffer competitive disadvantage from doing the right thing. Ultimately the public bears the brunt of the noncompliance impacts. Regulations shouldn't create the situation where a choice to violate seems like a viable idea. The Acid Rain Program—discussed at the beginning of this chapter—worked well not because it was a market program, but because it created a regulatory box so tight that compliance was the only way out. NSR is the opposite of a tight box.

Conclusion

Tolstoy famously said, "All happy families are alike; each unhappy family is unhappy in its own way."[127] So it is with environmental rules. If they work, it's because all the pieces fit together into a complete whole, resulting in good compliance that brings about the desired action in the real world. But if heavy force is applied to a rule that isn't resilient to that pressure, or a key piece is missing, collapse in compliance inevitably follows. Each poorly designed rule can collapse in a different way. But the lack of structural integrity eventually reveals itself in large numbers of violations.

The rules with excellent compliance records highlighted in this chapter were adopted under very different legal regimes and address very different types of compliance obligations. They use different strategies: performance standards,

[124] ELI, "Grandfathering Coal," 10544 (noting that states had the authority but lacked either the resources or the political will to control their existing sources. A number of states put nominally stringent limits on old power plants, but when the plants didn't comply, the states didn't enforce.) See also GAO, "EPA Needs Better Information on NSR," 17–18.

[125] ELI, "Grandfathering Coal," 10544.

[126] See Adam Babich, "Back to the Basics of Anti-Pollution Law," *Tulane Environmental Law Journal*, Vol. 32 (Winter, 2018): 41–42.

[127] Leo Tolstoy, *Anna Karenina* (1873), 1.

markets, requirements that apply equally to everyone. One rule is quite short, while the others are technically complex with extensive details on what is required and how it should be done.

They have one thing in common, however: their solution is simple, even though the underlying problem is complex. They make the most of advanced monitoring, electronic reporting, and data analytics. They rely on transparency to put pressure on the regulated and make the system operate smoothly. They use the power of uniform commands to overcome political and practical barriers. They make compliance more attractive than noncompliance. Their excellent compliance outcomes were achieved without the need for extensive enforcement pressure. In short, they meet Next Gen principles for rules with compliance built in.

The programs with dismal compliance records also have some common lessons. A rule with a "hope for the best" compliance theory is doomed. Compliance has to be designed in, not assumed. Much can be learned about a rule's weaknesses by hypothetically asking: If regulated parties want to avoid complying, how would they do that? This isn't a moral question, it's a practical one, like observing that water flows downhill. If the rule asks for water to flow uphill, and at the same time leaves open the downhill path, guess what?

2
Noncompliance with Environmental Rules Is Worse Than You Think

Serious noncompliance with environmental rules is common. It is common across all programs and industry types. Significant violations occur at 25 percent or more of facilities in nearly all programs for which there is compliance data. For many programs with the biggest impact on health, serious noncompliance is much worse than that. Significant violation rates of 50–70 percent are not unusual. These widespread violations have a direct effect on people's health.

What we want is less air and water pollution. We want people not to be at risk of dangerous exposures or catastrophic environmental accidents. We want safer chemicals in the market. We want kids to be able to drink water and to play outside without endangering their health.

Compliance is how we get there. Congress sets the lofty public health goals. Regulations translate those aspirations into concrete requirements. The rubber meets the road when firms do—or don't do—what those rules require. That's compliance: Are companies taking the necessary action to protect public health or not? Laws and rules are a fine start, but what we truly care about is whether they produce action in the real world.

The introduction of this book explains why rule design is the most important determinant of compliance. If a rule makes compliance the path of least resistance, compliance will be good. Otherwise, we can expect widespread serious violations. Chapter 1 of this book—"Rules with Compliance Built In"—provides detailed examples of successful and unsuccessful rules and explains how their structure determined the compliance outcome. You could be excused for thinking, as most environmental regulators do, that the bad examples are outliers, and that most rules have fairly good compliance performance. You could be excused, but you would still be wrong. Are the examples of rules with widespread violations anomalies in an otherwise great compliance record? Unfortunately, no. The poor outcomes discussed in chapter 1 are regrettably just the tip of the iceberg. The broader compliance record is the subject of this chapter.

Environmental compliance evidence is of four main types: (1) statistically valid compliance "rates," of which there are very few; (2) programs where the compliance status of a very large percentage of the sources is known, so

Next Generation Compliance. Cynthia Giles, Oxford University Press. © Cynthia Giles 2022.
DOI: 10.1093/oso/9780197656747.003.0003

something meaningful can be said about the compliance performance of the entire regulated community; (3) rules where we don't know what the compliance performance is, but there is compelling evidence suggesting it is probably bad; and (4) programs for which we have no idea about compliance, of which there are many. After reviewing evidence in these four categories, I describe two important regulatory programs for which the publicly stated evidence about compliance is flawed and serious violations are much more common than public reports claim.

My review of the noncompliance evidence could have presented a couple of illustrations in each category and moved on. Instead, I give multiple examples—nearly all very brief—because the number and diversity of the examples underscore the central point: serious noncompliance occurs everywhere. It's in all types of programs and in all kinds and sizes of companies. It cannot be dismissed as a problem confined to a few industry sectors or a small number of atypical regulations. The sheer number of examples is part of the rebuttal to skeptics, who may acknowledge that there are some rules with poor outcomes but who still cling to the belief that overall compliance is the norm.

A well-established paradigm is not easily knocked off its perch. The dual assumptions, that compliance overall is good and assuring compliance is the job of enforcers, have a tight grip on environmental policy.[1] That's not going to change until the paradigm's adherents accept that our current system isn't getting us there. The evidence presented here makes that case.

What Kinds of Violations Matter?

For most people, the idea of pollution conjures up an image of smoke rising from a tall stack or dirty water flowing from a pipe. Everyone understands why illegal discharges from those sources are a problem. And they realize that higher amounts of violating pollution are generally more troubling. But there are also significant health threats regulated by EPA that don't fit that model.

Much of the most serious pollution does not come from clearly defined sources like a stack or a pipe. For air pollution, significant amounts of dangerous air emissions come from much more dispersed sources, like the leaks from

[1] There are some scholars who have acknowledged the pervasiveness of serious environmental violations, but they expressly or implicitly assume that deficiencies in enforcement are the principal reason. See, e.g., David L. Markell and Robert L. Glicksman, "Dynamic Governance in Theory and Application, Part I," *Arizona Law Review*, Vol. 58 (2016): 563, 590–91; Victor B. Flatt and Paul M. Collins Jr., "Environmental Enforcement in Dire Straits: There Is No Protection for Nothing and No Data for Free, *Environmental Law Reporter New and Analysis*, Vol. 41 (2011): 10679; Daniel A. Farber, "Taking Slippage Seriously: Noncompliance and Creative Compliance in Environmental Law," *Harvard Environmental Law Review*, Vol. 23 (1999): 297.

valves, pipes, and tanks at industrial facilities; releases from oil and gas well sites across the country; and the emissions from millions of trucks, ships, and cars. And although discharge from wastewater pipes is still a serious problem, we face a growing threat from stormwater—rain that washes bacteria, nutrients, and chemical contamination from industrial facilities, pavement, and farms into the nation's waters.[2] Widespread violations of rules for these more diffuse types of pollution have a huge collective impact.

Many EPA rules are intended to prevent pollution from happening at all, not just allow it in limited amounts. For example, requiring that hazardous waste can only be sent to a licensed facility for treatment or disposal assures the public that dangerous wastes are handled by companies with the expertise and resources to prevent leaks. It's not that all violations of prevention rules lead directly to harm. It's that the more times regulated parties engage in unsafe practices, the more likely it is that dangerous incidents will happen. That's why compliance with prevention rules is important. You never know when a violation will combine with unpredictable events to cause serious damage: the unlicensed pesticide applicator dowses a condo with a chemical not approved for indoor use and causes severe and permanent damage to an entire family; the inadequately inspected tank explodes, exposing workers and neighbors to dangerous chemicals; or the cracked containment wall leaks, sending toxic chemicals into the drinking water supply. When these terrible incidents occur, government investigations usually reveal that the company failed to take the required preventive measures. In other words, there was a violation. By insisting on compliance with prevention requirements, we avoid creating the circumstances for these catastrophes to occur.[3]

Monitoring and reporting are also key compliance obligations. That's how companies and government know if the standards have been met. When companies don't monitor or don't report their activities, serious pollution problems can be happening unobserved. If waste manifests are not completed, no one knows that dangerous hazardous waste has gone missing. If a company skips monitoring, they are unaware that they have a serious leak spewing toxic chemicals into neighboring communities. These failures are not unimportant "paperwork"

[2] Toxic algal blooms are increasing in the United States, caused by nutrients from multiple sources including stormwater runoff. See "How Human Activities Increase the Occurrence of Cyanobacterial Blooms," EPA, https://www.epa.gov/cyanohabs/causes-cyanohabs; "Harmful Algal Blooms," EPA, https://www.epa.gov/nutrientpollution/harmful-algal-blooms.

[3] Many regulations are intended to prevent contamination from reaching the environment or threatening people. Here are just few illustrations: oil spill prevention and countermeasures (SPCC), pesticide labeling that includes use restrictions, corrosion control in drinking water systems to inhibit lead contamination, liner requirements for landfills to avert leaks, lead paint removal work practices, safe disposal requirements for PCBs, financial assurance obligations proving that companies have the resources to address the problems they cause, work practices for asbestos removal, and checking for corrosion in tanks and pipes holding dangerous chemicals.

violations. If a drinking water provider doesn't sample the water to make sure it is safe before sending it out to consumers, no one who drinks the water would consider that a minor problem. Because the accuracy and timeliness of self-monitoring and reporting is central both to compliance and to program integrity, government rightfully considers violations of these requirements as very serious.

Compliance with rules is not all or nothing. Companies often comply with some rules but not others. They might violate rarely or frequently. Their emissions could barely exceed the limit or surpass it by a factor of 20. Some firms do completely ignore environmental requirements, but it is more common that some actions are taken to comply, but they fall far short. The firm installs pollution controls but operates them intermittently or incorrectly. It has a program to inspect for corrosion, but the people implementing it miss obvious defects—with sometimes catastrophic consequences. Samples are taken, but not in the right places or at the right times, so the key pollution is missed. For these and many other reasons, noncompliance isn't a simple yes/no proposition.

We care about compliance overall, but we care most about the violations with the greatest potential impact. That's why I focus on serious noncompliance in this book. The question of greatest interest isn't whether a thorough examination can find any violation of any standard—although if many companies routinely have violations something is amiss—but how common it is to discover significant problems. In some programs, there is a well-established definition of what qualifies as significant, like the amount of violating pollution or the frequency of failure to monitor or report. When available, I focus on data about these most serious kinds of violations.

It is worth noting that noncompliance and compliance are not always flip sides of the same coin. It is possible for inspectors to confirm some violations without knowing whether a company is complying with everything else. An inspector can see that the stack is belching smoke without checking that every required report was filed on time. It often takes intensive effort to say with certainty that a facility is 100 percent compliant, and usually that's not an important question. That's why this book focuses on *noncompliance* and not compliance rates. Data about violations are more reliable than claims of full compliance, and the public health threat from serious violations is what we care most about.

Some people may wonder if the threats from widespread violations are less worrying than they appear because pollution measured at many ambient monitors has declined.[4] They hope that progress reducing some of the most

[4] Ambient monitors measure air and water pollutants in the community, which aggregates pollution from all sources. They are different from facility-specific monitors intended to measure the pollution from individual facilities.

troubling air and water pollution means we don't need to worry about pervasive, significant noncompliance. Unfortunately, ambient monitoring results don't give us that reassurance.

One reason is that despite progress, we still have very serious pollution problems. Ninety-seven million Americans live in areas that don't meet air quality standards.[5] Almost half of the nation's rivers and streams are in poor condition.[6] Some of these trends have recently been going in the wrong direction after years of improvement.[7] Widespread violations contribute to these ongoing health threats.

But an even more important reason is that many of the serious violations occurring across the country today result in exposures, or the risk of exposures, that will never be spotted by ambient monitors. Ambient monitors look at long-term trends for some air and water pollutants[8] in some places[9] some of the

[5] See "Air Quality—National Summary," EPA, https://www.epa.gov/air-trends/air-quality-national-summary (2020 data); "Nonattainment Areas for the Criteria Pollutants," EPA, https://epa.maps.arcgis.com/apps/MapSeries/index.html?appid=8fbf9bde204944eeb422eb3ae9fde765 (displaying nonattainment areas for each of the criteria pollutants). Furthermore, ambient monitors may undercount actual pollution. A recent study using satellite data concluded that 24 million people lived in areas that should have been, but were not, classified as nonattainment for $PM_{2.5}$, doubling the number that EPA reported as living in $PM_{2.5}$ nonattainment areas nationally. Daniel M. Sullivan and Alan Krupnick, "Using Satellite Data to Fill the Gaps in the US Air Pollution Monitoring Network," Resources for the Future, Working Paper RFF WP18-21, September 2018, at 2–3.

[6] EPA, "National Water Quality Inventory Report to Congress," EPA 841-R-16-011, August 2017, at 2, https://www.epa.gov/sites/production/files/2017-12/documents/305brtc_finalowow_08302017.pdf. The national water quality assessment is based on statistical sampling. The 2017 survey report concludes that 46% of the river miles are in poor condition, but it can't identify where those poor-quality river miles are because the conclusion is based on a random sample. Fifty-five percent of the river miles assessed by the states were deemed impaired (unable to support one or more of the uses designated for them by the states, such as fishing or swimming). EPA, "National Water Quality Inventory," 8.

[7] See, e.g., Seth Borenstein and Nicky Forster, "US Air Quality Is Slipping After Years of Improvement," *AP News*, June 18, 2019, https://www.apnews.com/d3515b79af1246d08f7978f026c9092b; "Water Quality Changes in the Nation's Streams and Rivers," USGS, https://nawqatrends.wim.usgs.gov/swtrends/ (an interactive map showing where water pollution is getting worse); Nadja Popovich, "America's Air Quality Worsens, Ending Years of Gains, Study Says," *New York Times*, October 24, 2019, https://www.nytimes.com/interactive/2019/10/24/climate/air-pollution-increase.html.

[8] The air monitoring network focuses on the National Ambient Air Quality Standards (NAAQS) criteria pollutants: ozone (O_3)—formed in the atmosphere from the interaction of nitrogen oxides (NO_x) and volatile organic compounds (VOCs) in sunlight, particulate matter (PM), nitrogen dioxide (NO_2), lead (Pb), sulfur dioxide (SO_2), and carbon monoxide (CO). There is some, but very limited, air toxic ambient monitoring: as of 2019 there were only 27 ambient toxic monitoring sites nationwide, each only required to check for 19 compounds, although most check for more. See "Air Toxics—National Air Toxics Trends Stations," EPA, https://www3.epa.gov/ttnamti1/natts.html. Note that there are 188 listed air toxics under the Clean Air Act. For the contaminants included in water quality monitoring, see "Water Quality in the Nation's Streams and Rivers—Current Conditions and Long-Term Trends," the United States Geological Survey, https://www.usgs.gov/mission-areas/water-resources/science/water-quality-nation-s-streams-and-rivers-current-conditions?qt-science_center_objects=0-qt-science_center_objects; EPA, "National Water Quality Inventory."

[9] The majority of US counties don't have ambient air pollution monitors. Sullivan and Krupnick, "Using Satellite Data," 2. See also "Interactive Map of Air Quality Monitors," EPA, https://www.epa.gov/outdoor-air-quality-data/interactive-map-air-quality-monitors (map showing where all the

time.[10] That's all they are intended to do. They tell us nothing about serious pollution that occurs far from any monitors, or toxic contaminants the monitors aren't looking for, or releases that happen when monitors aren't checking. And air and water monitors can't tell us how we are doing with programs to prevent exposure, such as lead paint, pesticides, drinking water contaminants, and dangerous chemicals in products.

Compliance is the first line of defense. And for many problems it's the only one. Some violations may eventually be observed in concerning results from long-term ambient monitoring. But most of the time they won't. People will be at risk, but ambient monitoring won't tell us that. Widespread serious violations of rules designed to protect public health are alarming, whether or not ambient monitors are raising a flag.

What Do We Know about Noncompliance?

This book looks primarily at noncompliance on a national scale. That's because federal rules set the bar for delegated programs across the county. They are intended to meet the charge from Congress that everyone be protected, regardless of where they live. Individual states might have much better or worse results, but we can only know if we are achieving the national goals of our federal laws by looking at the national picture. What does the evidence say?

Violations Are Common in the Few Programs with True Noncompliance Rates

For decades, a noncompliance rate has been the holy grail for compliance work: What percentage of the regulated firms have violations? If we reliably know that and have information on the types and seriousness of the violations, we would know if we are doing a generally good or terrible job protecting the public.

ambient air quality monitors are located). Less than a third of the nation's river miles are monitored over a multiple year period for EPA's national water quality survey. EPA, "National Water Quality Inventory," 8.

[10] Most ambient air monitors operate every three, six, or 12 days, depending on the pollutant. See "Ambient Monitoring Technology Information Center, Sampling Schedule Calendar," EPA, https://www.epa.gov/amtic/sampling-schedule-calendar (displaying annual monitoring schedules). The national statistical sample of US waterways occurs on a five-year cycle. EPA, "National Water Quality Inventory," 4.

There are two ways to identify a noncompliance rate with confidence. One is through statistically valid sampling; if we examine compliance at a randomly selected representative sample of regulated firms, the data on that sample can tell us what the rate is for the population as a whole. The other approach is to look at the compliance status of 100 percent (or very close to it) of the regulated facilities. This approach doesn't require randomized sampling because it looks at the entire universe.

There is surprisingly little information that could realistically be called a noncompliance rate. Because there has never been and will never be enough inspectors to inspect all or even a significant fraction of regulated facilities, figuring out a meaningful "rate" of noncompliance has been challenging.[11] State and federal inspectors are, for good reason, focused on the facilities that regulators have reason to believe might be in violation. Very few enforcement offices have the resources to inspect enough randomly selected facilities to be able to say anything with confidence about the rate of noncompliance. If there are 1,000 facilities in a state covered by a rule, and if in any year the state selectively inspects 100 of them and finds 25 are in violation, that is not a 25 percent noncompliance rate. Under this scenario the state doesn't know what's happening at 900 facilities. The true noncompliance rate could be anywhere between 5 percent and 90 percent; it can't be determined from 100 targeted inspections.[12]

In the early 2000s, EPA attempted to get a statistically valid noncompliance rate for some programs. This was EPA's response to continual pressure to say something conclusive about noncompliance rates. EPA worked with statisticians to develop reliable information about the rate of noncompliance for some sectors. So that the inspections could be largely random—and thus representative of the whole sector—EPA and the states had to forgo inspections that they would otherwise have done at facilities where there was reason to believe there were violations.

[11] For some programs, boots-on-the-ground inspections have been the most reliable way to determine compliance, as is discussed elsewhere in this book. That isn't always the case; sometimes violations can be determined off-site from document reviews or facility-run monitoring, and sometimes inspectors miss some of the most serious violations because it isn't possible to identify those violations through on-site inspections alone. "Inspections" is used here to mean whatever investigatory method is best for determining noncompliance.

[12] For many years, probably decades, some states have argued for calling the rate of violations discovered during inspections a noncompliance rate for the whole sector. The percentage of inspected facilities with violations—often called a "hit rate"—may tell you how common violations are at the inspected facilities, but it tells you nothing about compliance at facilities not inspected; pick a different group of facilities to inspect and you get a different rate. A true noncompliance rate can only be determined through data about the entire universe or a statistically representative sample. See EPA, "Expanding the Use of Outcome Measurement for EPA's Office of Enforcement and Compliance Assurance: Report to OMB," July 31, 2006 (2006), 12, https://archive.epa.gov/compliance/resources/reports/compliance/research/web/pdf/outcome-measurement.pdf.

For each of three major environmental laws—clean water, clean air, and hazardous waste—EPA looked at one sector's compliance with one regulation. EPA learned two things: noncompliance was common and figuring out noncompliance rates this way is prohibitively expensive.[13]

Here's what EPA found out about noncompliance rates:

The noncompliance rates EPA found during this exercise, shown in Table 2.1, range from bad to dismal. Thirty-four percent of the studied RCRA hazardous waste generators were in violation; 49 percent of the ethylene oxide manufacturers were violating rules about air toxics; and a whopping 61 percent of municipalities were violating the rules about discharge of raw sewage and contaminated stormwater.

In addition, EPA learned that figuring out noncompliance rates this way isn't practical. It costs too much money, takes too much time, and it reduces the inspections EPA and states can do at facilities likely to be violating. Statistical sampling cannot possibly be done for even a small number of the sectors that EPA regulates, most of which have so many regulated facilities that the number needed for a statistically representative sample is unaffordably large. And it fails on another score too: because taking a representative sample is designed to figure out the rate of noncompliance, it only tells EPA what the percentage of violators is, not who they are. It may show that 50 percent of facilities are violating, but it doesn't tell regulators which ones, so isn't useful for taking direct action.

An alternative way to figure out noncompliance rates is by looking at the compliance status of the entire universe of regulated facilities. No sampling is required. This kind of nearly complete universe information—so-called near-census data—is a better way to figure out noncompliance rates if the data are available as part of regular government operations because it gives useful rate information without all the costs and other downsides of sampling.

EPA has data on almost the entire universe of large, individually permitted discharges under the Clean Water Act (the National Pollution Discharge Elimination System, or NPDES, program). Under NPDES, individually permitted facilities are required to submit usually monthly reports about their discharges and self-disclose any violations. These are self-reported compliance levels, not verified by government, but because the requirement to report is universal, it is possible to find out from these reports what the self-reported rate of noncompliance is without the need to divert resources to investigating a representative sample.

[13] EPA, "Expanding the Use," 14–15. The project had both direct costs (additional costs to conduct the inspections and analyze the results—over $300,000 in 2018 dollars) and unquantified opportunity costs (the pollution or risk reductions EPA could have achieved by doing the same number of targeted—rather than random—inspections).

Table 2.1 Compliance data from EPA's 2006 report "Expanding the Use of Outcome Measurement."

Sector and regulation	Number of random inspections required	Noncompliance rate[a]
Organic Chemical Manufacturing small quantity generator hazardous waste requirements under RCRA[b]	112	34.3% (+/- 8.1%)
Ethylene Oxide Manufacturers Clean Air Act toxic air pollution requirements[c]	67	49.2% (+/- 5%)
Municipal Combined Sewer requirements under Clean Water Act[d]	214	61.4% (+/- 5%)

[a] The noncompliance rates cited here include all noncompliance, not just significant noncompliance, because that's the only information provided in the report.

[b] Wastes from organic chemical manufacturing are defined as hazardous under the Resource Conservation and Recovery Act (RCRA). "Defining Hazardous Waste: Listed, Characteristic and Mixed Radiological Wastes, the F and K lists," EPA, https://www.epa.gov/hw/defining-hazardous-waste-listed-characteristic-and-mixed-radiological-wastes#FandK. The small quantity generator rules define how those wastes should be stored, transported, and disposed to prevent releases of those hazardous wastes into the environment.

[c] Ethylene oxide is identified as a hazardous air pollutant under the Clean Air Act. It can cause harm to the brain and central nervous system, in addition to irritating eyes, skin, nose, throat, and lungs. See "Background Information on Ethylene Oxide," EPA, https://www.epa.gov/hazardous-air-pollutants-ethylene-oxide/background-information-ethylene-oxide#why. Ethylene oxide is also a carcinogen. EPA, "Background Information on Ethylene Oxide." Ethylene oxide has recently been in the news because of exposure concerns that started with a facility in Illinois. See Illinois EPA, "Illinois EPA Director Seals Portions of Sterigenics Due to Public Health Hazards from Ethylene Oxide Emissions," Press Release, February15, 2019, https://www2.illinois.gov/Pages/news-item.aspx?ReleaseID=19717. EPA is considering additional regulations for ethylene oxide. EPA, "Ethylene Oxide—Updates," https://www.epa.gov/hazardous-air-pollutants-ethylene-oxide/ethylene-oxide-updates.

[d] The term "combined sewers" refers to the situation where human sewage, stormwater runoff, and indirect discharges of industrial waste are funneled into the same pipes. When rainfall leads to high volumes of stormwater, treatment authorities often discharge this untreated or partially treated combined waste into surface waters. Regulations governing discharges from combined sewers are designed to protect the public from the serious health threats posed by pathogens and industrial contaminants in the nation's waters.

There are about 7,000 major NPDES water dischargers each year. "Major" dischargers include the largest facilities discharging pollutants into the nation's waters. These include large industrial facilities like refineries and chemical manufacturing plants as well as cities that run sewage treatment plants. The percentage of NPDES majors that have self-reported violations over the period 2013 to 2021 (the years shown on EPA's dashboard as of the end of 2021) has been between 72 percent and 75 percent a year.[14] The self-reported rate of more serious violations, labeled as significant noncompliance (SNC), has for many years been between 20 percent and 25 percent.[15]

Under a recent regulation, nonmajor water pollution dischargers are also required to submit their discharge reports to EPA and states electronically.[16] Nonmajors are usually smaller industrial facilities and cities, and their discharges include nutrients, sediments, and a host of chemical contaminants. The new requirement will give EPA compliance data on close to the entire universe of approximately 40,000 facilities that are significant enough to require individual permits but for which compliance information has often been largely inaccessible to EPA and the public. Prior to the new universal requirement, about 36 states and territories did have 75 percent or more of their nonmajor (sometimes called "minor") water dischargers report electronically to EPA.[17] That's not enough for a completely reliable rate of noncompliance, but it is pretty close.[18] The rate of

[14] See "Enforcement and Compliance History Online (ECHO)," EPA, https://echo.epa.gov/ (select topic Analyze Trends: State Water Dashboard, Facility type: Major, Box: Violations, Data: % Facilities with violations).

[15] See "Clean Water Act Action Plan," EPA, October 15, 2009, at 3, https://www.epa.gov/compliance/clean-water-act-cwa-action-plan (SNC for majors about 24%); EPA ECHO, screen capture on December 28, 2018 of SNC for major water dischargers for fiscal years 2011 through 2018 (on file with author) (showing SNC for majors varied between 20% and 25% for the years 2011 through 2018). I do not cite EPA's ECHO data for SNC majors as displayed on ECHO as of late 2021 because the historic data has changed without clear explanation. EPA has recently started allowing states to revise years old data and the percentage of facilities reporting electronically under the NPDES electronic reporting rule has also been changing, both of which make SNC trend data on ECHO less reliable. SNC is a defined term that includes violations that are more frequent, higher volume, or more serious.

[16] "National Pollutant Discharge Elimination System (NPDES) Electronic Reporting Rule," *Federal Register*, Vol. 80 (October 22, 2015): 64063, https://www.federalregister.gov/documents/2015/10/22/2015-24954/national-pollutant-discharge-elimination-system-npdes-electronic-reporting-rule. The NPDES e-reporting rule requires electronic submission by nonmajor individually permitted sources starting in December of 2016, so more reliable rates for this universe of water pollution dischargers will eventually become available, although the rule has still not been fully implemented. *See NPDES eRule Readiness and Data Completeness Dashboard*, EPA, https://echo.epa.gov/trends/npdes-erule-dashboard-public (presenting state-specific data on e-rule implementation and also trends in percent of permitted facilities that electronically report; as of November 2021, ECHO reports that about 75% of the individually permitted sources required to report electronically were submitting electronic reports). GAO recently called EPA to task for not being aboveboard about the significant missing and inaccurate water compliance data presented on the ECHO site. GAO, "EPA Needs to Better Assess and Disclose Quality of Compliance and Enforcement Data," GAO-21-290, July 12, 2021, https://www.gao.gov/products/gao-21-290.

[17] "U.S. EPA Annual Noncompliance Report (ANCR) Calendar Year 2015," EPA Office of Enforcement and Compliance Assurance, August 2016, at 9, https://echo.epa.gov/system/files/2015_ANCR.pdf. The most recent ANCR was for 2015; the dashboard using NPDES e-reporting rule data will eventually take the place of the ANCR. The states for which EPA had actual discharge data are labeled in the ANCR as "verified" states. States that only provided summary information were labeled "non-verified." EPA, "ANCR for 2015," 6.

[18] Note that because 36 states and territories provided this universe data for facilities in their states, it cannot be directly translated into a *national* noncompliance rate. The facilities in states where

serious noncompliance for facilities in these states in 2008 was 60 percent.[19] With a sustained EPA effort to call attention to these astonishingly high rates of serious violations—aided by a prominent article in the *New York Times*[20]—the rate has steadily declined; in 2015 the self-reported serious violation rate for the verified dischargers was an improved but still poor 32 percent.[21] These serious problems continue; EPA reports that in 2018 the significant violator rate for individually permitted water dischargers (major and nonmajor) in 2018 was a discouraging 29.4 percent.[22]

In contrast to the preceding discouraging outcomes, rules employing Next Gen strategies had excellent compliance results. Two of those rules are highlighted in chapter 1: the Acid Rain Program and the Greenhouse Gas Reporting Program. As a result of strong compliance design, both rules had noncompliance rates less than 2 percent. Not coincidently, these rules have near-census data as part of the program design, so it is possible to be confident about noncompliance rates. The same impressive results occurred for the Mercury and Air Toxics Standards (MATS), which employed Next Gen strategies akin to the Acid Rain Program.[23] Note that the same regulated sector—coal-fired power—had two impressive

electronic reporting was not required may have a noncompliance record that is better or worse than the reporting states.

[19] EPA, "ANCR for 2015," 6.

[20] Charles Duhigg, "Clean Water Laws Are Neglected, at a Cost in Suffering," *New York Times*, September 12, 2009.

[21] EPA, "ANCR for 2015," 6. States that did not require facility electronic reporting or provide that information to EPA gave EPA only summary information. That summary data provided no facility-specific information, just conclusions, like "10% of our non-majors had serious violations." For years, that summary data have suggested that these nonverified states had noncompliance rates that were dramatically lower than verified states, a conclusion that is not supportable and that EPA rejected in its ANCR in 2015. EPA, "ANCR for 2015," 6. For example, in 2008, states with verified data reported a serious noncompliance rate for nonmajors of 60%, while the states with nonverified summary data claimed a serious noncompliance rate of only 18%. EPA, "ANCR for 2015," 6. Note that the data from EPA's ANCR for 2015 does not match with the data displayed for 2015 on EPA's ECHO dashboard as of November 2021 for a host of reasons that make the ECHO data less reliable, some of which are described *supra* in note 20.

[22] "Data Quality Record for Long-Term Performance Goal," EPA, updated January 1, 2020, https://www.epa.gov/sites/default/files/2020-06/documents/dqr-3-1-environmental-law-compliance.pdf (support for objective 3.1: compliance with the law for EPA's strategic plan for 2018-2022); GAO, "EPA Needs to Better Assess," 30–31.

[23] See US Energy Information Administration, "Coal Plants Installed Mercury Controls to Meet Compliance Deadlines," *Today in Energy*, September 18, 2017, https://www.eia.gov/todayinenergy/detail.php?id=32952# (compliance record). For example, the MATS rule required continuous monitoring for mercury. 40 C.F.R. § 63.10000(c)(1)(vi). Compliance with MATS was also aided by external developments, like the reduced price of gas and technological innovation in mercury removal, which significantly reduced the costs of compliance. See Calpine Corporation, Exelon Corporation, and Public Service Enterprise Group, "Comment Letter on Proposed Supplemental Finding that it is Appropriate and Necessary to Regulate Hazardous Air Pollutants from Coal- and Oil-Fired Electric Utility Steam Generating Units," Docket ID No. EPA-HQ-OAR-2009-0234, January 15, 2016, at 3, https://www.regulations.gov/document?D=EPA-HQ-OAR-2009-0234-20549 (finding that the actual cost of complying with MATS was less than 25% of costs EPA estimated in the final rule).

compliance outcomes (acid rain, MATS) and one compliance disaster (NSR),[24] further evidence that it is rule design, not the sector being regulated, that drives compliance results.

Overwhelming Data in Many Programs Show that Serious Violations Are Widespread

EPA usually doesn't have statistically valid sampling or near-census data about compliance. So most of the time there is nothing that can credibly be called a noncompliance rate.[25] However, for some individual rules or programs, EPA has reliable compliance data on 70 percent or more of the universe. That's enough to estimate how common it is that large facilities in that sector have serious violations.

Here are some examples:

Coal-fired power plants. Coal-fired power plants have produced by far the largest volume of dangerous air pollution of any industrial sector in the United States.[26] Of the largest 25 coal-fired power companies, responsible for about 70 percent of the US coal-fired power production in 2005,[27] 18 were sued for violating the Clean Air Act's requirement to upgrade pollution controls when upgrading the plant.[28] That means at least 70 percent of the largest 25 coal-fired power companies were in serious violation of the Clean Air Act.

[24] See the discussion of NSR later in this chapter, and also in chapter 1 (comparing Acid Rain and NSR).

[25] Environmental policy practitioners may be wondering why the preceding discussion of noncompliance rates doesn't include rates for public drinking water systems and major stationary sources of air pollution. Doesn't EPA routinely claim to have noncompliance rates for these two important categories of regulated sources? It does. But those claimed rates are demonstrably unreliable, as is discussed later in this chapter.

[26] See Emanuele Massetti et al., "Environmental Quality and the U.S. Power Sector: Air Quality, Water Quality, Land Use and Environmental," Oak Ridge National Laboratory, ORNL/SPR-2016/772, January 4, 2017, vii, https://info.ornl.gov/sites/publications/files/Pub60561.pdf. See also GAO, "Wider Use of Advanced Technologies Can Improve Emissions Monitoring," June 2001, at 19; American Lung Association, "Toxic Air: The Case for Cleaning Up Coal-fired Power Plants," March 16, 2011, at 1, https://www.lung.org/getmedia/c3b2b744-7c7e-4941-b0cd-5a5e468515d1/toxic-air-report.pdf.pdf.

[27] See "Ownership of Existing U.S. Coal-fired Generating Stations," Center for Media and Democracy, https://www.gem.wiki/Existing_U.S._Coal_Plants (listing top 25 coal-fired utilities in 2005).

[28] Sixteen were sued by EPA, and two by Sierra Club (MidAmerican Energy and Entergy). See "Coal-Fired Power Plant Enforcement," EPA, https://www.epa.gov/enforcement/coal-fired-power-plant-enforcement (partial list of EPA coal-fired power plant cases). There are many other coal-fired power plant settlements with companies not on the top 25 list.

Petroleum refineries. Emissions from petroleum refineries include some of the same pollutants found at power plants, along with smog-causing volatile organic compounds and air toxics, including benzene, a known carcinogen. EPA has entered into 37 Clean Air Act settlements with US companies that refine over 95 percent of the nation's petroleum refining capacity. In other words, the companies responsible for virtually all of the nation's total production were in serious violation.[29]

Cement manufacturing plants. Cement manufacturing plants are the third largest industrial source of air pollution. All of the top five, and nine of the top 10 cement manufacturers in the United States—responsible for 82 percent of the total US production—entered into enforcement agreements with EPA for serious Clean Air Act violations.[30]

Combined sewer overflows. Just about every large city was in consistent and serious violation of the Clean Water Act and was eventually sued by EPA to fix the public health threat posed by discharges of raw sewage and contaminated stormwater into the nation's rivers. EPA and states have taken actions at 97 percent of large combined sewer systems, 92 percent of large sanitary sewer systems, and 79 percent of Phase 1 municipal separate stormwater systems.[31]

[29] These settlements cover 112 refineries in 32 states and territories. On full implementation, those cases will result in annual emissions reductions of more than 95,000 tons of nitrogen oxides and more than 260,000 tons of sulfur dioxide. See "Petroleum Refinery National Case Results," EPA, https://www.epa.gov/enforcement/petroleum-refinery-national-case-results. The defendants in these cases include BP, Chevron, CITGO, Conoco, ExxonMobil, Hess, Koch, Sunoco, Tesoro, Total, and Valero, among many others.

[30] The top 10 US cement manufacturers in 2010 were, in declining order: *CEMEX, Inc.; Holcim (US) Inc.; Lafarge North America Inc.; Lehigh Cement Co.; Buzzi Unicem USA Inc.; Ash Grove Cement Co.; Essroc Cement Corp.*; Texas Industries, Inc. (TXI); *Eagle Materials, Inc.; and St. Marys Cement Group.* In 2010, the top five companies produced nearly 60% of total US portland cement, and the top 10 accounted for 82% of total production. See "2010 Minerals Yearbook, Cement," USGS, 16.3, https://prd-wret.s3-us-west-2.amazonaws.com/assets/palladium/production/atoms/files/myb1-2010-cemen.pdf. The companies italicized in the preceding list entered into enforcement agreements with EPA. See "Cement Manufacturing Enforcement Initiative," EPA, https://www.epa.gov/enforcement/cement-manufacturing-enforcement-initiative; EPA, "EPA Reaches Agreement with Lehigh Cement on Clean Air Violations," Press Release, June 18, 2008, https://archive.epa.gov/epapages/newsroom_archive/newsreleases/68bb6c787b74b7968525746c004d4b66.html (settlement with Lehigh, not on EPA's partial list of settlements); EPA, "Nevada Cement Co. Facility in Fernley, Nev., to Reduce Emissions, Upgrade Pollution Controls," Press Release, May 12, 2017, https://www.epa.gov/newsreleases/nevada-cement-co-facility-fernley-nev-reduce-emissions-upgrade-pollution-controls (settlement with Nevada Cement, not on EPA's partial list). Nevada Cement is owned by Eagle Materials. See "Eagle Materials Cement," Eagle Materials, http://www.eaglematerials.com/products/cement.html.

[31] EPA, "Public Comment on EPA's National Compliance Initiatives for Fiscal Years 2020-2023," *Federal Register*, Vol. 84 (February 8, 2019): 2850. "Large" means serves a population of over 50,000 or has more than 10 million gallons a day wastewater discharge. Here are just some of the biggest cities whose sewer systems were sued by EPA for sewage and/or stormwater contamination violations: Atlanta, Baltimore, Boston, Chicago, Cincinnati, Cleveland, Dallas, District of Columbia, Houston, Indianapolis, Kansas City, Los Angeles, Miami, Nashville, New York, Philadelphia,

Mineral processing. Mineral processing facilities generate more toxic and hazardous waste than any other industrial sector. EPA's national enforcement initiative to reduce risk from this sector initially focused on compliance in the phosphoric acid industry.[32] Of the 20 facilities in this industry nationally,[33] 13 were covered by enforcement agreements as of 2016,[34] a serious violation rate of over 60 percent.

Sulfuric and nitric acid manufacturers. These acids are used in fertilizer, chemical, and explosives production. Acid production plants emit many thousands of tons of nitrogen oxides, sulfur dioxide, and sulfuric acid mist each year.[35] Complete data on noncompliance isn't publicly available, but EPA says this about violations in this sector: "EPA investigations have found a high rate of noncompliance with NSR/PSD in connection with plant expansions and process changes."[36]

Underground storage tanks (UST). There are over 540,000 regulated underground storage tanks at about 200,000 facilities in the United States.[37] These tanks store gasoline, oil, and chemicals. A leak from an underground tank, especially one that goes undetected for an extended period of time, can release dangerous substances into soil and groundwater that can be both a threat to drinking water and expensive to clean up. State reports

Pittsburgh, San Diego, Seattle, and St. Louis. Note that CSO and stormwater requirements discussed here are different from the secondary treatment regulation discussed in chapter 1.

[32] See "National Enforcement Initiative: Reducing Pollution from Mineral Processing Operations," EPA, https://19january2017snapshot.epa.gov/enforcement/national-enforcement-initiative-reducing-pollution-mineral-processing-operations_.html. Phosphoric acid facilities have a high risk of releases of acidic wastewaters, which also contain metals and can cause serious water contamination and fish kills. For example, a 2007 incident at the Agrifos phosphoric acid facility in Houston released 50 million gallons of acidic hazardous wastewater into the Houston Ship Channel. A 2009 sinkhole at the PCS White Springs phosphoric acid facility in north Florida released over 90 million gallons of hazardous wastewaters into the Floridian aquifer, the drinking water source for Florida and South Georgia. See "Mosaic Fertilizer, LLC Settlement," EPA, https://www.epa.gov/enforcement/mosaic-fertilizer-llc-settlement (look under the heading: Health Effects and Environmental Benefits).

[33] EPA, "National Enforcement Initiative Mineral Processing" (chart titled "EPA's progress toward inspecting and addressing phosphoric acid facilities).

[34] Innophos, Mosaic, PCS Geismar, Agrifos, and CF Industries settlements are all described on EPA's civil settlements web page. "Civil Cases and Settlements by Statute," EPA, https://cfpub.epa.gov/enforcement/cases/ (search for each case by company name). Some settlements covered more than one facility.

[35] "Acid Plant New Source Review Enforcement Initiative," EPA, https://www.epa.gov/enforcement/acid-plant-new-source-review-enforcement-initiative.

[36] "Air Enforcement, Stationary Sources," EPA, https://www.epa.gov/enforcement/air-enforcement.

[37] See "Semiannual Report of UST Performance Measures End Of Fiscal Year 2020 (October 1, 2019–September 30, 2020)," EPA, https://www.epa.gov/sites/default/files/2020-11/documents/ca-20-34.pdf. Federal UST regulations do not apply to septic tanks, smaller tanks, or residential or farm tanks. See "Learn About Underground Storage Tanks (USTs), Do All Tanks Have to Meet Federal EPA Regulations?," EPA, https://www.epa.gov/ust/learn-about-underground-storage-tanks-usts#regs.

reveal that the rate of significant violations by USTs is somewhere between 28 percent and 50 percent.[38]

In Many Programs, Compliance Evidence Is Spotty, but the Signs Aren't Good

There are a much larger number of rules, sectors, and programs for which there isn't enough information to even approximate a noncompliance rate. Nevertheless, for many such programs there are troubling signs that serious noncompliance is widespread. Some examples:

Oil and gas wells. Oil and gas wells and the storage tanks located near the wellheads frequently emit volatile organic compounds (VOCs) and benzene that directly pose a threat to health and collectively contribute to the formation of ground-level ozone, a known serious health issue. There are about 1 million active wells in the United States.[39] Even Trump's EPA admitted that there have been significant excess emissions and Clean Air Act noncompliance at these wells, although the full extent of the problem is not known.[40]

[38] EPA has national regulations designed both to prevent such leaks and to detect them quickly if a leak does occur. States inspect about 45% of the facilities with USTs per year. See "UST Performance Measures," EPA, https://www.epa.gov/ust/ust-performance-measures (posting EPA annual reports). States submit summary information to EPA about inspected facilities' compliance. "Significant operational compliance" was the traditional compliance metric, and EPA is now transitioning to the more complete "technical compliance measure." See "Significant Operational Compliance (SOC) Performance Measures," EPA, https://www.epa.gov/ust/significant-operational-compliance-soc-performance-measures. These compliance measures focus on the most serious kinds of violations—noncompliance with the requirements designed to prevent releases from underground tanks, the obligation to quickly detect releases should they occur and the new testing requirements. The rate of serious violation using the older metric hovered around 28%; the rate using the recent and more complete TCR measure is worse—about 50%. See EPA, "UST Performance Measures" (end-of-year reports for FY2020 and FY2019).

[39] "U.S. Oil and Natural Gas Wells by Production Rate," U.S. Energy Information Administration, December 2020, https://www.eia.gov/petroleum/wells/.

[40] See "New Owner Clean Air Act Audit Program for Upstream Oil and Natural Gas Exploration and Production Facilities, Questions and Answers," EPA, March 29, 2019, at 1, https://www.epa.gov/sites/production/files/2018-06/documents/qaoilandnaturalgasnewownerauditprogram.pdf. For example, in an enforcement case with Noble Energy, EPA found that emissions controls were not properly designed or sized to control VOC emissions. See "Noble Energy, Inc. Settlement," EPA, April 22, 2015, https://www.epa.gov/enforcement/noble-energy-inc-settlement. EPA issued a compliance alert in 2015 to address the widespread air violations states and EPA were observing in the field. "Compliance Alert: EPA Observes Air Emissions from Controlled Storage Vessels at Onshore Oil and Natural Gas Production Facilities," EPA, September 2015, https://www.epa.gov/sites/production/files/2015-09/documents/oilgascompliancealert.pdf. Some states have inspection programs for some wells, but neither the inspection methods nor the number of inspections is sufficient to determine how common serious violations are. Serious emission problems in natural gas gathering operations are also common, as evidenced by EPA's recent Enforcement Alert about violations during pigging operations: "EPA Observes Air Emissions from Natural Gas Gathering Operations in

Animal agriculture. EPA estimates that there are about 20,000 large animal agricultural operations in the United States; confined animal operations produce more than *three times* the sewage produced by the entire US human population.[41] EPA does not have reliable data about exactly how many of these sources there are, or whether they are complying with the regulatory limits on pollution.[42] We know the problems are significant though, because water quality studies routinely cite industrial animal agriculture as a major contributor to serious water quality degradation.[43]

Stormwater. Stormwater is runoff from rain falling on city streets, industrial plants, and construction sites, which adds chemicals, nutrients, pathogens, and other contaminants to the nation's waters.[44] People can be exposed to all of these contaminants when they drink water, eat fish, or swim or boat in rivers, lakes, and beaches, as millions do. Hundreds of thousands of sources are regulated by stormwater rules.[45] Compliance with federal stormwater requirements is unknown, but the huge number of river miles impaired by stormwater suggest that compliance is poor.[46] One study in

Violation of the Clean Air Act," September 2019, at 1–2, https://www.epa.gov/sites/production/files/2019-09/documents/naturalgasgatheringoperationinviolationcaa-enforcementalert0919.pdf.

[41] See "NPDES CAFO Permitting Status Report: National Summary, Endyear 2020, completed 05/11/21," EPA, https://www.epa.gov/npdes/npdes-cafo-regulations-implementation-status-reports (number of CAFOs); "National Pollutant Discharge Elimination System Permit Regulation and Effluent Limitation Guidelines and Standards for Concentrated Animal Feeding Operations (CAFOs), Final Rule," *Federal Register*, Vol. 68 (February 12, 2003): 7176, 7180 (amount of waste). Manure in such large quantities carries excess nutrients, chemicals, and microorganisms that find their way into waterways, lakes, groundwater, soils, and airways. See "Putting Meat on the Table: Industrial Farm Animal Production in America," Pew Commission on Industrial Farm Animal Production in America, April 2008, at 9, http://www.pewtrusts.org/~/media/assets/2008/pcifap_exec-summary.pdf; "Animal Waste and Water Quality: EPA Regulation of Concentrated Animal Feeding Operations (CAFOs)," Congressional Research Service, RL31851, February 16, 2010, at 4, https://nationalaglawcenter.org/wp-content/uploads/assets/crs/RL31851.pdf.

[42] See "The EPA's Failure to Track Factory Farms," Food and Water Watch, August 2013), at 4–5, https://foodandwaterwatch.org/wp-content/uploads/2021/03/EPA-Factory-Farms-IB-Aug-2013_0.pdf. See also Aman Azhar, "Pollution from N.C.'s Commercial Poultry Farms Disproportionately Harms Communities of Color," *Inside Climate News*, October 13, 2021 (example in one sector in one state).

[43] Food and Water Watch, "EPA's Failure," 2; EPA, "National Water Quality Inventory" (see n.6), 8 (citing crop production and animal agriculture as leading causes of water quality problems).

[44] EPA's water quality reports document the strong link between stormwater and water quality impairment. These wet weather discharges contain sediments, oil and grease, chemicals, nutrients, metals, and pathogens, all of which are among the biggest contributors to degraded water quality and some of which can endanger human health. EPA, "NPDES E-Reporting Rule," 64068.

[45] There are about 95,000 industrial facilities covered by stormwater regulations. Another about 250,000 construction sites per year are required to control stormwater runoff. About 5,000 nonmajor municipal systems collect stormwater and are required to meet the federal standards. EPA, "NPDES E-Reporting Rule," 64081.

[46] See "Limited Knowledge of the Universe of Regulated Entities Impedes EPA's Ability to Demonstrate Changes in Regulatory Compliance," EPA OIG, September 2005, at 16, 18: "According to EPA staff, there is a high level of noncompliance with stormwater regulations." As a result of the NPDES e-reporting rule finalized by EPA in 2015, nationwide data on stormwater sources are scheduled to become available in 2021. EPA, "NPDES E-Reporting Rule," 64087.

North Carolina found that only 36 percent of regulated locations fully complied with stormwater standards.[47]

Worker Protection Standard (WPS). EPA Worker Protection Standard regulations are designed to protect agricultural workers and pesticide handlers by requiring owners to provide workers with information about pesticide safety, to limit their potential exposure to pesticides, and to quickly address any exposures that do occur.[48] EPA reports that only 3,407 of the 346,000 regulated entities were inspected in 2019, which is less than 1 percent. From these inspections, 1,903 violations were reported.[49]

Small quantity hazardous waste generators. The purpose of rules governing the roughly 40,000 firms generating large quantities of hazardous waste is to prevent releases of hazardous waste into the environment.[50] The rules are less strict for the about 375,000 small and very small quantity generators.[51]

[47] The 1993 study attempted to measure compliance with construction stormwater runoff controls in the state of North Carolina. See Raymond J. Burby and Robert G. Paterson, "Improving Compliance with State Environmental Regulations," *Journal of Policy Analysis and Management*, Vol. 12, No. 4 (Autumn 1993). Unlike many Clean Water Act evaluations that rely primarily on industry self-reported data, this study did field investigations to make an independent determination of compliance. The report was dismal: developers failed to install 27% of the control measures specified, and 51% of the installed measures were not properly maintained. The study found that more than 20% of the plans were deficient, so even full compliance would not have achieved the pollution-reduction standard. In total, only 36% of the sites fully complied with the standard to retain all sediment on site. Burby and Paterson, "Improving Compliance," 759.

[48] See "Agricultural Worker Protection Standard (WPS)," EPA, https://www.epa.gov/pesticide-worker-safety/agricultural-worker-protection-standard-wps. Among other things, these rules require keeping workers out of areas being treated with pesticides. Exposure to pesticides can be a very serious matter. In one 2018 case, EPA found that a company failed to notify workers to avoid fields recently treated with pesticides, resulting in exposure and hospitalization of workers. See EPA, "EPA Reaches Agreement with Syngenta for Farmworker Safety Violations on Kauai," News Release, February 12, 2018, https://archive.epa.gov/epa/newsreleases/epa-reaches-agreement-syngenta-farmworker-safety-violations-kauai.html.

[49] See EPA Enforcement and Compliance History Online (ECHO), https://echo.epa.gov/ (select Analyze Trends: Pesticide Dashboard, WPS Dashboard, Box 3 (Inspections of WPS Regulated Facilities and Box 4: Violations by WPS Regulated Facilities)). For all the reasons previously discussed, the percent of violations found at such a limited and targeted number of inspections does not indicate a rate of noncompliance. Nevertheless, the data from these limited inspections are not encouraging.

[50] The rules mandate storage, labeling, and transportation requirements, and require that hazardous waste only be sent to appropriately licensed facilities for treatment or disposal. See EPA, "ECHO" (select Analyze Trends: Hazardous Waste, Box 1—Facilities, LQG (Large Quantity Generators)) (number of large quantity generators).

[51] See "Guide to Regulated Facilities in ECHO," EPA, https://echo.epa.gov/resources/guidance-policy/guide-to-regulated-facilities (showing number of RCRA regulated facilities under heading "Resource Conservation and Recovery Act (RCRA) Designations"). Although the small quantity and "conditionally exempt" (even smaller quantity) generators generate less waste, there are a lot more of them, so the collective impact of the smaller generators can still be large. Under RCRA there are no federally mandated state inspection requirements for small quantity generators, although states are supposed to have a program for periodically inspecting these facilities. See "Compliance Monitoring Strategy for the Resource Conservation and Recovery Act Subtitle C Core Program," EPA, September 2015, at 21, https://www.epa.gov/compliance/compliance-monitoring-strategy-resource-conservation-and-recovery-act.

EPA does not know the number or compliance status of smaller quantity generators, and a significant percentage of small quantity generators have never been inspected.[52] However, EPA has frequently found examples of firms inaccurately claiming to be small—and thus less regulated—and one statistically valid sample found a 34 percent noncompliance rate by one type of small quantity generators.[53]

Vehicle emissions. Cars and trucks are a major source of some of our most serious air pollution problems.[54] Mobile sources are responsible for more than half of the total nitrogen oxides (NO_x) emissions in the United States.[55] Even if cars and trucks meet emission standards when manufactured, which we know is not universally the case (see Volkswagen[56]), emission controls can deteriorate over time, resulting in vehicles that unlawfully emit many times the allowable amount of pollution.[57] Owners of some vehicles also illegally tamper with emissions controls, significantly contributing to pollution in communities across the country.[58] Although

[52] See EPA OIG, "Limited Knowledge," 18–19.

[53] See Timothy A. Wilkins, "EPA's 'Next Generation' Enforcement Hitting Region 6 Facilities Now," *Bracewell blog*, June 15, 2012 (note that the blog discusses events in 2015 so appears to be incorrectly dated), https://bracewell.com/insights/epas-next-generation-enforcement-hitting-region-6-facilities-now (describing EPA enforcement concerning the "common problem" of generators underreporting their hazardous waste volumes). For the statistically valid rate of noncompliance, see EPA, "Expanding the Use of Outcome Measurement," 14.

[54] See generally EPA, "Our Nation's Air (2018)," https://gispub.epa.gov/air/trendsreport/2018/#sources. See also Phillip Brooks, EPA Air Enforcement Director, "Presentation at the Association of Air Pollution Control Agencies: Tampering and Aftermarket Defeat Devices," August 27, 2019, at 2–4, https://cleanairact.org/wp-content/uploads/2019/09/Tampering-and-Aftermarket-Defeat-Devices-Phil-Brooks.pdf.

[55] Brooks, "Tampering and Aftermarket Defeat Devices," 2. Among other things, NO_x pollution contributes to the formation of ozone (smog), a serious health threat.

[56] In September of 2015, EPA commenced an enforcement action against Volkswagen for installing defeat devices on diesel vehicles sold in the United States, causing significant excess NO_x pollution. See "Volkswagen Clean Air Act Civil Settlement," EPA, https://www.epa.gov/enforcement/volkswagen-clean-air-act-civil-settlement. The case ultimately resulted in the resignation of the CEO, over $20 billion in payments by the company, and criminal convictions. See Hiroko Tabuchi, Jack Ewing and Matt Apuzzo, "6 Volkswagen Executives Charged as Company Pleads Guilty in Emissions Case," *New York Times*, January 11, 2017.

[57] Shaohua Hu et. al., "Development and Establishment of a Monitoring Network using Portable Emissions AcQuisition System to Quantify Heavy-Duty In-Use Vehicles Emissions in California," Presentation at Air Sensors International Conference, September 12–14, 2018, slide 11, https://asic.aqrc.ucdavis.edu/sites/g/files/dgvnsk3466/files/inline-files/Shaohua%20Hu%20-%20UPDATED%20-2018%20ASIC%20Conference_PEAQS_Hu%20S_Final_0.pdf (1.4% of trucks emitted 50% of the soot, and 3.9% of trucks emitted 50% of NO_x from trucks at one location in California). See also Chelsea V. Preble, Troy E. Cados, Robert A. Harley, and Thomas W. Kirchstetter, "In-Use Performance and Durability of Particle Filters on Heavy-Duty Diesel Trucks," *Environmental Science and Technology*, Vol. 52 (2018): 11913–21, https://pubs.acs.org/doi/10.1021/acs.est.8b02977.

[58] See "National Compliance Initiative: Stopping Aftermarket Defeat Devices for Vehicles and Engines," EPA, https://www.epa.gov/enforcement/national-compliance-initiative-stopping-aftermarket-defeat-devices-vehicles-and-engines. Software and hardware intended to change emissions control performance from its condition when new are referred to as "aftermarket defeat devices" (as contrasted with defeat devices that are built into the vehicles as originally sold, as happened with Volkswagen). See also Brooks, "Tampering and Aftermarket Defeat Devices," 8–10 (noting that

the rate of these serious violations is not known, the evidence so far shows that the problem is widespread.

Oil spill prevention. Oil spills into surface waters present a significant risk. Preventing spills and responding quickly to limit harm when they do occur is the mission of the Spill Prevention, Control, and Countermeasure regulations (SPCC). Over 460,000 facilities that store oil are regulated, including oil production facilities and farms. In a 2008 proposed revision to the SPCC rule, EPA admitted that it doesn't know the extent of noncompliance with this rule but believed that there might be zero (!) compliance with the SPCC rule in the farming community.[59]

For Many Programs, Compliance Is Unknown

In 2005, EPA's Office of Inspector General (OIG) did a review of universe size and compliance across the programs EPA is charged with implementing.[60] That report included EPA's best estimate that there were 41.1 million entities regulated through the programs established under federal environmental laws.[61]

The OIG found that the Office of Enforcement and Compliance Assurance (OECA) concentrates most of its compliance monitoring and enforcement activities on large facilities and knows little about the identities or cumulative pollution effects of smaller entities.[62] This book has already mentioned a long list of industrial sectors and rules that EPA is charged with administering, but there are many more. In every case, Congress directed EPA to adopt regulations to address risks to health and the environment. What follows is a partial list of additional EPA programs and the number of firms regulated under each. It isn't important to grasp the full list or understand each example and why it's important. Instead, the purpose is to show how extensive the total number of programs—and the

heavy-duty trucks with deleted emissions controls emit NO_x at over 300 times the allowable amount, and that 10% or more of trucks may have had emission controls deleted). EPA enforcement has also found large-scale sales of passenger vehicle aftermarket defeat devices. Brooks, 19.

59 EPA, "Regulatory Impact Analysis for the Final Amendments to the Oil Pollution Prevention Regulations (40 C.F.R. PART 112), Volume II—Technical Appendices," November 11, 2008, Appendix B, 17–18. https://www.regulations.gov/document/EPA-HQ-OPA-2007-0584-0172.

60 EPA OIG, "Limited Knowledge."

61 EPA OIG, "Limited Knowledge," 3, 10. There are 45 programs under nine statutes listed in the appendix of the OIG report. Note that I include here fewer than the total claimed in the report because some categories are self-evidently inappropriately listed as statutes. See EPA OIG, "Limited Knowledge," 22. The number of regulated entities is likely higher today. For example, of the six areas that the OIG focused on for detailed analysis in its report, the OIG found that between 2001 and 2005, the size of the regulated universe increased by 35%. EPA OIG, "Limited Knowledge," 6.

62 EPA OIG, "Limited Knowledge," 6, 14.

number of regulated firms—are for which EPA does not have reliable compliance information.

Additional examples not already touched on elsewhere include:[63]

- Over 110,000 minor and "synthetic minor" stationary sources of air pollution,[64]
- 580,000 firms regulated under the Emergency Planning and Community Right-to-Know Act (EPCRA), which requires industry to report on the storage, use, and releases of hazardous substances,[65]
- Over 3 million chemical facilities regulated under the so-called "core" Toxic Substances Control Act (TSCA), which regulates the manufacture, distribution, and use of chemicals,[66]
- More than 6 million establishments that have PCBs, which are regulated under TSCA to prevent the release of PCBs (compounds with both cancer and non-cancer health effects) into the environment,[67]
- More than 700,000 federal government, private nonresidential, and residential apartment buildings that contain friable asbestos and are subject to regulations under the Clean Air Act that protect against release of asbestos (a carcinogen) during demolition and renovation activities,[68] and
- About 320,000 renovators who do 18 million renovation projects a year in homes with lead paint that are subject to the Lead Renovation, Repair and Painting (RRP) Rule (designed to prevent exposure to dangerous lead contamination, particularly for children). EPA conducts about 1,130 targeted RRP inspections a year.[69]

[63] Superfund—the shorthand name for the law and program that cleans up contaminated sites in the United States—is the largest program that is not included in this book. That's not because Superfund isn't important but because it is not a regulatory compliance program. Superfund is a cleanup program for contaminated properties, with liability provisions designed to ensure that cleanups are funded not by the taxpayer but by the companies that created the problem. The regulatory program that governs how hazardous waste is treated today, with a goal of preventing the kind of contamination seen at Superfund sites, is the Resource Conservation and Recovery Act (RCRA). That's why RCRA is discussed in this book, but Superfund is not.

[64] EPA OIG, "Limited Knowledge," 23 (size of regulated universe) and 17 (lack of compliance information). "Synthetic minor source" means a source that has the potential to emit regulated pollutants in amounts that are at or above the thresholds for major sources but has agreed to an enforceable restriction so that its potential to emit is less than major source levels.

[65] EPA OIG, "Limited Knowledge," 24.

[66] EPA OIG, "Limited Knowledge," 24 (size of regulated universe) and 18 (lack of compliance information).

[67] EPA OIG, "Limited Knowledge," 24. PCBs are found in transformers, capacitors, and other electrical equipment, as well as oil used in motors and hydraulic systems, fluorescent light ballasts, and caulk. See "Learn about Polychlorinated Biphenyls (PCBs)," EPA, https://www.epa.gov/pcbs/learn-about-polychlorinated-biphenyls-pcbs.

[68] See "Asbestos Fact Book," EPA Office of Public Affairs (A-107), February 1985, at 4.

[69] See EPA OIG, "EPA Not Effectively Implementing the Lead-Based Paint Renovation, Repair and Painting Rule," Report No. 19-P-0302, September 9, 2019, at 2 (number of renovators and projects

Most of the examples in this book are programs for which the evidence strongly suggests serious violations are widespread, but the exact percentage of noncompliance isn't definitively known; in the fraction that represents the noncompliance "rate," EPA doesn't have the numerator. The preceding abbreviated list reminds us that for many programs EPA also doesn't know the denominator.[70]

For Some Important Programs, EPA's Understanding of Noncompliance Is Wrong

In addition to the many areas where EPA doesn't know how bad the noncompliance picture is, there is good reason to believe that some of what EPA thinks it knows is incorrect. The following are two examples.

Drinking water

Most people understand that compliance with standards to protect the safety of drinking water is vitally important. Exposure to contamination in drinking water can cause serious health problems, like acute health distress for infants from nitrates and waterborne disease outbreaks that affect millions in the United States each year.[71] Contaminants in drinking water, such as arsenic, lead, and disinfection byproducts, can also contribute to long-term chronic health problems, especially for children.[72]

There are about 150,000 regulated public drinking water systems in the United States. Approximately 50,000 of these are community water systems, responsible for providing safe drinking water to roughly 94 percent of the people living in the United States.[73]

subject to the rule), at 11 (average number of inspections per year, noting that inspections are at less than one-half percent of the estimated universe of renovators).

[70] EPA OIG, "Limited Knowledge."

[71] A 2006 study estimated there were between 4.3 million to 11.7 million annual cases of acute gastrointestinal illnesses in the United States attributable to drinking water from community drinking water systems. John M. Colford Jr. et al., "A Review of Household Drinking Water Intervention Trials and an Approach to the Estimation of Endemic Waterborne Gastroenteritis in the United States," *Journal of Water and Health*, Vol. 4, Suppl. 2 (2006): 71, cited in GAO, "Unreliable State Data Limit EPA's Ability to Target Enforcement Priorities and Communicate Water Systems' Performance," GAO-11-381, June 2011, at 5.

[72] GAO, "Unreliable State Data," 5–6.

[73] "Population Served by Community Water Systems with No Reported Violations of Health-Based Standards," EPA, Exhibit 1 (fiscal years 1993–2019), https://cfpub.epa.gov/roe/indicator.cfm?i=45 (percent of population served by community water systems); "Background on Drinking Water Standards in the Safe Drinking Water Act (SDWA)," EPA, https://www.epa.gov/dwstandardsregulations/background-drinking-water-standards-safe-drinking-water-act-sdwa (number of PWS and CWS in the United States). The remaining about 6% of the population are supplied by private drinking water wells, which are not regulated at the federal level. A word about nomenclature. EPA regulates public water systems. There are three types of public water systems: about 50,000

EPA's knowledge about systems' compliance with the drinking water standards is entirely dependent upon information from states. Drinking water systems are required to treat drinking water, test for signs of contamination, and provide that information to states. States are required to tell EPA about violations.[74] Using the state-reported data, EPA issues annual reports on the noncompliance record of the nation's drinking water systems. Based on information provided by the states, in 2016 EPA reported that 34 percent of public water systems had at least one violation, 8 percent violated health-based standards, and 26 percent violated monitoring requirements.[75]

Those numbers are troubling, but the actual number of violations is unfortunately much worse. Among other things, there are loopholes in the monitoring requirements, incentives to avoid admitting serious health-based violations, and huge gaps in the information the states provide to EPA.[76] The impact of loopholes and misaligned incentives is hard to quantify, but overwhelming evidence documents one thing: states are not telling EPA about all violations. Multiple assessments over many years have found the same thing. In audits of 38 states between 2002 and 2004, EPA found that states didn't report 38 percent of public systems' health-based violations and 71 percent of their monitoring and reporting violations.[77] A 2011 Government Accountability Office (GAO) review of community water system data from 14 states found that those states did not report 26 percent of health-based violations, and 84 percent of the monitoring violations.[78] Other reports by EPA and the EPA OIG had similar findings.[79] A recent National Academy of Sciences study concluded that an estimated 26 percent

community water systems (serving the same population year-round), about 85,000 transient noncommunity water systems (supplying water in transient locations like gas stations or campgrounds where people don't stay for long periods), and about 18,000 nontransient, noncommunity systems (supplying water to the same people at least six months a year but not all year, like schools or factories that have their own drinking water systems). EPA, "Background on Drinking Water Standards."

[74] 40 C.F.R. § 142.15(a)(1) (2011).

[75] See "Providing Safe Drinking Water in America: National Public Water Systems Compliance Report," EPA, https://www.epa.gov/compliance/providing-safe-drinking-water-america-national-public-water-systems-compliance-report (2016 National Snapshot).

[76] See chapter 1 for a discussion about the compliance problems for two particularly concerning drinking water contaminants: pathogens and lead. The structure of those rules contributes both to the violations and the failure to accurately report them.

[77] See EPA, "2006 Drinking Water Data Reliability Analysis and Action Plan," EPA 816-R-010, March 2008, at 18. Community water systems—the ones that supply drinking water to people's homes—had an even worse record in EPA's 2006 analysis; 49% of health-based violations by community systems were not reported. GAO, "Unreliable State Data," 14.

[78] GAO, "Unreliable State Data," Highlights Summary.

[79] See, e.g., GAO, "Unreliable State Data," 22–24 (describing prior EPA analyses of data reliability); EPA OIG, "EPA Claims to Meet Drinking Water Goals Despite Persistent Data Quality Shortcomings," March 5, 2004, at 4–6.

to 38 percent of health-based and 77 percent to 91 percent of monitoring and reporting violations were either not reported or inaccurately reported.[80]

Violations of health standards are of course deeply concerning, but monitoring violations can be just as serious. When a drinking water system doesn't monitor or monitors incorrectly, it can very easily miss contamination that causes health problems. A GAO review confirmed that conclusion, finding that monitoring violations were a strong and statistically significant predictor of health-based violations.[81] The real violation numbers revealed in repeated audits are therefore even more alarming than they appear; large numbers of additional violations with a direct impact on health are hiding in the extensive (and unreported) monitoring noncompliance.[82]

The net effect of these giant holes in reporting by states is that EPA's official record about drinking water system compliance dramatically undercounts violations. Somewhere between 25 percent and 50 percent of the health-based violations, and up to 90 percent of monitoring violations, are not counted in EPA's reports.[83] This dismal performance doesn't even include the additional violations obscured by monitoring loopholes and incentives to avoid discovering problems. The actual number of systems violating drinking water standards isn't known, but it is likely twice, or more, what is stated in EPA's public reports.[84]

[80] Maura Allaire, Haowei Wu, and Upmanu Lal, "National Trends in Drinking Water Violations," *Proceedings of the National Academy of Sciences*, Vol. 115, No. 9 (February 27, 2018): 2078, 2083 (health-based), and 2079 (monitoring and reporting). The PNAS study focused on Total Coliform Rule (TCR) violations because it described those as "more accurately reported than other types of violations." Allaire, "National Trends," 2079. That is an understatement. In a thorough 2000 data quality review, EPA found that TCR violations were reported to EPA 68% of the time (i.e., 32% were not reported). Allaire, 19, citing EPA, "Data Reliability Analysis of the EPA Safe Drinking Water Information System, Federal Version (SDWIS/Fed)," EPA 816-R-00-020, October 2000. For the other health-based standards underreporting was much worse: 85% of other Maximum Contaminant Level violations, and 93% of Surface Water Treatment Technique violations were not reported. EPA, "Data Reliability Analysis," 6.

[81] GAO, "Unreliable State Data," 16.

[82] GAO, "Unreliable State Data," 17. GAO also examined the effect of this inaccurate state reporting on EPA's enforcement prioritization system, which is designed to identify the most serious violators and ensure quick enforcement action to return violators to compliance. That prioritization system, called the EPA Drinking Water Enforcement Targeting Tool, looks at the state reported violation data for systems serving more than 10,000 people and gives each system a score based on the violations reported to EPA; if that score is above the cutoff level, it triggers an obligation for an enforcement response by the state or EPA. The GAO found that 73% of drinking water systems would have received a different score if EPA had known about the unreported violation data. GAO, "Unreliable State Data," 23–24.

[83] More recent data about state failure to report violations to EPA aren't presented here because there aren't any. EPA stopped doing data verification reports for the drinking water program in 2010. See GAO, "Unreliable State Data," 29. See also GAO, "Drinking Water; Additional Data and Statistical Analysis May Enhance EPA's Oversight of the Lead and Copper Rule," GAO-17-424, September 2017, at 37 (EPA reports that it has not conducted another data verification audit since they were discontinued in 2011).

[84] See EPA OIG, "EPA Claims," 8: "EPA has reported to Congress and the public that it met an important annual performance goal when available evidence indicates it did not."

Stationary sources of air pollution

Most of EPA's knowledge about violations at stationary sources of air pollution comes from states.[85] States do most of the inspections, and states receive the reports from facilities about their performance. States are supposed to identify high priority violations (HPVs) and enter that information into a database maintained by EPA. EPA then uses this information to confer with states about how to address serious violations within the time frames set out in national guidance.[86] The national average state-reported rate of significant violations for major air sources has been about 3 percent in recent years.[87] That seems like a fairly good performance record—certainly much better than the on average 20 percent to 25 percent significant violation rate self-reported by major Clean Water Act facilities over many years. But is it correct?

An OIG investigation into significant air violators in the late 1990s found that states failed to report the vast majority of serious violations to EPA. In Pennsylvania, where the investigation started, the state reported that only six of its about 2,000 major stationary sources were in significant violation—an incredible rate of less than one-third of 1 percent. When the OIG looked at just 270 state files—a small fraction of the total—it found another 64 facilities that should have been reported as having significant violations: a rate of 24 percent.[88] The OIG then expanded its investigation to five additional states and found that the same failure to report was widespread. The actual rate of significant violation was about 25 percent for those states too.[89] Other states not included in the OIG investigation showed similar worrying data. Ohio reported that over two years only four of its 1,700 major sources were significant violators. New York, with 2,300 major sources, reported zero significant violations.[90] All 10 EPA regions told the OIG that states were underreporting significant violators.[91] One state candidly admitted to the OIG that it didn't list significant violators because it did

[85] Stationary sources are distinguished from mobile sources of air pollution, such as cars and trucks, which are also regulated under the Clean Air Act.

[86] In 2014, EPA changed the High Priority Violator policy, effective in fiscal year 2015. The guidance about addressing serious violators is typically called "timely and appropriate" guidance because it sets standards for the speed and manner in which serious violations are addressed. See US EPA, "Revised Timely and Appropriate (T and A) Enforcement Response to High Priority Violations (HPVs) Policy," August 25, 2014, https://www.epa.gov/enforcement/revised-timely-and-appropriate-t-and-enforcement-response-high-priority-violations-hpvs.

[87] See EPA Enforcement and Compliance History Online (ECHO), https://echo.epa.gov/ (select topic: Analyze Trends: State Air Dashboard, select Classification: major, Box 4 (High Priority Violations, select % Facilities (Majors) with HPVs) (data 2015 through 2021).

[88] See "Validation of Air Enforcement Data Reported to EPA by Pennsylvania," EPA OIG Region 3, February 14, 1997, at 11–12.

[89] See "Consolidated Report on OECA's Oversight of Regional and State Air Enforcement Programs," EPA OIG, September 25, 1998, at 7–10.

[90] EPA OIG, "Consolidated Report," 7–10.

[91] EPA OIG, "Consolidated Report," 3, 8–10.

not want EPA involved in the resolution of a violation, saying that EPA's involvement "delayed the process."[92]

Twenty years have passed since the EPA OIG documented that over 85 percent of significant violations by major air sources went unreported,[93] but the problem persists. For 2014, the last year before EPA changed the HPV definition, 14 states reported zero high-priority violators[94] and an additional four states reported high priority violation rates for major sources of less than 1 percent.[95] In 2014, these 18 states together reported only six HPVs for their collective 2,484 major air sources.[96] That's a serious violation rate for large air pollution emitters in those 18 states of just 0.24 percent.[97] EPA's data says that nationwide in 2014 for the roughly 14,000 largest air pollution sources in the country, the state-reported high priority violator rate was an incredible 3 percent.[98] It has stayed close to the 3 percent mark for every year since.[99]

To judge how suspiciously low this state-reported serious air violator information is, compare it to what water dischargers reported for the same year. Major water pollution dischargers, unlike most major air emitters, report their actual pollution levels directly to both EPA and states, so there is data to evaluate claims of compliance. The self-reported rate of significant violation by the largest water pollution dischargers in 2014 was 22 percent.[100] The comparison between air and water polluters buttresses the OIG's prior conclusion and shows how improbable it is that major air pollution sources—under regulations that are much more complex than those that apply to water dischargers—had the low rates of serious violation that states report.

The underreporting found by the OIG was a result of states not notifying EPA of the significant violators that the states had identified. An additional problem is that often states themselves don't know about serious violations. Of the state files reviewed by the OIG, 35 percent either failed to conduct the required tests

[92] EPA OIG, "Consolidated Report," 10–11.

[93] See EPA OIG, "Consolidated Report," 8 (the percentage unreported is the total number of unreported violations the IG discovered as a percentage of all violations).

[94] The policy for tracking more serious violations—called High Priority Violations—was revised in 2014. HPVs are a subset of all violations and intended to focus attention on the most important problems.

[95] The 2014 data presented in this paragraph was obtained from EPA's public ECHO data system in November 2019. That data is no longer available on the public website, but an excel spreadsheet presenting that data by state is on file with the author ("2014 EPA HPV data").

[96] 2014 EPA HPV data.

[97] Twelve additional states reported HPV rates below 2%; the 30 states with claimed HPV rates between 0% and 2% reported just 51 majors as HPVs, a collective HPV rate of just 0.9%. 2014 EPA HPV data.

[98] 2014 EPA HPV data.

[99] EPA, ECHO (select topic: Analyze Trends: State Air Dashboard, select Classification: major, Box 4 (High Priority Violations, select % Facilities (Majors) with HPVs) (data 2015 through 2021).

[100] See the sources cited *supra* in notes 14 and 15, and the explanation for the difference between historic data and what is displayed on EPA's public ECHO data site today.

or failed to document the inspection sufficiently for the OIG to determine if the proper inspection was conducted.[101] So not only were states failing to tell EPA about detected violations, but some were also failing to conduct inspections of sufficient rigor to find out which facilities were violating. Adding to this problem is the fact that some kinds of serious violations cannot be discovered through the kinds of inspections that states normally do, even if they are properly done.[102]

Not only are states underreporting what they know are serious air violators, facilities are not accurately reporting their emissions to states, so facilities are not flagged as violators in the first place. Part of the reason is companies' use of "emission factors" to estimate their air pollution releases. Emission factors are long-term, industry-wide averages of air pollution from a source or process.[103] They were never intended to predict actual emissions at an individual location.[104] By definition, even if the emission factors were perfect, as many as half of the facilities would emit more.[105] And they are far from perfect. EPA itself identifies 62 percent of its emission factors as "below average" or "poor."[106]

Not surprisingly, field investigations frequently uncover actual emissions that are substantially higher than emission estimates. Time and again, monitoring data have revealed that estimated pollution levels significantly underreport actual pollution amounts, sometimes by an order of magnitude or more.[107] At

[101] EPA OIG, "Consolidated Report," 14 (faulty inspections as a percentage of total inspection files reviewed).

[102] GAO, "EPA Should Improve Oversight of Emissions Reporting by Large Facilities," GAO-01-46, April 2001, at 10.

[103] EPA, "AP-42, Fifth Edition Compilation of Air Pollutant Emissions Factors, Volume 1: Stationary Point and Area Sources," January 1995, Introduction, 1, https://www.epa.gov/air-emissions-factors-and-quantification/ap-42-compilation-air-emissions-factors#5thed: "In most cases, these factors are simply averages of all available data of acceptable quality and are generally *assumed* to be representative of long-term averages for all facilities in the source category (i.e., a population average)" (emphasis added). See also "Basic Information of Air Emissions Factors and Quantification," EPA, https://www.epa.gov/air-emissions-factors-and-quantification/basic-information-air-emissions-factors-and-quantification.

[104] See Rachel Leven, "Most of EPA's Pollution Estimates Are Unreliable. So Why Is Everyone Still Using Them?," *Center for Public Integrity*, January 29, 2018, https://publicintegrity.org/environment/most-of-the-epas-pollution-estimates-are-unreliable-so-why-is-everyone-still-using-them/. See also EPA, "Compilation of Air Pollutant Emissions Factors," Introduction, 2: "Use of these factors as source specific permit limits and/or as emission regulation compliance determinations is not recommended by EPA."

[105] EPA, "Compilation of Air Pollutant Emissions Factors," Introduction, 2: "Because emission factors essentially represent an average of a range of emission rates, approximately half of the subject sources will have emission rates greater than the emission factor."

[106] See Leven, "Pollution Estimates are Unreliable"; Associated Press, "Emissions Often Underestimated, EPA Standards Old," *Cleveland.com*, April 22, 2010, https://www.cleveland.com/business/index.ssf/2010/04/emissions_often_underestimated.html. See also EPA OIG, "EPA Can Improve Emissions Factor Development and Management," March 22, 2006, at 8.

[107] See, e.g., Daniel W. Hoyt and Loren H. Raun, "Measured and Estimated Benzene and Volatile Organic Carbon (VOC) Emissions at a Major U.S. Refinery/Chemical Plant: Comparison and Prioritization," *Journal of Air and Waste Management*, Vol. 65 (July 25, 2015): 1020, https://doi.org/10.1080/10962247.2015.1058304; GAO, "EPA Should," 12. See also Leven, "Pollution Estimates Are

refineries, for example, actual emissions have been discovered to be four times, 25 times, 132 times, and even 448 times the estimated amount.[108] EPA recently issued an enforcement alert warning companies that AP-42 emission factors are not accurate predictors of emissions from individual facilities and that use of those factors for emissions reporting can result in unlawful underestimation of emissions.[109]

Underreporting emissions by estimating is a big issue; EPA projected in 2001 that about 80 percent of facilities used emission factors for their emissions reporting.[110] The title of one 2018 investigative report says it all: "Most of the EPA's Pollution Estimates are Unreliable, so Why is Everyone Still Using Them?"[111]

Some kinds of direct monitoring aren't necessarily much more reliable. A recent EPA OIG report found errors in over half of the stack test reports it reviewed in one state.[112] In addition, over 80 percent of the stack test reports lacked key

Unreliable"; Ann E. Carlson, "The Clean Air Act's Blind Spot: Microclimates and Hotspot Pollution," *University of California, Los Angeles Law Review*, Vol. 65 (2018): 1036, 1041, 1059.

[108] See GAO, "EPA Should," 12 (fugitive emission leaks were four times estimates); David Hindin (EPA Office of Compliance), "The Future of Environmental Monitoring: Making the Invisible Visible," Presentation at National Environmental Monitoring Conference, August 2014, at 14 (flare emissions at Marathon were 25 times estimate), https://nemc.us/docs/2014/Presentations/Wed-Plenary-26.1-Hindin.pdf; Loren H. Raun and Dan W. Hoyt, "Measurement and Analysis of Benzene and VOC Emissions in the Houston Ship Channel Area and Selected Surrounding Major Stationary Sources Using DIAL (Differential Absorption Light Detection and Ranging) Technology To Support Ambient HAP Concentrations Reductions in the Community (DIAL Project)," City of Houston, Bureau of Pollution Control & Prevention, 2011, at 1, 92, http://www.greenhoustontx.gov/reports/dial20110720.pdf (true emissions underestimated by a factor of as much as 93 for benzene and 132 for VOCs); Hoyt, "Measured and Estimated Benzene" (floating tank emissions 448 times estimate). Note that EPA has updated some refinery emissions estimates. See Emissions Estimation Protocol for Petroleum Refineries, EPA, April 2015, https://www3.epa.gov/ttn/chief/efpac/protocol/Protocol%20Report%202015.pdf.

[109] EPA, "EPA Reminder About Inappropriate Use of AP-42 Emission Factors," Enforcement Alert, November 2020, https://www.epa.gov/compliance/epa-reminder-about-inappropriate-use-ap-42-emission-factors.

[110] GAO, "EPA Should," 14 (citing EPA as saying that in 2001 only 4% of reporting facilities used direct measurement, and about 80% used emission factors). See also EPA OIG, "EPA Can Improve," 4, 8, and 10 (three industries undercontrolled as a result of emission estimates' understating actual emissions); Environmental Integrity Project, "Toxic Shell Game," March 26, 2018 (reasons why emission factors understate actual emissions), 6, https://www.environmentalintegrity.org/wp-content/uploads/2017/02/Toxic-Shell-Game.pdf.

[111] Leven, "Pollution Estimates are Unreliable." See also David Hasemyer, "EPA Agrees Its Emission Estimates from Flaring May Be Flawed," *Inside Climate News*, October 13, 2016, https://insideclimatenews.org/news/12102016/epa-natural-gas-oil-drilling-flaring-emissions-estimates-flawed-fracking. See also Lisa Song, "They Knew Industrial Pollution Was Ruining the Neighborhood's Air. If Only Regulators Had Listened," *ProPublica*, November 29, 2021, https://www.propublica.org/article/they-knew-industrial-pollution-was-ruining-the-neighborhoods-air-if-only-regulators-had-listened.

[112] EPA OIG, "More Effective EPA Oversight Is Needed for Particulate Matter Emissions Compliance Testing," Report No. 19-P-0251, July 2019, at 11. Stack tests are measurements of air pollution done at the stack—the chimneys or smokestacks located at industrial facilities. The tests are done by companies selected and paid by the polluting facility and can take days to complete.

data necessary to evaluate the reliability of the results.[113] EPA admits that the problems the OIG found were not limited to one state or region.[114] Because stack tests can be as infrequent as once every five years, a mistake means that unlawful pollution can go unnoticed for years.[115]

When you put all this data together, it is obvious that the official national report substantially understates the extent of serious air violations. The last thorough look concluded that states were informing EPA about less than 15 percent of the significant violations—and that's just for violations the states knew about. Incorrect emissions reporting accounts for untold additional violations. State reporting practices have not appreciably changed. Nor have the political dynamics, which discourage states from revealing violators to EPA. It would be nice if the program with the biggest public health impacts, and also the most complex regulatory requirements, had by far the best compliance record. Unfortunately, the evidence shows that isn't credible.

The Challenges of Federalism

In the preceding examples of unreliable data—drinking water and large stationary air sources—there is a common theme: states not informing EPA about violations. For all the reasons discussed previously, states frequently don't know when there is a violation. But repeated audits show that states are often not informing EPA of known violations, despite the obligation to do so. Why? A central factor is that states don't want the scrutiny—from EPA or the public—that comes with raising their hands. If EPA knows about serious violators, it might insist that the matter be addressed more quickly or more aggressively. If the public knows how extensive the violations really are, they are likely to be upset and put even more pressure on the overburdened and underresourced states. There are many other factors too, like antiquated information technology (IT) systems and too-confusing rules. Fortunately, Next Gen offers the opportunity to bypass this historic gulf, which I will explain in chapter 10.

[113] EPA OIG, "More Effective EPA Oversight," 15. Twenty-five of thirty stack test reports reviewed were missing at least one element of calibration information; EPA's training says that without calibration, stack test results are meaningless. EPA OIG, "More Effective EPA Oversight," 15.

[114] EPA OIG, "More Effective EPA Oversight," 12.

[115] EPA OIG, "More Effective EPA Oversight," 11. Continuous emission monitoring systems (CEMS), on the other hand, have a quite good record for reliability. See EPA OIG, "EPA Effectively Screens Air Emissions Data from Continuous Monitoring Systems but Could Enhance Verification of System Performance," EPA Report No. 19-P-0207, June 2019.

How the Data Add Up

The evidence presented here shows that violations are common. When we narrow the focus to just the most serious violations, we find noncompliance rates of 25 percent or more. That's true even in programs that have had persistent and focused attention for decades. Rates of serious violation that are much higher—up to 70 percent or more—occur far too frequently.

EPA has promulgated regulations intended to improve our air and water and to reduce our risk from hazardous pollutants. But all these serious violations reveal the large gap between the goals of those rules and the situation on the ground.

As I mentioned in the introduction to this book, observing that many companies do not comply is not a moral statement. Trying to make it one distracts from the central point. There certainly are companies that are reckless or criminal, and our rules need to make that irresponsible conduct harder to do and easier to detect. But many companies don't decide to violate, they just don't make compliance a priority and so fall short. The people who bear the brunt of the violations don't care about the reasons. They just want it to stop.

That's the goal of Next Gen too. The point isn't to pass judgment. It's to make the rules work. Once we accept that violations happen all the time under the traditional model, we can put our effort into designing rules to make that far less likely.

Next Gen is a paradigm shift. It presents a way to dramatically improve compliance, and thereby reduce risks to health, but it requires letting go of the fiction that most companies comply. Policy makers' guesstimate that only about 5 percent to 10 percent of facilities violate is wrong. Serious violations are widespread and happen in companies of all sizes and all sectors and all programs.

Dislodging the belief that most companies comply is not easy. It has been the accepted wisdom for so long that people who have that view are not aware that the evidence doesn't support it. Summary statements of the facts meet skepticism. That's why extensive recitation of the evidence is presented in this chapter.

Enforcement will continue to play an essential role in boosting compliance. Many of our nation's most important environmental advances have depended on enforcement, and that will continue to be the case. Some of the alarming noncompliance problems discussed in this book have been the focus of consistent enforcement effort that has helped to turn the tide, however expensive and time consuming—and avoidable—that may have been.

But even the most committed and smart enforcers cannot achieve the impossible. A handful of enforcers at EPA and the states can't force compliance on millions of regulated entities. We will always need civil and criminal enforcement. Enforcement will always be central to the environmental protection mission. But the most important thing we can do to get better compliance is write rules with compliance built in. Give the enforcers a fighting chance by improving compliance out of the gate. Here's what we would all like: rules for which compliance is pretty good even if enforcement never comes knocking.

Environmental laws in the United States have brought us a long way. The traditional paradigm was the basis for significant progress, but that paradigm is getting in the way now. The beliefs that most companies comply, and that enforcement can take care of the rest, cannot be squared with the facts. Continuing to believe that will make it impossible to deliver on the promises that Congress made 50 years ago and the urgent problems we face today. When we look the facts in the eye and acknowledge that we need a change, it opens the door to solutions that will work. That's what the rest of this book is about.

3

Rules about Rules

Agencies don't write regulations when the fancy strikes them. They do so under directions from Congress, and subject to a host of executive orders, government-wide guidance, and agency requirements. Regulations are also subject to review in court. In this book I argue that our approach to environmental regulations needs to change. Understanding that argument, and how it would work in practice, requires some familiarity with agency regulatory authority—the rules about rules. This chapter provides that very quick overview.

Every subject touched on here in a few paragraphs is complicated and nuanced. Scores of books, probably thousands of articles, and countless academic, legal, policy, and advocacy conferences—not to mention briefs filed in court—have explored these topics, which are subject to fierce debate and political maneuvering. Scholars and practitioners would be horrified by the simplistic rendition offered here. But my purpose is not to present the complexities of these topics. It is to show how many hoops—both visible to the public and not—EPA has to jump through to promulgate a regulation. It describes the playing field for EPA's rules to provide necessary context for the changes I propose in the rest of the book.

Agency Authority to Regulate Comes from Congress

Agencies' authority to write regulations depends entirely on what Congress gives them.[1] That authority is contained in laws passed by Congress that delegate authority to the agency. For EPA, those are laws about clean air and water, hazardous waste, clean drinking water, safe chemicals, and a host of other subjects. Such so-called "enabling legislation" contains both the direction to regulate and the boundaries on the agency's authority. If a regulation strays from that Congressional directive, a court can invalidate some or all of the rule.[2] That's why

[1] Environmental Law Institute, "Environment 2021: What Comes Next?," *Environmental Law Institute* (July 2020): 4, https://www.eli.org/research-report/environment-2021-what-comes-next. This ELI report contains a summary of the rules about rules covered in this chapter.

[2] ELI, "What Comes Next," 6; 5 U.S.C. § 706(2).

Next Generation Compliance. Cynthia Giles, Oxford University Press. © Cynthia Giles 2022.
DOI: 10.1093/oso/9780197656747.003.0004

you sometimes see Supreme Court cases about EPA regulations that spill a lot of ink about the meaning of a few words in the statute under which the rules were written. EPA can only promulgate regulations within the confines of the authority conferred by Congress.

Sometimes Congress relies on EPA's expertise to design the best approach to achieve the goals Congress has set. Other laws provide quite specific instructions about the content of agency rules. There is room for Next Gen ideas in regulations under every statute, but some laws allow a wider band of potential options than others. The key takeaway is every regulation can include Next Gen provisions, but, as with every aspect of regulations, they have to fit within the limits set by Congress.

The Administrative Procedure Act Sets the Process Boundaries

Through enabling laws Congress tells agencies what to do. Another statute—the Administrative Procedure Act (APA)—tells agencies how to do it. The APA sets out the broad contours of the process agencies must follow in writing regulations.[3] The part of the APA that is relevant for the rules that are the subject of this book is Section 553.[4] It's only 357 words, but it packs a big punch. It says that notice of proposed rules has to be published in the Federal Register, and that people must be allowed to submit their views on proposed rules. Hence the common term "notice and comment rulemaking." Section 553 requires that the notice state the terms or substance of the proposed rule and include in the final rule a concise statement of the rule's basis and purpose.

Regulations covered by APA Section 553 are considered "informal" rulemaking, although anyone who has been involved with such rules may marvel at the word informal being applied to this process. Over the years, a body of law and practice has grown up around Congress's regulatory process mandates. Some of that law is about what constitutes reasonable notice and meaningful opportunity to comment. So that people have a fair chance to know what they are commenting on, agencies have to provide notice of data or studies on which

[3] 5 U.S.C. § 551 et seq.; ELI, "What Comes Next," 4; Jack M. Beerman, *Insde Administrative Law* (Wolters Kluwer, 2011)199–209.

[4] 5 U.S.C. § 553. The APA rulemaking requirements are described as accommodating three influences in agency rulemaking: openness and democracy, agency expertise, and political involvement in agency decision-making. Beerman, *Inside Administrative Law*, 201.

the agency relies, and the final rule has to be reasonably related to what was proposed.[5]

The APA also calls for a "concise general statement" of the rule's basis and purpose. The final regulation and its explanatory preamble can run hundreds of pages in the Federal Register, and there are often accompanying documents that are hundreds of pages more. That length is in response to court decisions requiring that the agency explain the basis for its decisions and respond to substantial comments. Many EPA rules generate significant interest; some proposed rules draw hundreds of thousands, and some even millions, of public comments.[6] So EPA often goes into quite a lot of depth in explaining how it made decisions and what changed or didn't in response to comments.

It does little good to promulgate a regulation that is overturned because the agency failed to follow the law. That's why EPA—at least in most administrations—is scrupulous in making sure that it adheres to both the enabling statute and the APA. If Congress didn't authorize it, or it can't be done under the process rules, then the answer is no.[7]

Notice and comment is an obligation, but it is also a benefit for rule writers. Hearing from the affected industry, other knowledgeable experts, advocacy groups, and the affected communities can help refine ideas and uncover valuable information. It's not window dressing; comments inform and improve regulations.

The lesson of the APA for Next Gen is that major structural ideas for a regulation have to be included early, so that there is meaningful notice and an opportunity to comment. That's best practice totally apart from the APA, for reasons explained in the next chapter. Good policy—rules designed for strong implementation—thus aligns perfectly with the legal obligations of the APA. By the same token, if Next Gen isn't built into a rule during the early design, the APA will be an additional barrier to adding it later.

[5] Beerman, *Inside Administrative Law*, 202–09. Some changes to the proposal are expected of course. That's the purpose of taking public comment. If the final rule is a "logical outgrowth" of the proposal, that's fine under the APA. Beerman, *Inside Administrative Law*, 204–07.

[6] See, e.g., the Mercury and Air Toxics Standards (more than 900,000 public comments), "EPA Fact Sheet, Mercury and Air Toxics Standards," https://www.epa.gov/sites/default/files/2015-11/documents/20111221matsadjustmentsfs.pdf; Clean Power plan (4.3 million public comments), "Fact Sheet: Clean Power Plan by the Numbers," https://archive.epa.gov/epa/cleanpowerplan/fact-sheet-clean-power-plan-numbers.html.

[7] Lack of interest in following the rules about rules is one reason that the Trump EPA's regulations have had such a sorry record in the courts. Cary Coglianese and Daniel E. Walters, "Litigating EPA Rules: A Fifty-Year Retrospective of Environmental Rulemaking in the Courts," *Case Western Reserve Law Review*, Vol. 70, No. 4 (2020): 1033.

Presidential Executive Orders and Directives from the Office of Management and Budget Control Regulatory Development

It isn't just Congress that tells the agencies what to do. The president does too. Executive orders (EOs) are the principal vehicle for binding instructions from the president about regulations. The Office of Management and Budget (OMB) is the president's enforcer for these presidential mandates.

Executive Order 12866, issued by President Clinton, lays out the principal obligations for agencies in developing regulations.[8] That order builds on the centralized review of agency regulations that was started by President Reagan.[9] Presidents have tinkered with the system, but the basic frame remains the same: all significant regulations, including an analysis of the potential costs and benefits, must go to OMB for review before they can be published. The specific office within OMB that conducts these reviews is the Office of Information and Regulatory Affairs (OIRA). OIRA is the gatekeeper. As anyone who has encountered a gatekeeper knows, that's a powerful position. If OIRA wants changes to the rule, it is usually in a position to press the point, because without OIRA sign-off, the rule isn't going anywhere.

OIRA's written guidance about benefit-cost analysis goes by the catchy name Circular A-4.[10] The basic idea of this requirement is to promote the seemingly noncontroversial idea that regulations should only be promulgated when the benefits justify the costs. Hard to argue with that. But what counts as costs and benefits and how should they be weighed? There's the rub.

As will become more evident in the next chapter, one of the key challenges of such benefit-cost calculation (or cost-benefit, same thing) is its heavy focus on quantification. At the heart of this analysis, as the name implies, is the comparison between costs and benefits. After all, the overall goal is regulations where benefits are higher than costs. To allow comparison, benefit-cost analysis requires first that costs and benefits be identified and quantified. Then those

[8] Executive Order 12866 of September 30, 1993, "Regulatory Planning and Review," *Federal Register*, Vol. 58, No. 190 (October 4, 1993), https://www.reginfo.gov/public/jsp/Utilities/EO_12866.pdf, as amended Executive Order 13563 of January 18, 2011, "Improving Regulation and Regulatory Review," *Federal Register*, Vol. 76, No. 14 (January 21, 2011): 3821, https://www.reginfo.gov/public/jsp/Utilities/EO_13563.pdf. Daniel A. Farber, Lisa Heinzerling, and Peter M. Shane, "Reforming Regulatory Reform: A Progressive Framework for Agency Rulemaking in the Public Interest," *Advance: The Journal of the ACS Issue Briefs*, Vol. 12 (2018): 11; Beerman, *Administrative Law*, 162–63.

[9] Farber, "Reforming Regulatory Reform," 11; Lisa Heinzerling, "Inside EPA: A Former Insider's Reflections on the Relationship between the Obama EPA and the Obama White House," *Pace Environmental Law Review*, Vol. 31, No. 1 (Winter 2014): 325–69.

[10] Office of Management and Budget, "Circular A-4," September 17, 2003, https://www.reginfo.gov/public/jsp/Utilities/a-4.pdf.

costs and benefits have to be converted to a common metric—dollars—so they can be compared. Often the costs are relatively easy to quantify in dollars. For EPA rules, costs are just what they sound like, usually what industry would have to spend to implement the actions proposed in the rule. There is a lot to disagree about, and understanding the full range of costs isn't simple, but usually EPA can figure out what information it needs and how to get it.[11] Most costs naturally come in dollars, so there isn't a huge amount of translation necessary. Benefits are another story. The effect of reduced pollution on human health or the benefits of cleaner water for natural ecosystems can be much harder to figure out, and it can be exceedingly difficult to translate identified benefits into dollars.

Proponents of benefit-cost analysis argue that reasoned presentation of the costs and benefits of any proposed course of action leads to much better, and better informed, rules. Detractors suggest that the mismatch between the ease of identifying costs and the challenges of quantifying benefits create an anti-regulatory bias.[12] This persistent debate is part of the reason for the executive memorandum issued by President Biden, directing OMB to revisit Circular A-4 and take a fresh look at OMB's standards for regulatory action.[13]

EPA Has Its Own Guidance on Writing Rules

EPA has agency-specific guidance for its own staff to implement the many requirements of federal law and executive directives as the agency develops regulations. The agency's internal practice is defined in two guidance documents: one describing the internal process steps required to move from congressional instruction to final regulation and the other directing how benefit-cost analysis should be done. Both adopt the prevailing—and wrong—articles of faith that compliance with rules is generally good and it is up to enforcement to ensure that it is. As such, these EPA policies reinforce the barriers to Next

[11] One of the main places to get information about costs is the regulated industry itself. That introduces some bias, because companies have reasons to object to new rules and to claim they will cost too much. That, among other reasons, is why costs in rules are often overstated. Nathaniel O. Keohane, "The Technocratic and Democratic Functions of the CAIR Regulatory Analysis," in Winston Harrington, Lisa Heinzerling, and Richard D. Morgenstern eds., *Reforming Regulatory Impact Analysis* (Resources for the Future, April 2009), 37; Elizabeth Kopits et al., "Retrospective Cost Analyses of EPA Regulations: A Case Study Approach," *Journal of Benefit Cost Analysis*, Vol. 5, No. 2 (2014): 176–80.

[12] Harrington, *Reforming Regulatory Impact Analysis*, 8; Farber, "Reforming Regulatory Reform," 12; Heinzerling, "Inside EPA," 329; David M. Driesen, "Is Cost-Benefit Analysis Neutral?," *University of Colorado Law Review*, Vol. 77, No. 2 (2006): 335–404.

[13] Presidential Memorandum, "Modernizing Regulatory Review" (January 20, 2021), https://www.whitehouse.gov/briefing-room/presidential-actions/2021/01/20/modernizing-regulatory-review/; Richard L. Revesz, "A New Era for Regulatory Review," *The Regulatory Review*, February 16, 2021, https://www.theregreview.org/2021/02/16/revesz-new-era-regulatory-review/.

Gen. Why that happens, and how to fix it, is the topic of the next chapter. In the following I describe them very briefly just to round out the picture of the rules about rules.

EPA's Action Development Process

It won't surprise you to learn that EPA doesn't make up the process steps anew every time it writes a regulation. EPA wants to be consistent, provide opportunities for management direction at key points, and ensure that the process is efficient, predictable, and compliant with all the applicable requirements. The system for achieving that is called the Action Development Process, or ADP.[14]

EPA regulations vary widely in scope, complexity, and cost. Rules that get the most scrutiny are rules that are very complicated, have significant costs or benefits, affect more than one agency program, or attract a lot of public interest. Routine and noncontroversial regulations get less. For higher profile rules (labeled Tier 1 and Tier 2 under the ADP) EPA forms an intra-agency workgroup, consisting of the lead program office and representatives from the other interested offices, which is charged to move through the four major steps in the process: early guidance (preliminary take on options to consider and data to be collected), analytic blueprint (workgroup plan for conducting the analysis), options selection (choice of preferred option), and final agency review (last chance look at the written final action).

All four steps are taken to issue a proposed rule. After public comment is received on the proposal, EPA reviews the comments and starts the same four steps over again to work toward a final rule.

The essential point from a Next Gen perspective is that the possibilities for significant changes go down as the process moves forward. Considerable time, effort, and money are put into each step, and the options under consideration narrow as the rule moves toward final agency review. It is never well received, or effective, to raise a new issue or problem toward the later stages of the process. Therefore, any foundational issues—like Next Gen structural design changes necessary to ensure effective implementation—need to be flagged at the initial stages. As is true for the APA, this structure means that the earliest choices in rule design matter most.

[14] EPA, "EPA's Action Development Process: Guidance for EPA Staff on Developing Quality Actions," EPA Office of Policy (Revised: March 2011), https://yosemite.epa.gov/sab/sabproduct.nsf/5088B3878A90053E8525788E005EC8D8/$File/adp03-00-11.pdf. At EPA, the ADP applies to more than just regulations, which is why it's called the action, not the regulation, development process.

EPA's Economic Analysis Guidelines

Every agency is charged with taking OMB's Circular A-4—the guidance on doing benefit-cost analysis—and applying it to their own agency's work. At EPA, that's done through the Guidelines for Preparing Economic Analyses ("Guidelines).[15] The Guidelines contain lots of detail about how to figure out the costs and benefits of a rule. The current version of the Guidelines is 302 pages long, with lots of charts, math, and discussion of equilibrium models. Almost no one who isn't an economist reads the Guidelines. Including inside of EPA. So the fine print of those guidelines isn't top of mind even for rule writers.

Nevertheless, this internal EPA instruction on benefit-cost analysis illustrates why Next Gen has been a steep uphill climb: the Guidelines enshrine in written agency policy the incorrect beliefs that compliance may be assumed, and it is up to enforcement to make sure it is good. Benefit-cost analysis is central to the regulatory narrative as a result of the EOs and the OMB Guidance described earlier. And as currently practiced it is also a significant barrier to Next Gen. That doesn't have to be true: a robust benefit-cost analysis could be consistent with strong implementation design. But as applied today, it isn't. Because the Guidelines inject the false compliance assumptions into the benefit-cost math, they stack the deck against better-designed rules, as the next chapter discusses in more depth.

Who Gets to Say? Judicial Review of Agency Regulations

The rules about rules, from legislation through presidential directives to internal agency guidance, provide the structure within which rule writers operate. As should be very apparent by now, it is far from the wide-open field that some anti-regulatory advocates claim. An additional constraint comes from judicial review.

A final agency rule can be contested in federal court under the APA on the grounds that it is "arbitrary and capricious, an abuse of discretion or otherwise not in accordance with the law."[16] There are many caveats and details that are important, but what's relevant here is that every rule is potentially the subject of litigation. The regulated entities might sue saying the rule is too stringent, environmental advocates and community groups might sue claiming it isn't stringent

[15] EPA, "Guidelines for Preparing Economic Analyses," EPA National Center for Environmental Economics, December 17, 2010 (updated May 2014), https://www.epa.gov/environmental-economics/guidelines-preparing-economic-analyses. Proposed revisions to the 2010 Guidelines were published in 2020, but no changes have been adopted yet.

[16] 5 U.S.C. § 706(2)(A); ELI, *What Comes Next*, 6; Beerman, *Administrative Law*, 109–12.

enough. Often both do, with states also suing for good measure. Lawsuits about regulations have become a permanent feature of EPA rule-writing.[17]

The fact that a regulation can, and if it is a significant rule probably will, be challenged in court has a powerful impact on rule writers. It urges them to be conservative in assumptions and regulatory choices to reduce the chances that a regulation fails judicial review. It can take years to develop, propose, take comment on, then finalize a rule. No agency wants to follow that with years of litigation, only to be told by a court to go back to the drawing board. Therefore, rule writers are usually interested in making rules as defensible as possible.[18]

Next Gen can align with a defensively oriented posture about rules. After all, its main purpose is making sure rules achieve the purpose that Congress directed. However, sometimes reaching for a better implementation result requires an innovative regulatory strategy. That's not a problem unless the agency allows it to be. EPA just has to carefully document the basis for its decisions, something it has a great deal of practice doing.

Don't Believe Everything You Read

The very brief description of the context for regulations in this chapter is mainly about what's written down. What the laws, executive orders and guidance say. But that's not the whole story. For laws passed by Congress, judicial interpretation can vary by court and also over time. And for the directives that are entirely internal, as the saying goes, your mileage may vary.

For example, the executive orders on regulatory review state that only certain economically significant rules are subject to the benefit-cost obligation and OIRA review. In fact, OIRA does what it likes. These facts of life are revealed in Lisa Heinzerling's eye-opening, behind-the-scenes analysis of EPA's interaction with OIRA.[19] There is little brake on OIRA's insistence on reviewing additional rules, demanding changes, or interposing lengthy delays, because without

[17] Coglianese, "Litigating EPA Rules," 1025. While lawsuits challenging major EPA rules are common—over 50% of the most significant EPA rules were contested in court—the common perception of nearly universal challenge to EPA rules is overblown. Coglianese, 1012, 1015, 1021. However, EPA rules are a more frequent entry in the judicial docket than are challenges for other federal agencies. Coglianese, 1018.

[18] A 50-year retrospective on judicial challenges to EPA rules finds that EPA has a strong record of withstanding judicial scrutiny through professional analysis and internal management processes but noting that the Trump EPA was a notable departure from that otherwise successful judicial record. Coglianese, 1010, 1030.

[19] Heinzerling, "Inside EPA." For example, Prof. Heinzerling notes that some 80% of the Obama EPA rules reviewed by OIRA as of 2013 were not economically significant. Heinzerling, 347.

OIRA's sign-off, that rule will never see the light of day.[20] EPA is the agency most at the mercy of these attitudes: it receives more sustained attention from OIRA than does any other federal agency.[21] OIRA's pre-promulgation role in rule oversight insulates it from judicial review.[22]

EPA's practice in developing regulations also varies from the official version. The written ADP process guidance may say that affected offices participate in both early rule design choices and briefings where key decisions are made. But in practice, who's going to make them? If rule writers don't want to hear from other offices, or really don't understand why compliance has anything to do with rule design, it is easy to skip that consultation. And they do. There is no recourse for these informal and common adjustments to the written directives.

Next Gen is about the messy, complicated, and expectation-busting facts of real life. Those realities apply to the internal governmental process for rules too. So while it is important to know what the formal guardrails are, which I very briefly outlined in this chapter, it is just as vital to understand how things really work. What matters isn't tinkering with the written-down standards, it's delivering the public health benefits in the actual world. That's why the next chapter takes a deep dive into the seemingly arcane but actually all-important world of rule-writing guidance, and what has to change to get those to push for, rather than against, more effective regulations.

[20] Heinzerling, "Inside EPA," 359 ("OIRA's actual practice in reviewing agency rules departs considerably from the structure created by the executive order governing OIRA's process of regulatory review"). Farber, "Reforming Regulatory Reform," 12–13.

[21] Heinzerling, 348.

[22] Beerman, *Administrative Law*, 163.

4

Getting in Our Own Way

How EPA Guidance Reinforces Faulty Compliance Assumptions

The mistaken beliefs that compliance is generally good, and it is up to enforcement to ensure a strong compliance outcome aren't just the common wisdom.[1] They are reflected in formal EPA policies and practice for writing rules. That turns difficult-to-change assumptions into nearly insurmountable obstacles. The overwhelming data and the compelling arguments of Next Gen won't carry the day as long as EPA guidance provides cover for rule writers already inclined to ignore compliance. The solution lies not just in removing the barriers these policies erect—although that's essential—but revamping the guidance to put implementation center stage.

The two EPA guidance documents that turn compliance folklore into policy are the directive outlining process steps for writing regulations, and the instructions for doing benefit-cost analysis. The playbook for writing rules is known as the "EPA Action Development Process Guidance" (2011 ADP).[2] The official name of the guidance for benefit-cost analysis is "EPA Guidelines for Preparing Economic Analyses" (2010 Guidelines).[3] Plenty of people who work at EPA have never read either of these. They are almost never mentioned outside of EPA. But they loom large for anyone trying to shift the agency toward rules that will work in the real world.

This chapter starts by explaining the compliance impacts of the not-helpful ADP process guidance, which is worse in practice than the written directive

[1] Academics reinforce this common wisdom. See Martha G. Roberts, "Integrating Compliance and Regulatory Design in EPA Rulemaking," *New York University Environmental Law Journal*, Vol. 20, No. 3 (2014): 546 (noting that the significant body of literature on EPA regulation and compliance often focuses exclusively on strategies to increase compliance once a rule has been promulgated and that some studies on regulatory reform explicitly exclude compliance topics as unrelated).

[2] EPA, "EPA's Action Development Process: Guidance for EPA Staff on Developing Quality Actions," EPA Office of Policy (Revised: March 2011), https://yosemite.epa.gov/sab/sabproduct.nsf/5088B3878A90053E8525788E005EC8D8/$File/adp03-00-11.pdf ("2011 ADP"). The 2011 ADP is the most recent version available to the public; there have been some minor changes since then.

[3] EPA, "Guidelines for Preparing Economic Analyses," EPA National Center for Environmental Economics, December 17, 2010 (updated May 2014), https://www.epa.gov/environmental-economics/guidelines-preparing-economic-analyses ("2010 Guidelines"). Proposed revisions to the 2010 Guidelines were published in 2020, but no changes have been adopted yet.

Next Generation Compliance. Cynthia Giles, Oxford University Press. © Cynthia Giles 2022.
DOI: 10.1093/oso/9780197656747.003.0005

suggests. I then describe how the guidance for benefit-cost analysis endorses the faulty assumption of strong compliance. Using examples from actual EPA rules, I show how the traditional benefit-cost calculation method misunderstands the compliance challenge and compounds the already daunting problem of improving rule compliance design. The chapter ends with a proposed solution for revamping policy directives and putting the emphasis where it belongs: better rules, not better accounting.

EPA's Process Guidance for Writing Regulations Could Support Next Gen, but as Practiced It Doesn't

As was described in chapter 3, EPA's Action Development Process Guidance (ADP) proscribes the steps involved in writing a regulation and who is supposed to be involved at each step. The lead office for each rule (air, water, chemicals, waste) chairs the cross-agency workgroup that shepherds the rule to completion. The chair of the workgroup guides the team through the key steps in the process: preliminary analytic blueprint, early guidance, detailed analytic blueprint, options selection, then final agency review.

A lot of irreversible design choices are made in the initial stages. The workgroup's preliminary analytic blueprint is presented to senior managers at an early guidance meeting. That's when the basic strategy and the additional research needed to support that strategy are chosen. Once the rule has started down a path, it is virtually impossible to change it.[4] If compliance issues are not considered then, they likely miss their chance to be included, even if an eventual compliance analysis shows that the chosen strategy is an inevitable compliance disaster. It isn't just the internal dynamics that will strongly resist a late game change of direction; external stakeholders will also apply the brakes. Direction signaling during outreach can help build outside support, but it can also make it difficult to change plans.

EPA's enforcement and compliance office (Office of Enforcement and Compliance Assurance—OECA) in theory participates in the choices for these early steps. The good news is that OECA is listed as a core member of the workgroup for every significant rule.[5] That role is strengthened by the assertion in the 2011 ADP that the preliminary analytic blueprint is a collaborative effort of

[4] Winston Harrington, Lisa Heinzerling, and Richard D. Morgenstern, "What We Learned," in Winston Harrington, Lisa Heinzerling, and Richard D. Morgenstern eds., *Reforming Regulatory Impact Analysis* (Resources for the Future, April 2009), 225 (key elements of rule design are decided early in the regulatory process; strong internal pressures discourage change even if the regulatory impact analysis subsequently finds the originally preferred approach isn't the best one).

[5] EPA, "2011 ADP," 29.

the whole workgroup and not just a product of the lead office,[6] and that other offices should have meaningful opportunities to contribute to early guidance decisions.[7] Views of all workgroup members are supposed to be heard and considered during workgroup deliberations.[8] In other words, the formal ADP process suggests that OECA should be able to raise compliance design issues at these early decision-making stages.

The reality isn't quite as good. The role envisioned for OECA in the 2011 ADP is considerably more narrow than Next Gen–style compliance design. It uses the by now familiar compliance-is-about-enforcement frame.[9] And in practice, the lead office owns the rule and feels free to obtain input and direction from its own management without advising—never mind consulting—other workgroup members. By the time OECA staff learn of the first workgroup meeting, it is usually already too late.

As the rule proceeds through the process, the options narrow and are refined. At each stage there are opportunities to build in strong compliance design. Or to ignore the aggravating voices of the compliance representatives who are apparently unaware that most companies comply. Compliance is not seen as central, or the rule writers' responsibility, so it often isn't included in either the analysis or the briefings as compliance-defeating ideas get baked in or compliance-promoting ideas fall by the wayside.[10] As a participant in briefings for the administrator in high-profile rules, I can tell you how extremely uncomfortable it is, and how deeply unpopular you are—not to mention unsuccessful—when you raise compliance objections to ideas that have built up momentum. Formal nonconcurrence by a nonlead office in the final rule is the last gasp method for raising serious concerns, but it is considered the nuclear option, so is rarely exercised.[11] Plus it doesn't usually change anything.

The informal choices to pay little attention to compliance at the initial stages of rule development are the logical outgrowth of the assumptions that have dominated rule-writing since forever. The dual beliefs that compliance will be good just because and it is enforcement's job to accomplish that mean that compliance and design aren't spoken in the same sentence. The lead rule writers often

[6] EPA, "2011 ADP," 33.
[7] EPA, "2011 ADP," 35.
[8] EPA OIG, "EPA Does Not Always Adhere to Its Established Action Development Process for Rulemaking," Report No. 21-P-0115 (March 31, 2021), 4.
[9] EPA, "2011 ADP," 17: "OECA assigns a workgroup member to all Tier 1 and 2 actions when necessary to ensure that actions are clear and concise, that compliance measures can be understood and that final actions can be enforced."
[10] Other ADP requirements that are not regularly observed: AAs/RAs from all offices participating on the workgroup should be invited to the options selection meeting, EPA, "2011 ADP," 38; workgroup members should be included in any pre-briefs of senior managers, in particular when important issues affecting their office's interests are discussed. EPA, "2011 ADP," 39.
[11] EPA, "2011 ADP," 41–43 (describing nonconcurrence option).

don't think compliance is relevant—if at all—until the final stages of the rule-writing process, when decisions will be made about wording of the regulation, recordkeeping, and the like. You know, enforcement-type issues. The horse has long since left the barn.

Apart from equating compliance with enforcement, the 2011 ADP guidance isn't on its face hostile to Next Gen. But in practice it is. Most rule-writing offices don't think the compliance staff should be sticking their nose into rule design, and rule writers can ensure they don't by not including enforcement staff until it is far too late. Yes, the 2011 ADP guidance says compliance representatives are supposed to be included at the earliest stages, but that almost never happens, in part because rule writers don't see how that's relevant.

The ADP didn't create the feel-free-to-ignore-compliance attitude that pervades rule-writing. But as implemented, it locks that attitude in. The lead rule-writing office is in charge, and if it wants to pay no attention to compliance, that's easy to do. Nothing requires that implementation be front of mind when choosing what options to pursue. The ADP cements business as usual, and in so doing, it adds to the already high barriers to Next Gen. The good news is that the ADP could be transformed from a problem to a solution, and that's what I propose later in this chapter.

Unlike the ADP, which is at least superficially neutral toward Next Gen, the guidance on economic analysis goes all in on the compliance assumption. That's what I turn to next.

Putting the Rabbit in the Hat: Benefit-Cost Analysis and the 100 Percent Compliance Assumption

As explained in chapter 3, economic analyses are required by presidential executive orders (EOs) and by the instructions from the Office of Management and Budget (OMB) about meeting the requirements of the EOs.[12] Under these directives, EPA tallies up the benefits and the costs of major proposed rules, to ensure not only that benefits justify the costs but that benefits exceed costs by as much as possible.[13]

[12] Executive Order No. 12291, *Federal Register*, Vol. 46 (February 17, 1981): 13193; Executive Order No. 12,866, *Federal Register*, Vol. 58 (October 4, 1993): 51735; Executive Order No. 13,563, *Federal Register*, Vol. 76 (January 21, 2011): 3821; Office of Management and Budget, "Circular A-4" (September 17, 2003), https://obamawhitehouse.archives.gov/omb/circulars_a004_a-4/.

[13] OMB, *Circular A-4*, 10 (identify the alternative that maximizes net benefits). EPA, "2010 Guidelines," 11-1 (seeking outcome that yields the largest possible net benefits). See also Alan J. Krupnick, "The CAMR: An Economist's Perspective," in Harrington, *Reforming Regulatory Impact Analysis*, 143. By this metric, EPA rules hit the ball out of the park. See, e.g., US EPA: "Benefits and Costs of the Clean Air Act 1990–2020, the Second Prospective Study," finding that the benefits of the Clean Air Act have exceeded costs by a factor of more than 30 to 1. https://www.epa.gov/

The 2010 Guidelines take the assumption that compliance is generally good to a new level; it directs rule writers to assume 100 percent compliance when they do economic analysis.[14] I am not making this up. Here's the exact language of the 2010 Guidelines:

> As a general rule, when preparing analyses of regulations analysts should develop baseline and policy scenarios that assume full compliance with existing and newly enacted (but not yet implemented) regulations.

In other words, it roundly endorses the idea that rule writers don't have to worry about how to assure compliance. They can just assume it will happen. Problem solved.[15]

We get to this unsupportable position as a result of the mistaken beliefs about compliance described throughout this book. Rule writers fall prey to the unfounded but widely held assumption that compliance is strong. And they accept the unwarranted but universally embraced belief that noncompliance is enforcement's responsibility.[16] The 100 percent compliance assumption is the math statement of these articles of faith. It articulates in economist-speak the view that compliance is not the rule writers' problem. Out of sight, out of mind.

The Guidelines didn't invent this dismissive attitude toward implementation challenges. That perspective predates both the ADP and the Guidelines, both of which reflect rather than establish agency views about compliance. The ADP doesn't set up either obligations or process check points to ensure that robust implementation is front of mind in defining a rule's structure. Economic analysis adopts the same policy choices. Ill-conceived benefit-cost analysis isn't the cause

clean-air-act-overview/benefits-and-costs-clean-air-act-1990-2020-second-prospective-study. Surprisingly, regulations about surface water pollution seem to be exempt from this otherwise powerful mandate. David A. Keiser, Catherine L. Kling, and Joseph S. Shapiro, "The Low but Uncertain Measured Benefits of US Water Quality Policy," *Proceedings of the National Academy of Sciences*, Vol. 116, No. 12 (March 2019): 5262–69; https://doi.org/10.1073/pnas.1802870115.

[14] EPA, "2010 Guidelines," 5–9. EPA issued proposed revisions to the Guidelines in 2020, which have not been issued in final. The 2020 proposed changes revise some of the verbiage around compliance but retain the same approach: 100% compliance is the default, and a different assumption can only be used when there is "strong evidence" to support it. EPA, "Guidelines for Preparing Economic Analyses, Review Copy Prepared for EPA's Science Advisory Board's Economic Guidelines Review Panel" (April 3, 2020), 5–18 ("2020 Proposed Guidelines"), https://yosemite.epa.gov/sab/sabproduct.nsf//LookupWebProjectsCurrentBOARD/30D5E59E8DC91C2285258403006EEE00/$File/GuidelinesReviewDraft.pdf.

[15] The Guidelines establish a presumption of 100% compliance but allow a deviation from that if the rule writers clear a very high bar. See, e.g., EPA, 2010 Guidelines, 5–9. Because the 100% compliance default confirms the bias that rule writers and analysts already have, and it is by far the easier approach since no explanation is required for using it, that's what nearly every rule does.

[16] OMB accepts that view too. See OMB, "Circular A-4," 7–8 (compliance and enforcement are viewed as the same idea).

of shortsighted or incomplete rule design that will lead to compliance failure. But by justifying those choices, the Guidelines make it much harder to get Next Gen ideas into rules.

Here's an oversimplified example to show how that happens. Let's say the rule is going to direct a certain sector of industry to control pollutant X to a defined limit. To calculate costs, EPA will look at how much it costs a typical company in that sector to meet the pollution limit. Then EPA will multiply that per-facility cost by the number of regulated sources. The resulting number is considered the cost of the rule. For benefits, EPA will calculate how much of pollutant X will be removed from the environment if every firm installs the controls and meets the limit. Then EPA figures out the health improvements and other benefits that would result from that pollution reduction and tries as much as possible to quantify those benefits and convert them to dollars. That number is the total monetized benefits. Then they are compared: Are total $ benefits more than total $ costs? If yes, the rule has potential. The greater the amount by which benefits exceed costs, the better the rule's chances.

That's all well and good, says the Next Gen proponent to the rule writers, but why do you think the regulated companies will do it? How about including real-time monitoring? More detailed reporting? Third-party auditing? Let's talk about adding some strategies that would make the rule more self-implementing and improve the chances that most of the companies do what the rule requires.

Here's where the Next Gen options run headfirst into the 100 percent compliance assumption. The Next Gen ideas likely will increase costs, even if only slightly. Which is lower cost: strategies that will improve compliance, or assuming 100 percent compliance is free? The rule writers may think Next Gen is a great idea. But since Next Gen can only increase costs (it doesn't usually cost zero) and by definition never increases benefits (you have already assumed you get all the benefits), including Next Gen is a tough sell.[17] The same thing happens when rules are considered at OMB; compliance driving provisions that industry doesn't want can be axed without hurting the benefit-cost ratio, at least on paper.

The 100 percent compliance assumption doesn't just brush aside compliance concerns, it inadvertently encourages strategies that are likely to make the problem worse. Because the benefits are fixed via the magic of an assumption, every reduction in industry costs improves net benefits. Rule writers are therefore motivated to include complicated exemptions, exceptions, and qualifications that reduce industry costs but are anathema to strong compliance.[18] There is no

[17] Roberts, "Integrating Compliance and Regulatory Design," 572 (the 100% compliance assumption makes compliance-driving measures appear to be all cost and no benefit).

[18] This concern was raised in a 2002 National Academies report, which recommended changing the 100% compliance assumption because it is likely to result in the agency's "neglecting the important issue of the relative cost and effectiveness of alternative implementation and enforcement

accountability for inserting these politically convenient compliance nightmares into rules, because compliance doesn't enter into it. You have an enforcement office, right? No further inquiry needed.

In reality, rule writers usually don't directly cite the 2010 Guidelines' 100 percent compliance assumption to defend their decisions. The Guidelines don't need to instruct rule writers to take compliance for granted; they are already doing that on their own. But by enshrining that assumption in formal written policy, the Guidelines crystallize what is wrong with the just-ignore-it position, and help illustrate how we should, and should not, address the problem. That's why this chapter goes into some depth about what the 100 percent compliance assumption gets wrong and how it should be re-envisioned: to press for, rather than against, rules that deliver in the real world.[19]

Overly Simplistic Ideas about Compliance Make This Harder to Solve

The very few references to compliance in the benefit-cost literature usually employ an oversimplistic assumption that compliance is binary: all or nothing. Either a facility complies completely and fully, or it does nothing at all. In this view, 70 percent compliance means 70 percent of facilities achieve full compliance (incur all costs, achieve all benefits), while 30 percent are at zero (no costs, no benefits). That means it doesn't matter to benefit-cost analysis if you are wrong about the rate of compliance. If you assume 70 percent compliance, you also assume you get 70 percent of the benefits and incur 70 percent of the costs. So of course—because you have assumed it—any reduction in compliance will be perfectly mirrored by a reduction in assumed benefits and costs.[20] Using this approach, while a lesser compliance rate may change the absolute benefits from a rule, it won't change the benefit-cost ratio.[21] You know it won't, because you

measures." National Research Council, "Estimating the Public Health Benefits of Proposed Air Pollution Regulations" (The National Academies Press 2002): 59–60, https://doi.org/10.17226/10511.

[19] Note that this discussion of benefit-cost analysis does not apply to all rules. Benefit-cost evaluation isn't required for every rule and is expressly prohibited by statute in some cases.

[20] See, e.g., EPA, "2010 Guidelines," 5–10; Institute for Policy Integrity, "Comments on Proposed Revision to the Hours of Service Regulation for Property Carrying Commercial Motor Vehicles, 75 Fed. Reg. 82170 (Dec. 29, 2010), Docket No. FMCSA-2004-19608, March 4, 2011," 14 ("The agency's assumption of 100 percent compliance in conducting its cost-benefit analysis is methodologically sound because changes in the compliance rate should have an equal effect on costs and benefits."), https://policyintegrity.org/documents/Policy_Integrity_Final_Comments_on_HOS_Rule.pdf.

[21] Institute for Policy Integrity, "Comments on Proposed Revision," 14. Two EPA Regulatory Impact Analyses discussed in this chapter made this same assumption. See discussion later in this chapter of the Lead Renovation, Repair and Painting rule and the Oil Spill Prevention, Control, and Countermeasure rule.

decided in advance to adopt an assumption that mandates it won't. On this basis, even the few who concede that a 100 percent compliance assumption doesn't align with reality dismiss it as not relevant to the task of benefit-cost analysis.

EPA's rule to prevent lead in drinking water illustrates how far from binary the compliance problem really is. The drinking water regulation that lays out the obligations for avoiding lead requires sampling across the distribution system to detect the presence of lead, further action if lead is found above a certain level, and multiple recordkeeping and reporting requirements so government can know what's going on.[22] Here are just some types of violations that occur under that rule. Drinking water systems don't sample where lead is most likely to be found, despite a requirement to do so, so government has no idea what the lead risks in that system are.[23] Water suppliers engage in "sampling out," a common practice in which systems collect more known-to-be-clean samples and thereby dodge the percentage-based trigger for additional action, also making it impossible to assess the extent of lead contamination.[24] They fail to monitor or report, thereby masking more serious problems; GAO found that monitoring and reporting violations are statistically significant predictors of violations of health-based standards.[25] States also don't tell EPA about lead in drinking water violations, including a whopping 92 percent of health-based violations and 84 percent of monitoring and reporting violations, so nationwide we are blind to the extent of the problem.[26]

When you stack up the compliance picture for just this one rule, it shows how unsophisticated it is to portray compliance as a yes/no proposition. It is nowhere close to that. The goal of the rule—its bottom line—is preventing exposure to lead in drinking water. There are a host of obligations in the rule intended to achieve that, including sampling, monitoring, recordkeeping, treatment, and reporting requirements. Violating any one of these obligations doesn't necessarily mean that people were exposed to lead, but it might. What all these violations

[22] Code of Federal Regulations, Title 40, Part 141 Subpart I (1991). For a summary, see EPA, "Lead and Copper Rule," https://www.epa.gov/dwreginfo/lead-and-copper-rule#rule-summary.

[23] One investigation found that almost half of the drinking water systems in studied in Georgia incorrectly claimed to be sampling where lead was likely to be high, although they weren't actually doing that. Brenda Goodman, Andy Miller, Erica Hensley, and Elizabeth Fite, "Lax Oversight Weakens Lead Testing of Water," a joint investigation by *WebMD* and *Georgia Health News* (2017), https://www.webmd.com/special-reports/lead-dangers/20170612/lead-water-testing.

[24] In another drinking water rule where this practice was studied, researchers found that it allowed almost one-third of violations to go undetected. Lori S. Bennear, Katrina K. Jessoe, and Sheila M. Olmstead, "Sampling Out: Regulatory Avoidance and the Total Coliform Rule," *Environmental Science & Technology*, Vol. 43, No. 14 (2009): 5176–77.

[25] U.S. Government Accountability Office, "Drinking Water: Unreliable State Data Limit EPA's Ability to Target Enforcement Priorities and Communicate Water Systems' Performance," (GAO 11-381, 2011), 16.

[26] See the discussion of drinking water rules in chapter 1.

definitely mean is that it is impossible to say how we are doing with preventing lead exposure.[27]

That's not the only kind of complexity that makes use of a compliance rate unilluminating in benefit-cost analysis. EPA's data isn't about compliance, it's about violations. These are not flipsides of the same coin. When a company self-reports noncompliance, it's likely that there actually was a violation. But when a facility doesn't raise its hand to admit a violation, that doesn't necessarily mean that it is complying. EPA routinely finds violations that were not self-disclosed. All of the huge New Source Review (NSR) violations, for example, were ferreted out by EPA, not self-reported by the companies.[28] A compliance rate cannot be inferred from the rate of reported violations.

Violations also vary widely in scale. Within the self-reported violations there will be some horrendously bad violators and some pretty close to the line. There will also be serious problems among the many who fail to self-report violations, including probably all of the intentional violators, as program after program has shown.[29] And many programs don't have a self-reporting requirement, so EPA doesn't know how bad the violations are.[30] In those cases there is literally no way to even guesstimate a violation rate.

An aggregate "compliance rate" perspective also says nothing about the distributional impacts of the violations. Polluting and higher risk facilities are more likely to be located in communities of color and low-income areas, so even completely randomly distributed violations would have disproportionate impacts on already overburdened communities. And of course, violations are not randomly distributed. Studies increasingly show that pollution burden falls most heavily on Black communities and other communities of color; violations are certainly contributors to that problem.[31] As government grapples with including

[27] For another example of the many complexities of compliance and the multiple ways compliance can run off the rails, see EPA, "Compliance Tips for Small, Mechanical Wastewater Treatment Plants," EPA Compliance Advisory, April 2021 (listing over 30 "common issues" that can lead to serious water discharge violations).

[28] See the discussion of the compliance disastrous NSR rule in chapter 1.

[29] See chapter 2. See also Yingfei Mu, Edward A. Rubin, and Eric Zou, "What's Missing in Environmental (Self-)Monitoring: Evidence from Strategic Shutdowns of Pollution Monitors," National Bureau of Economic Research, Working Paper 28735 (April 2021), DOI 10.3386/w28735 (presenting statistical evidence of strategic shutdowns of air monitors when air quality is expected to deteriorate); Daniel Nicholas Stuart, "Strategic Non-Reporting Under the Clean Water Act," chapter in "Essays in Energy and Environmental Economics," PhD diss., Harvard University 2021, https://nrs.harvard.edu/URN-3:HUL.INSTREPOS:37368502 (finding statistical evidence of strategic nonreporting by water pollution dischargers when violations are more likely).

[30] See discussion in chapter 2.

[31] See, e.g., Hiroko Tabuchi and Nadja Popovich, "People of Color Breathe More Hazardous Air. The Sources Are Everywhere," *New York Times*, April 28, 2021, https://www.nytimes.com/2021/04/28/climate/air-pollution-minorities.html.

environmental justice in regulatory analyses, let's not hide a disparate impact in a superficially neutral metric like a compliance rate.[32]

Put all this together and you see the error of the all-or-nothing compliance assumption and its corollary that costs and benefits vary in exact tandem. It is not accurate to assume that some firms do everything all the time, some firms do nothing, and violating facilities are randomly distributed around the country. That isn't reality.

The way it works now—when anyone pays attention to compliance in writing a rule, which is unusual—is all this complexity is reduced to a single number, the mythical compliance rate. Then that already oversimplified number is dumbed down again to an all-or-nothing complies-every-minute or doesn't-ever-comply-with-anything. Real life is distorted beyond recognition.

The reason to recite the many complexities of compliance isn't to argue for a different compliance rate in benefit-cost analysis. It is to explain why the conventional wisdom that it is safe to ignore compliance in regulatory impact analysis is misguided. The baseline is wrong; no industry anywhere has 100 percent compliance. And if we could know exactly how companies will respond to a new rule—which we are nowhere close to being able to do—we would not find that compliance is all or nothing and thus has no impact on the comparison of benefits with costs.

Another Bias: Only Industry Costs Count

One more fiction requires explanation. That is what counts as costs. Industry costs matter, government costs do not. Regulatory options are compared using the metric of expense to companies.[33] The fact that some regulatory choices make government's compliance work much harder doesn't make it on to the ledger.

Some compliance disasters can look fairly good using this one-sided accounting system. For example, it might cost industry less to allow every company

[32] The same concern arises when choices are made about exempting some facilities from regulation. The firms making theoretically neutral claims for exemption like the age of their facility or its size/revenue/employees are also likely to disproportionately affect environmental justice communities. Harrington, "What We Learned," in Harrington, *Reforming RIA*, 233. See also Richard L. Revesz, "A New Era for Regulatory Review," *The Regulatory Review*, February 16, 2021, https://www.theregreview.org/2021/02/16/revesz-new-era-regulatory-review/; Elinor Benami et al., "The Distributive Effects of Risk Prediction in Environmental Compliance: Algorithmic Design, Environmental Justice, and Public Policy," *FAccT'21*, Virtual Event, Canada (March 3–10, 2021): 95–97, 99, https://doi.org/10.1145/3442188.3445873 (government choice to focus attention on the rate of noncompliance shifts government attention away from polluters located in minority communities).

[33] See discussion in chapter 3; Daniel H. Cole and Peter Z. Grossman, "Beyond Compliance Costs: Comparing the Total Costs of Alternative Regulatory Instruments, in Kenneth R. Richards and Josephine van Zeben eds., *Policy Instruments in Environmental Law* (Edward Elgar Publishing, 2020), 33.

to develop its own system for monitoring or measuring compliance. Or industry costs can be cut by saying companies with site-specific factors have lesser standards or are exempt. Not considered is the fact that such provisions provide lots of opportunity for confusion and mistakes, not to mention room to maneuver for companies interested in avoiding compliance obligations. Investigating these situations to determine what's a legitimate claim and what isn't would be a big and often impossible task. Counting only industry costs allows rule writers to disregard the additional burden these kinds of provisions create for regulators.[34] EPA's 2010 Guidelines say that rule writers should consider government costs,[35] but that almost never happens.

I am not arguing that including government burdens in the costs column is the way to fix to compliance flaws in economic analysis. It isn't. But the failure to consider government costs is another bias that prevents rule writers from acknowledging the compliance implications of their choices. If rule writers carried their 100 percent compliance, leave-it-to-enforcement assumption to its logical conclusion, and included in their analysis the actual costs to government of achieving close to 100 percent compliance that way, it would bring them up short. That doesn't happen though. The practice of focusing only on industry costs avoids a reckoning with the totally unrealistic expectation that enforcement will handle it. As a result, the costs of regulatory options are improperly accounted for in the benefit-cost analysis, with many of the compliance-defeating strategies appearing far more economically attractive than they actually are.[36]

Suppose that rule writers had been told when the compliance-catastrophic NSR program was built that the only way to achieve compliance by coal-fired power plants was to have EPA sue virtually every coal-fired power plant in America with cases that would cost millions per annum and consume a large fraction of national enforcement capacity for the next 25 or 30 years. And, by the way, that benefits would be delayed until those cases were completed. Would that NSR rule design have passed a benefit-cost test?[37] Perhaps the rule writers would have decided that indefinitely exempting all existing coal-fired power plants from modern pollution control rules and including lots of confusing fact-specific exceptions wasn't such a terrific cost-saving idea after all.

My point isn't that enforcement costs should be included in a benefit-cost analysis. That would do nothing to dislodge the unfounded and counterproductive assumption that enforcement can be counted on to salvage all manner of

[34] See discussion in chapter 6.
[35] EPA, "2010 Guidelines," § 9.2.4.2.
[36] See discussion in chapter 6.
[37] See discussion of NSR in chapter 1. Note that a Next Gen analysis of NSR would have predicted the disastrous compliance outcomes that actually happened. The regulation contained many loopholes and places to hide, a big compliance gray zone that made hiding easy, and powerful financial incentives to find a way out.

poor regulatory designs. And even more to the point, it wouldn't change anything about the real government budget, which is set by Congress without regard to what rule writers assume in obscure benefit-cost analyses. Figuring in enforcement costs as though money for that purpose would actually be forthcoming would just create a new illusion to pile on top of the existing one.

The point is that regulatory impact analysis as performed today sweeps compliance under the benefit-cost rug. The goal of benefit-cost analysis is better policy.[38] We can debate—and many do—whether it generally achieves that objective.[39] But benefit-cost's approach to environmental compliance is not driving better outcomes. Just the opposite.

Assuming Compliance Undermines Both Purposes of Next Gen: Getting the Design Right and Taking a Hard Look at the Big Picture

Next Gen plays two important roles in effective rules. Both are relevant to a proposal to fix the currently broken system for assuring compliance in environmental regulations.

The first is to help design a rule that has all the necessary components for a compliance-resilient program. The current overly simplistic, all-or-nothing idea of compliance interferes with the more sophisticated work necessary to drive better compliance behavior. Even for traditional pollution rules that have a way to reliably measure the end point EPA cares most about, measurement alone isn't a robust compliance assurance mechanism. You also need a strategy to inspire continuous use of the monitors and adequate quality control, as well as a way for the monitoring to be available to the government, the public, or both, and consequences for bad performance so that the data drive a better outcome, not simply better documentation of compliance failure.

Next Gen isn't a single idea inserted into a rule. It is an array of strategies. In the impressively effective Acid Rain Program, excellent compliance was achieved because of continuous emissions monitoring, data-substitution requirements that inspired companies to operate the monitors properly, clear and simple compliance requirements, electronic reporting that identified violations quickly and made them hard to miss, and automatic consequences that made violation more expensive than compliance. Take out any of these pieces and you might have had a different outcome.[40]

[38] EPA, "2010 Guidelines," 1–2 ("a thorough and careful economic analysis is an important component in informing sound environmental policies.").

[39] See Harrington, *Reforming Regulatory Impact Analysis*.

[40] See discussion of the Acid Rain Program in chapters 1 and 6.

That is the norm. Usually there isn't one big problem where all the noncompliance happens, so there isn't a single change that will solve it all. It is more common to find a collection of places where things can run aground, requiring a suite of structural drivers. Rule writers need to think through each significant element of the rule and consider what implementation drivers are necessary for each. Strategies for ensuring use of monitoring equipment are different from designs to ensure accuracy in reporting, which aren't the same as motivators to accomplish necessary training. It's not one or the other, it's all of the above. Next Gen isn't one compliance-driving idea, it is a compliance system.

The recently finalized rule for big reductions in hydrofluorocarbons (HFCs) provides an up-to-date illustration of what a Next Gen compliance system looks like.[41] The rule addresses the urgent need to cut back on these intensively climate-altering compounds. HFCs are used in a variety of applications, but the most common are in refrigeration and air conditioning. The rule covers HFCs that are hundreds to thousands of times more damaging to the climate than carbon.[42] Congress directed EPA to develop regulations to reduce the emissions of HFCs to 15 percent of the 1990 levels by 2036.[43] The climate benefits from controlling these super-pollutants are therefore enormous; EPA projects that the annual net benefits from the rule are $1.7 billion in 2022 and will increase over time.[44]

EPA's rule was informed by the experience of HFC phasedown in Europe, which has been plagued by pervasive illegal imports. By some estimates, Europe is missing its target by more than 30 percent, and that is likely understating the problem.[45] Without deliberate design to prevent that, the same thing would happen in the United States, pulling the rug out from under the rule's climate ambition.[46]

EPA's rule adopts a suite of compliance drivers to both block illegal imports at the border and cut off demand for them within the United States. There are too many creative compliance-enhancing provisions in the proposed rule to list here, but a few of the key ones are:

[41] EPA Final Rule, "Phasedown of Hydrofluorocarbons: Establishing the Allowance Allocation and Trading Program Under the American Innovation and Manufacturing Act," *Federal Register*, Vol. 86, No. 190 (October 5, 2021): 55116 ("Final Rule for HFC Phasedown").

[42] EPA, "Final Rule for HFC Phasedown," 55123.

[43] EPA, "Final Rule for HFC Phasedown," 55116.

[44] EPA, "Final Rule for HFC Phasedown," 55118.

[45] Environmental Investigation Agency, "Doors Wide Open," *Eia-International.org*, April 2019, https://reports.eia-international.org/doorswideopen; King & Spaulding, on behalf of the Alliance for Responsible Atmospheric Policy, "Alliance for Responsible Atmospheric Policy—Refrigerant Imports Committee," Side Event presentation at COP12/MOP32 (November 23, 2020), available at https://www.alliancepolicy.org/site/usermedia/application/10/Bradford%20KS%20HFC%20Presentation%2023%20Nov%202020%20v4.pdf; see also discussion in EPA, "Final Rule for HFC Phasedown," 55166–168.

[46] EPA, "Final Rule for HFC Phasedown," 55167.

1. Automated real-time checking to be sure that a company has the necessary allowances *before* they can import HFCs.[47] No allowing import of illegal HFCs and then counting on government to check records and chase them down afterward, as has proven so disastrous in Europe.
2. Ban on the use of disposable containers, which is the principal way unlawful product is shipped in other countries.[48] Simple, direct, and easy to spot.
3. Every importer and producer must have an independent third-party audit of its reports to EPA.[49]
4. Every container must have a QR code that links to the website at which a potential buyer can determine if the seller and the container are legit, and every transaction is tracked in the system.[50] Accountability up and down the chain; there is no way to buy unlawful HFCs without knowing they are unlawful.
5. Tough administrative consequences for violations, including attempting to produce or import without necessary allowances. EPA has the authority to withhold, revoke, or retire allowances and impose a ban on holding future allowances, in addition to the penalties imposed in enforcement actions.[51]

There are a lot more, but this gives you the idea. Every direction the potential violator turns, the way is blocked. There are many places that unlawful activity is obstructed before the product can get into commerce, powerful incentives for buyers to only purchase lawful product, and no way for any party to a transaction to plausibly claim they didn't know. Good luck making it over this succession of hurdles. And, if you get caught trying—which the electronic system makes it much easier to detect—the jig is up. Lots of companies will decide this just isn't worth it. Exactly.

The second essential role for Next Gen—and this is the more challenging one—is to provide the insight that some regulatory designs will never work. Every rule could benefit from a stronger compliance foundation and fewer ways around. But sometimes no amount of Next Gen tinkering will work. A Next Gen analysis says there is no way to get there from here. This particular strategy will never be successful.

[47] EPA, "Final Rule for HFC Phasedown," 55186–187.

[48] EPA, "Final Rule for HFC Phasedown," 55173 (illegal trade in other countries primarily in disposable containers), 55172–175 (preamble discussion explaining why a ban is necessary to ensure compliance). Note that Europe in theory bans disposable containers, but Europe's ban contains so many exceptions that it is impossible to tell through observation alone if any observed disposable container is unlawful or not, making Europe's ban ineffective. EIA, "Doors Wide Open," 24. EPA's rule closes that loophole by making disposable containers universally prohibited after a fixed date. EPA, "Final Rule for HFC Phasedown," 55173.

[49] EPA, "Final Rule for HFC Phasedown," 55179–181.

[50] EPA, "Final Rule for HFC Phasedown," 55183–186.

[51] EPA, "Final Rule for HFC Phasedown," 55168–171.

In the lead in drinking water rule, for example, a Next Gen analysis might suggest that this is a program with more holes than cheese, as the expression goes. If EPA were to step back from all the challenges and look again, it might conclude that the rule is just too complicated. If every drinking water system's path has 10 different steps where there is opportunity for serious noncompliance due to misunderstanding, bad training, insufficient funds, or intentional gaming, and at every step we are counting on overworked and underresourced state regulators to ensure everything is done right, maybe that's just not a system that can be designed to succeed. The complexity itself might be the problem.

Maybe it would be better to cut through all the complication and adopt a simple requirement: replace the lead pipes. That's not cheap, but when it's done, we know a large share of the lead problem is addressed. One thing that's clear is that the system we have now isn't achieving the public health protection we need. I am not saying that replacing the lead pipes is the right answer—that would take a lot more analysis to figure out. What I am saying is that sometimes Next Gen suggests that we are going about solving the problem the wrong way. There are times when the nuanced, complicated, and involved approach that on paper seems more efficient doesn't—and can't be made to—deliver in real life.

Benefit-cost analysis as currently structured is not well suited to the needs-an-entirely-different-strategy situation. On one hand you have an expensive but certain-to-succeed strategy. On the other, you have a theoretically cheaper alternative that has zero chance of strong implementation. The 100 percent compliance assumption lets the agency treat these two as equally effective without ever acknowledging that it would take a huge infusion of government resources, plus some miracles, to make the supposedly cheaper one deliver. In the current state of play, the higher cost (to industry) option will lose every time.

The purpose of this deep dive into these foundational elements of the compliance assumption is to show how engrained the bias against Next Gen is in EPA's current approach to rules. Everything stacks up against it. The agency is allowed, even encouraged, to take out all the Next Gen compliance drivers, while still pretending it will get all the benefits. The fact that in real life all those benefits won't actually happen can't compete with the make-believe world of perfect compliance.

Should We Dump the 100 Percent Compliance Assumption?

You might think that because the 100 percent compliance assumption creates an impediment to Next Gen in rules that I would argue to get rid of it. Actually, no. Ditching that assumption would make the problem worse. Here's an example of a rule that tried, which illustrates how approaching it that way leads us astray.

The Lead-based Paint Renovation, Repair and Painting rule (commonly called lead RRP) finalized in 2008 was designed to protect the public from the hazards associated with lead paint.[52] Disturbing lead-painted surfaces can create hazardous lead dust and lead paint chips. A principal goal of the RRP rule was to protect children, who famously put so many things in their mouths, from the severe neurological and cognitive impacts of swallowing lead paint and dust.[53] The rule requires the roughly 320,000 renovators nationwide to follow standards and to be certified and trained in the use of lead-safe work practices for the estimated 18 million projects involving lead paint that happen each year.[54]

Unlike most rules, which say nothing about compliance but implicitly adopt the 100 percent compliance assumption, the lead RRP used a 75 percent compliance rate.[55] Not only did the rule expressly acknowledge that 100 percent compliance wasn't going to happen, but it also included a sensitivity analysis, checking to see what impact a lower compliance rate might have.[56] Sounds great, right? But no. A closer look at the economic analysis for this rule, which addresses the issue exactly as the Guidelines recommend, helps to illustrate why that approach is the wrong way to go about it.

There usually isn't reliable evidence to support a different compliance assumption

EPA's 75 percent compliance assumption in the lead RRP rule wasn't based on EPA compliance data but on OSHA's evaluation of compliance with OSHA safety

52 EPA, "Lead; Renovation, Repair, and Painting, Program," Final Rule, *Federal Register*, Vol. 73 (April 22, 2008): 21692 ("lead RRP final rule").

53 EPA, "lead RRP final rule," 21694.

54 EPA, "lead RRP final rule," 21694; EPA, "Lead Renovation, Repair and Painting Program Rules," https://www.epa.gov/lead/lead-renovation-repair-and-painting-program-rules (overview of requirements); EPA Office of the Inspector General, "EPA Not Effectively Implementing the Lead-Based Paint Renovation, Repair and Painting Rule" (Report No. 19-P-0302, September 9, 2019), 2 (number of renovators and projects annually).

55 EPA, "Economic Analysis for the TSCA Lead Renovation, Repair, and Painting Program Final Rule for Target Housing and Child-Occupied Facilities," 4–74 (March 2008) ("Economic Analysis Lead RRP Final Rule"), https://nchh.org/resource-library/EPA-HQ-OPPT-2005-0049-0916_Final_Economic_Analysis_3-08.pdf; EPA, "Economic Analysis for the Renovation, Repair, and Painting Program Proposed Rule" (February 2006), ch. 7, at 5 ("Economic Analysis Lead RRP Proposed Rule"); David Weil, "Assessing OSHA Performance: Evidence from the Construction Industry" (July 1999), available at https://ssrn.com/abstract=171406. The EPA Inspector General has challenged the economic analysis for the lead RRP rule on other grounds. See EPA OIG, "Review of Hotline Complaint Concerning Cost and Benefit Estimates for EPA's Lead-Based Paint Rule," Report No. 12-P-0600 (July 25, 2012).

56 This discussion is based on the economic analysis for the *proposed* rule. EPA, "Economic Analysis Lead RRP Proposed Rule." The sensitivity analysis is in chapter 7 of the Economic Analysis for the proposed rule, alternative estimates 2, 3, and 4 (ch. 7, at 5-10). The economic analysis for the final rule retained the 75% compliance assumption but didn't include a sensitivity analysis for compliance rates.

standards at construction sites.[57] There are many reasons why the compliance rate for the lead RRP rule is likely to be worse than the safety record at large construction sites regulated by OSHA. The most important is that both construction companies and their employees have self-interested motivation to prevent accidents and fatalities.[58] In contrast, the lead paint renovation firm isn't the principal beneficiary of the lead paint rules; the harm from RRP noncompliance lands on the people exposed to lead dust at the property once the contractor's work is over. For this and many other reasons, the already concerning 25 percent rate of violations in the OSHA studies likely understates the violation risk for RRP.

Lack of solid evidence about compliance is common for environmental rules. There are many programs—probably most—for which EPA has insufficient support for a specific number as a projected compliance rate.[59] It is telling that EPA's lead RRP analysts were not able to find in EPA's own compliance record relevant data to inform their compliance analysis.

While EPA's current Guidelines do acknowledge that 100 percent compliance might be wrong and that a different compliance rate may be allowable if there is substantiated evidence to support it, compliance data that would stand up to the rigors of an economic analysis will usually not be available.[60]

The idea that there is a single number—a compliance rate—that tells us whether we are achieving all the goals is an illusion

The desire for a single number that can tell us how good a job we are doing on implementation, however understandable, leads to dangerous oversimplification. It contributes to the all-or-nothing thinking that dominates consideration of compliance. Either the company does everything perfectly or it does nothing at all. That's not how it actually works.

Every regulation has a suite of provisions that might be violated, which can include work practice standards, emission limits, monitoring and reporting, training requirements, and notification obligations. Violations can be sporadic or continual, just over the limit or an order of magnitude above, minor, or egregious. Monitoring and reporting violations can make it impossible to know how

[57] EPA, "Economic Analysis Lead RRP Proposed Rule," ch. 4, at 14.

[58] This is cited in the 1999 Weil study as a principal reason for compliance by construction companies. Weil, "Assessing OSHA Performance," 13.

[59] For a survey of compliance information, or the lack thereof, for a wide variety of EPA administered programs, see chapter 2.

[60] See EPA, "2010 Guidelines," 5-9 to 5-10. The proposed 2020 revisions to the Guidelines are similar: "[A]nalysts should assume full compliance with regulations unless there is strong evidence to support an alternative assumption." EPA, "2020 Proposed Guidelines," 5-18.

bad it is. Then there are the straight-ahead fraudsters, who deliberately conceal their activities to prevent government from finding out.

The lead RRP rule illustrates the range of requirements that is typical in environmental rules. Firms and individual renovators have to be certified and take training from an EPA-accredited training provider. Everyone has to follow specified work procedures (e.g., containment, warning signs) and is prohibited from using other practices (no heat guns or power sanding). The site has to be cleaned up and waste handled appropriately.[61] A firm that does power sanding with no containment or cleaning is an obvious compliance fail posing high risk. But how should we measure the risk created by firms that do fake accreditation training so allegedly trained renovators actually don't know what safe practices are?[62] Or that falsely assure owners that their properties do not contain lead paint, so they fail to take any precautions?[63] A single-number compliance rate for the entire rule does not begin to capture the risks these kinds of violations unleash on the world.[64]

Some EPA programs condense a range of violation types into a single compliance metric—called serious, significant, or high-priority violations.[65] Such classifications attempt to consolidate what is known about reporting, monitoring, emissions, and work practice violations and single out the worst problems as a means of focusing enforcement resources.[66] Many of these serious violator metrics suffer from major shortcomings, including that they are almost entirely based on company- or state-reported data, which for some programs seriously underreport violations, and generally are constructed using a single threshold only, so don't differentiate between just-over-the-limit and an order of magnitude above.[67] For all their known flaws and cautions, a serious noncompliance

[61] See EPA, "lead RRP final rule," 21692 (summarized 21703–704).

[62] DOJ, "Environmental Training Company Owner to Serve Prison Time for Falsely Certifying Lead Abatement Course Completion," Press Release, December 7, 2017, https://www.justice.gov/usao-ct/pr/environmental-training-company-owner-serve-prison-time-falsely-certifying-lead-abatement.

[63] EPA "EPA Enforcement Actions Help Protect Vulnerable Communities from Lead-Based Paint Health Hazards—2020," News Release, October 29, 2020, https://www.epa.gov/newsreleases/epa-enforcement-actions-help-protect-vulnerable-communities-lead-based-paint-health-2 (see summary of Walter H. Clews case).

[64] See, e.g., EPA, "EPA's Lead-based Paint Enforcement Helps Protect Children and Vulnerable Communities—2018," https://www.epa.gov/enforcement/epas-lead-based-paint-enforcement-helps-protect-children-and-vulnerable-communities-2018#criminal, with links to scores of civil and criminal lead-based paint enforcement actions. EPA's website has similar lists of lead-based paint enforcement cases for each year.

[65] For the air program these are called High Priority Violations (HPV). Serious water point source discharge violations are called Significant Non-Compliance (SNC). In the drinking water program, the most problematic are labeled Serious Violators.

[66] Serious violators are just a fraction of the total number of all violators. In the water discharge program, for example, 60% to 75% of facilities self-report noncompliance each year. The rate of Significant Noncompliance is about 30%. Benami, "Distributive Effects," 92.

[67] See Benami, "Distributive Effects," for a description of how using such single threshold standards can bias government strategies.

metric is at least an attempt by knowledgeable experts to sort violations into informative categories. Most programs don't have that.

The nuances that EPA attempts to cram into a single number through serious noncompliance definitions are lost with the all-or-nothing "compliance" rate. As previously mentioned, what EPA has is some violation information. Lack of reported violations isn't a finding of compliance. There are multiple programs where EPA knows that violations are underreported.[68] Plus, what does a single number rate mean? If 25 percent of the lead certification trainers were fraudulent, sending thousands of workers into the world who know nothing about safely dealing with lead paint, that's a vastly different problem for controlling risk—and for the likelihood of achieving the benefits—than if 25 percent of contractors sometimes don't wet the containment sheeting before disposing of it.

The single compliance rate measure for benefit-cost analysis fails on two counts: EPA usually doesn't have anything close to the data necessary to develop one, and the single metric idea perpetuates the illusion that compliance is all or nothing and that every kind of violation counts the same for evaluating the rule's effectiveness/benefits/costs.

Sensitivity analysis—the recommended approach for analyzing assumptions that are uncertain—is not useful here

OMB's Circular A-4 and EPA's Guidelines recommend using sensitivity analysis to examine how assumptions about factors that are uncertain affect the bottom line.[69] A sensitivity analysis allows analysts to consider the effect of changing an assumption—in this case, the compliance rate—to see what effect that change has on the output of the model.[70] That's what the lead RRP rule did. EPA analysts tried the analysis assuming 100 percent, 75 percent, 60 percent, and 30 percent compliance to see how much those different assumptions affected the benefits and costs.

The bias built into the assumptions made that an uninformative exercise. By applying the compliance percentage uniformly to both costs and benefits, the sensitivity analysis decreed that benefits and costs move in lock step. That's the logical result of the all-or-nothing assumption contained in the single compliance rate construct. So guess what: the benefit-cost ratio stays pretty much the same under every compliance rate tried, because that's the logic embedded in the

[68] See, e.g., the discussion in chapter 2 of the data quality and completeness for the drinking water and stationary sources of air pollution programs.

[69] OMB, "Circular A-4," 41; EPA, "2010 Guidelines," 5-9, 5-10. The proposed 2020 revisions to the EPA Guidelines are similar. EPA, "2020 Proposed Guidelines," 5-18.

[70] EPA, "2010 Guidelines," 11-11.

assumption.[71] As compliance declines so will total benefits, but the benefit-cost ratio will be pretty stable.[72] This is the same flawed assumption that underlies the common expert view that varying the compliance rate doesn't really change the outcome.[73]

Using sensitivity analysis in this way hinders, rather than helps, exploration of options to improve compliance. Compliance isn't some external variable that magically occurs, or not, outside the agency's influence. Compliance is an end point that the rule has to make happen. Instead of running a sensitivity analysis that applies identical compliance assumptions for every option, the options should actively and intentionally include strategies intended to make compliance better.

Furthermore, the one-number-describes-everything approach of a single variable sensitivity analysis obscures the reality that will occur in nearly every rule: noncompliance will not be uniform everywhere. The economic analysis for the lead RRP rule, for example, is careful to point out that the sensitivity analysis only works if you assume that the rate of noncompliance isn't worse for some types of facilities.[74] If violations were more common in renovation jobs where small children live or poverty rates are higher, for example, then the benefits from higher compliance would far outstrip the additional costs. That's probably the case for RRP. These kinds of nonuniform impacts are common across all rules. Differential impacts are ignored in a single number all-or-nothing compliance rate sensitivity analysis.

[71] EPA's Guidelines say the same thing. Where noncompliance "occurs uniformly (or at random) across industry, changing the compliance rate assumption will not affect the benefit-cost ratio or the sign of net benefits." EPA, "2010 Guidelines," 5-10.

[72] The lead RRP makes additional assumptions about the baseline that result in net benefits dropping faster than the compliance rate. See EPA, "Economic Analysis Lead RRP Proposed Rule," ch. 7, at 24, Table 7-21, and ch. 7, at 5, note 5 (explaining the baseline assumptions that affect net benefits). The range of projected net benefits for this rule is so wide that they significantly overlap across compliance assumptions.

[73] See, e.g., EPA, "2010 Guidelines," 5-10; Institute for Policy Integrity, "Comments on Proposed Revision," 14; See also the compliance rate sensitivity analysis done for the Spill Prevention, Control, and Countermeasure (SPCC) rule: EPA, "Regulatory Impact Analysis for the Final Amendments to the Oil Pollution Prevention Regulations (40 CFR PART 112), Volume II—Technical Appendices" (November 11, 2008) ("SPCC RIA"), https://www.regulations.gov/document/EPA-HQ-OPA-2007-0584-0172. OMB challenged EPA's economic analysis for the SPCC rule, which was intended to relax regulatory standards, on the basis that compliance with the baseline existing rule was likely not 100%. EPA then did a sensitivity analysis, which made the familiar all-or-nothing assumption: costs to industry were assumed to be exactly twice as high at 50% compliance as they were at 25%. See EPA, "SPCC RIA," Appendix B, 21, Exhibit B-1. The SPCC compliance analysis is notable for stating that EPA does not know the extent of noncompliance in the SPCC regulated community (EPA, "SPCC RIA," Appendix B, 17) and that EPA believes that there might be zero (!) compliance with the rule in the farming community (EPA, "SPCC RIA," Appendix B, 22).

[74] This phrase is repeated throughout the lead RRP sensitivity analysis: "If it is assumed that noncompliance is independent of household LBP [lead-based paint] likelihoods, the type of RRP activities performed, and occupant composition (e.g., number, age and sex of occupants), then." See, e.g., EPA, "Economic Analysis Lead RRP Proposed Rule," ch. 7, at 6.

Someday EPA might have the depth of data to estimate with more certainty how well different regulatory strategies improve compliance, how that compliance might vary across each rule's variety of regulatory obligations, and how to account for that variability in a sensitivity analysis. We are nowhere near that point now. Doing it right would require a separate sensitivity analysis for each of the key provisions of the rule, including monitoring, reporting, training, and treatment requirements, each of which can have profound, and often unobserved, effects on benefits without reducing costs. Using sensitivity analysis with the paucity of knowledge we have about compliance today serves mainly to perpetuate the incorrect assumptions that hinder the search for more powerful compliance drivers. It creates the illusion of exactitude by restating incorrect assumptions in the impenetrable language of economics. Not only does it not add anything, it makes the problem worse by inadvertently reinforcing the idea that compliance is not a significant determinant of the rule's bottom line. Plus, it sends analysts down the rabbit hole of analytic complexity, diverting attention from what really matters: making the rule better.

Picking a different compliance rate and calling it done is worse than keeping the 100 percent compliance assumption

The lead RRP rule illustrates how picking a different compliance rate doesn't escape the compliance assumption, it just pegs it at a different percentage. Picking a different number doesn't inspire rule writers to change rule design to make compliance better. It doesn't change anything except the math in the economic analysis that will become irrelevant once the rule is promulgated. A 75 percent compliance assumption is just 100 percent compliance with a different hat.

And what does it mean to throw in the towel on compliance before you even get started? "Yep, compliance is going to be pretty bad" is quite the message for an agency adopting a new rule. What does a 75 percent—or 50 percent, or lower?—predicted compliance rate say about the agency's seriousness of purpose? How about reducing costs in your economic analysis by assuming that only 20 percent of firms will comply? That way lies madness.[75] If it's not worth trying to make it work, don't do it.

[75] That way also lies manipulation. Almost no one reads the fine print of rule economic analyses. How many people commenting on the lead RRP rule knew that it assumed that 25% of the people for whose protection the rule was written wouldn't be protected after all? Some administrations have proved willing to distort assumptions in an economic analysis to try to arrive at a preordained conclusion. See Michael A. Livermore and Richard L. Revesz, *Reviving Rationality: Saving Cost-Benefit Analysis for the Sake of the Environment and Our Health* (Oxford University Press, 2020). See also Catherine A. O'Neill, "The Mathematics of Mercury," in Harrington, *Reforming RIA*, 116. Let's not make that easier to do.

While 100 percent compliance is wildly unrealistic, it at least has the virtue of aspiring to a good outcome. Where it goes south is in taking that aspiration as fact, thereby ignoring the need to take action to make it (mostly) so. Using another compliance rate in exactly the same way does nothing to change the outcome, it simply adds the insult of low expectations. The only thing that seems less attractive than the 100 percent compliance assumption is endless debates about what different compliance rate to use instead. That would sidetrack compliance into an irrelevant accounting exercise away from the main action, which is what's going to be included in the rule.

The purpose of benefit-cost analysis, it bears repeating, is to make the rule better. To force consideration of options that will maximize public benefit. No assumption about the rate of compliance will do that. The important question is what are you going to do to assure strong compliance? Plugging a compliance assumption with all the flaws noted here into a model doesn't address that question. Some compliance rate assumptions are more cringeworthy than others, but they all suffer from the same fatal flaw: the agency is not the passive recipient of an externally determined compliance rate, it is in charge of making sure it's good. The only sensible benefit-cost strategy is one that makes this essential shift in thinking.

A serious look at government costs can be illuminating

The EPA Guidelines say that government costs should be considered in economic analysis, but that's widely ignored.[76] The RRP rule was an exception. If it had been done accurately, such an analysis might have shown the RRP rule writers how hopeless the rely-on-enforcement strategy actually was.

The lead RRP, which regulates about 320,000 renovators who perform approximately 18 million regulated projects annually,[77] estimated that 16 people would be needed to enforce the law.[78] Nationwide. *Sixteen*. You don't need to know anything about enforcement to understand how insufficient that is. Despite the ingenuity of enforcement staff in doing their best to pack the biggest punch they can, obviously that would never work.[79] That's what the EPA Inspector General found

[76] EPA, "2010 Guidelines," 9-13 to 9-14 (government costs should be considered); see also the discussion in chapters 3 and 6.

[77] EPA OIG, "EPA Not Effectively Implementing the Lead RRP," 2; EPA, "Economic Analysis Lead RRP Final Rule," ch. 4, at 24–29 (estimates of regulated "events").

[78] EPA, "Economic Analysis Lead RRP Final Rule," ch. 4, at 119. That analysis includes this trenchant observation: "Given the limited government resources expected to be available for enforcement and compliance assurance, EPA does not anticipate achieving full compliance with the rule."

[79] EPA has tried lots of innovative ideas in lead enforcement to punch above its weight. See, e.g., EPA "Corporate-wide Settlement with Lowe's Protects Public from Lead Pollution During Home Renovations," Press Release, April 17, 2014, https://www.epa.gov/enforcement/reference-news-rele

in a recent report, conveying the dog-bites-man conclusion that EPA is not effectively implementing the lead RRP.[80]

Had the real costs of attempting to achieve compliance exclusively through the expensive tool of enforcement been considered, it would have revealed two things. First is that the costs of achieving compliance that way would be unaffordable. At a bare minimum, EPA would need something like 60 times its present enforcement investment to have a credible enforcement presence.[81] Second is that Congress isn't going to give EPA the resources to increase its effort for this one rule by 6,000 percent. Congress doesn't consider benefit-cost analyses in writing budgets. It gives the agency what it gives, and that's it.

I don't favor estimating the costs of trying to force compliance on the regulated community exclusively through the most expensive tool of enforcement.[82] We need instead to consider Next Gen design strategies that are both more likely to work and far more cost-effective. But, if a rule is going to adopt a let-enforcement-do-it approach, it needs to be realistic. In virtually every case, such an analysis will show how completely impractical that strategy is. The benefit-cost numbers will be dramatically altered. EPA isn't going to get those additional resources, so this is a thought experiment, not a plan, but it might help persuade the rule writers that it is time to dust off the Next Gen playbook.

So What Can We Do?

Here's the situation. We need to make something happen (build Next Gen into rules), but nearly all of the incentives cut the other way: it doesn't fit the existing structure, it requires a change of deeply embedded beliefs, and it makes more work for people already overtaxed. All the existing guidelines say you don't have to do it, and even suggest you shouldn't. By one major metric—net benefits—you will do better by leaving it out. The only thing in its favor is that it will make the rules stronger and more effective. That's pretty important, but difficult to sustain in the face of all the pressures against.

ase-corporate-wide-settlement-lowes-protects-public-lead-pollution (settlement requiring Lowe's to ensure that all of its contractors nationwide comply with RRP).

[80] EPA OIG, "EPA Not Effectively Implementing Lead RRP."

[81] EPA presently does about 1,100 lead-based paint inspections a year. EPA OIG, "EPA Not Effectively Implementing Lead RRP," 11. Inspecting 20% of the renovators a year would mean 64,000 inspections, roughly 60 times the current number.

[82] Among other things, this kind of analysis presents a huge risk of a favorite Washington pastime: tell EPA to reallocate its existing resources to the one topic under consideration. You can do that seriatim through every program, thereby creating a shell game where EPA can be blamed for the result one program at a time, while Congress takes no responsibility for public health damage that results from insufficient resources.

It will therefore come as no surprise that for the most part this hasn't happened. There are certainly some notable exceptions, and plenty of people at EPA dedicated to effectiveness in the real world. But just like the changes the agency is trying to make through a rule, when everything is aligned against it, and we are relying primarily on good faith to overcome deeply rooted inertia, it will be a tough go.

During the Obama administration, EPA developed a lot of tools to encourage rule writers to embrace Next Gen. There was extensive training for rule writers across the agency, explaining the facts about widespread noncompliance and the factors that contribute to those implementation failures, and providing tools to address those in rules. There were workbooks for rule writers and checklists for assessing likely compliance. Specific rules were selected for extensive Next Gen consultation, in hopes of providing examples of how it looks in practice. Cross-agency workgroups were tasked with developing models and best practices. Compliance staff participated in rule-writing workgroups, senior executives launched persuasion campaigns and sometimes appealed to the administrator to intervene. EPA tried every trick in the institutional change book.[83]

Just as a Next Gen analysis would predict, all that effort had some effect in some places, but overall, it didn't take. And the already heavy lift became impossible when the Trump administration turned out to be not just uninterested but actively hostile to effective rules. With the advent of the Biden administration there are encouraging signs of openness to Next Gen approaches, and at least one recent rule that is the best example of Next Gen strategies since the Acid Rain Program.[84] That's great but doesn't reflect an agency-wide change of heart.

With the years of experience during the Obama EPA under our belts, it is time to face the fact that this institutional change—just like any other change that pushes against deeply engrained practices and beliefs—requires institutional drivers. We know that from Next Gen but have been reluctant to apply the same objective thinking to our own behavior. We have to make putting Next Gen in rules the path of least resistance, instead of the path of most resistance as it is today.

That's why Next Gen has to be mandated. And we need to insert institutional checkpoints that make it impossible to avoid. Those changes require revisions to the internal EPA guidelines that direct rule-writing process and economic analysis, and a charge to OMB to press the point. I recognize that this idea will be wildly unpopular. But that doesn't necessarily mean it is wrong.

[83] See Government Accountability Office, "Federal Regulations: Key Considerations for Agency Design and Enforcement Decisions," GAO-18-22, October 2017, at 11, 15–16.

[84] EPA, "Final Rule for HFC Phasedown."

What follows is an outline of what a Next Gen–driving institutional structure might look like. This is more a sketch than a developed proposal, meant to provoke discussion and communicate the direction and scale of the change that is necessary.

Step one: create a new framework for considering compliance in rules

Instead of leaving compliance to the end of both the options selection and economic analysis work—assuming it is considered at all—move it to the beginning and give it a prominent role in rule design from the outset. The most common response to Next Gen is to cast about for a less-than-100 percent compliance rate to use, as if tallying up predicted bad performance is any kind of an answer. Same for options selection; usually the most you can expect is discussion about ways to make the already-settled rule design more enforceable. These staple-it-on-at-the-end ideas are the wrong way around. We need to come at it from the opposite direction: How can the rule be designed to prevent violations?

That starts by not committing to a particular approach in advance. We should focus on the desired outcome and be less attached to the means to get there. Being more open to creative alternatives can lead to better results at lower cost than rigidly fixating on a single strategy. How easily the rule can assure a strong compliance outcome should be part of that calculus from the start. Sometimes Next Gen ideas can supplement traditional approaches. Sometimes Next Gen will show that a desired strategy is a nonstarter. Adherence to ideological preferences is at odds with the how-things-really-work pragmatism of Next Gen.

Having narrowed the range of options to the ones where compliance drivers are potentially feasible, rule writers should think about how each alternative under consideration can be set up so that implementation is good even if there were no enforcement. There is no recipe that works for every situation; the best answers depend heavily on the characteristics and pressure points for that particular industry/sector/problem. But we know a lot, and there are already-developed resources to guide that analysis. Some often-effective strategies are outlined in chapter 5. Whenever rule writers find themselves thinking that the regulated parties will do it because it is required or it is the right thing to do, they should think again.

Remember that it is common for a solid compliance design to include 10 or more design elements intended to inspire better performance across a range of compliance obligations, including monitoring, work practices, recordkeeping, reporting, pollution controls, risk-reduction protocols, and transparency. Some Next Gen ideas might involve real money—like effective monitoring—and some will be virtually no cost—like a presumption for what will happen with missing data in a report.

Rule writers should be able to ask this question: "Are there ways for regulated parties to obfuscate, avoid or ignore the most important actions the regulation is requiring?"—and be able to truthfully answer, "Not really." This isn't suggesting firms want to violate. That's usually not the case. It's saying that regulators need to make compliance the path of least resistance. When you get to the "not really" point, you have the suite of Next Gen strategies that should be included in the rule. It will almost never be just one thing. It will be a collection of measures that work together to make compliance much closer to the default setting. When this exploration concludes that there isn't any affordable way to assure strong results, that's telling you that you need to reconsider your basic approach. Start over and try something else.

The goal of this design exercise is to develop a compliance system for each option that will get as close as possible to the aspirational 100 percent compliance goal. That's the package of Next Gen measures that will accompany each option during its trip through options selection and proposal to final. Those compliance drivers are part of the rule. They are what justifies the assumption about benefits and are just as essential to the rule option as the standard. The Next Gen provisions can't be separated from the rest of the proposal. If you want to count the benefits described in the rule's economic analysis, this is what it is going to take. They are attached at the hip.

Extra credit for exploring multiple options—each with associated provisions for assuring compliance—to see how that affects the benefit-cost analysis. Once the full costs of assuring compliance are baked into each option, it might be more feasible to meaningfully compare than it is now, when the truth is obscured by the 100 percent compliance assumption unhindered by any obligation to try to achieve that. Including the real costs of assuring strong implementation for every considered alternative allows options to compete on level ground. We might discover that some popular strategies look less attractive when we stop allowing them to hide the fact that compliance will likely be bad.

Step two: make it mandatory

Next Gen is a heavy lift within the rule-writing infrastructure of the agency. Inertia is against it. It's unfamiliar. It goes against the established grain of "not my problem." Theory would tell you that under these circumstances, urging, exhorting, and explaining will not make it happen. And years of experience within EPA say the same thing. We tried that.

An additional challenge of my recommended approach for including Next Gen in rules—you can assume 100 percent compliance but only if you build in strategies to make that likely—requires judgment. Whether your rule clears that threshold or not isn't an objective or numeric standard. The big gray zone of a judgment-dependent requirement provides opportunity for confusion, legitimate differences, and attempts to avoid. We have seen how nonquantitative issues can be shoved aside and ignored in the quantification-heavy process of economic analysis. Structural features designed to prevent rule writers from doing that for Next Gen need to be built in.

This is why the formal ADP Guidance has to explicitly say that compliance design must be part of the rule-development process throughout, from the earliest stages to the last, with an obligatory sign-off from the compliance office at the analytic blueprint/early guidance stage, and again at option selection. That concurrence can include conditions of agreement, for example, the market strategy is fine from a compliance perspective, but only if real-time monitoring is mandatory and the rule includes provisions to ensure monitoring actually happens, like data-substitution requirements. This is not to imply that enforcement and compliance staff are the holders of all wisdom about compliance. They aren't. But requiring a checkpoint with people who know that compliance can't be taken for granted will underscore the obligation to consider implementation outcomes in rule design.[85] There are downsides to such a mandatory sign-off; it can feel heavy-handed when collaboration is what's needed. But cutting the other way is the powerful inertia against change. One of the most common ways to deal with new requirements is to go through the motions and appear to be doing what's required without actually changing anything. When it becomes clear people can get away with that, almost everyone will do it. Sign-off from another office that isn't going to succumb so readily to status quo compliance thinking is one way to stop that downward slide.

At the same time that the ADP guidance is changed to require consideration of compliance in drafting rule options, the Guidelines should be revised to explain how compliance design should be embedded in the economic analysis. It should require that analysts cannot use the 100 percent compliance assumption unless rule writers conduct the compliance analysis required under the Step One framework and build robust compliance drivers into the rule options.

When a rule is proposed, agencies have to look at the most current science. They are expected to explain how the rule is consistent with the available evidence, and why it is justified based on today's knowledge. Evidence about

[85] Another EPA colleague also recommends this approach. See Roberts, "Integrating Compliance and Regulatory Design," 576.

widespread noncompliance should be part of that evidentiary record too.[86] Under the new framework, analysts will not guess how bad compliance will be and call it a day. Instead, rule writers will strive to make compliance better. They will explore the reasons that compliance might be poor and develop strategies to counteract that. Strong compliance, and the expense of achieving it, will be part of the rule package benefits and costs.

Requiring a robust exploration of compliance challenges, and specific strategies to make it better, would force rule writers to grapple with the reality that some problems and some regulatory strategies are likely to have far worse compliance outcomes than others. It puts regulatory design center stage. An option with strong compliance design shouldn't be measured against an option that is a predictable compliance disaster as though these are both accomplishing the same thing. It isn't possible to meaningfully compare the protectiveness and costs of options unless each option is designed to ensure strong implementation.

That exploration could well change conventional wisdom on the comparative costs of options. For example, let's say the agency is writing a rule to control a harmful chemical, and among the options being considered are constraints on use of the chemical to protect the public, or a complete ban. The option that imposes constraints on use will almost always do better on a conventional benefit-cost analysis because that's probably lower cost for industry. A compliance analysis might present those options in a different light. A ban might cost a little more, but it may be close to a sure thing from an implementation perspective. The use constraints may be next to impossible to track without a robust new strategy for potentially intrusive monitoring and reporting. The cost of those requirements—an investment that is necessary for the option to pass the test of predicted strong compliance—may well shift the comparative net benefits of the two choices. If every option is required to include such necessary compliance drivers, the benefit-cost comparison occurs on equal footing: it compares the actual costs of achieving the benefits claimed and doesn't allow some options to claim benefits that the rule design hasn't earned.

A requirement to include compliance drivers will also encourage rule writers to explore Next Gen–type solutions that can dramatically improve compliance outcomes, sometimes without costing very much. If every option has to consider its likely compliance impact, and include strategies to make compliance better, rule writers will start looking around for ideas. That alone would be huge, because it inspires rule writers to welcome innovative compliance strategies, rather than resist ideas offered by Next Gen interlopers. It would put Next Gen

[86] The agency should not rely exclusively on industry's self-reported violations for this analysis, especially where those are known to be flawed. Studies showing that the self-reported data is incorrect or incomplete should be part of the analysis as well. See, e.g., chapter 2; Mu, "What's Missing in Environmental (Self-)Monitoring"; Stuart, "Strategic Non-Reporting."

proponents and rule writers on the same team. The currently strong incentives to create compliance-defeating exemptions, exceptions, and complexity will have a counterweight of accountability. The hidden compromise all too common in rules today—improve the benefit-cost ratio at the expense of implementation in real life—will be dragged into the open.

A spin-off benefit of this approach is that it could inspire more experimental methods for figuring out what compliance strategies work best. Today compliance is vaguely attributed to enforcement. The advanced thinking mentions general deterrence. Almost no one wonders why the rule has so many violations to begin with. If rule writers have to account for expected compliance, such questions become important. How powerful is transparency as a compliance strategy? Does shifting the burden of proof make a substantial difference? How much does real-time monitoring change behavior? These are all important questions, for which our current evidence is largely anecdotal. Once the answers matter to option selection in rules, agencies might be motivated to engage in field experiments to develop more robust understanding of compliance drivers.

Rule writers won't embrace the new compliance framework just because it is put into the ADP and economic analysis guidelines. This is Next Gen 101. Let's-find-a-work-around thinking will dominate. It would be easy to devolve into a boilerplate cut-and-paste recitation of blah blah blah compliance rhetoric. Rule writers will continue to be attracted to ways to trim costs without changing the benefits, as well as political compromises, leading to last-minute changes to the rule that add compliance-defeating exceptions and ambiguity or that delete Next Gen provisions while still claiming all the benefits, that is, exactly what happens now. One of the defining features of the rule-writing process today is that Next Gen proponents are not in the room when the final changes are being made and Next Gen provisions sink without a trace. How can we avoid that?

1. Require a written accounting of the evidence, thinking process and conclusions in the rule preamble and/or the regulatory impact analysis. That writing should examine the evidence about noncompliance, identify the provisions of the proposed rule that are most essential for accomplishing the goals, and devise a suite of Next Gen strategies for assuring that they happen. In other words, describe the strategy for getting close to 100 percent compliance that meets the requirements of the new framework. This will both help to keep the process honest and list the rule provisions that are essential to achieving the benefits.
2. Tie the Next Gen ideas to the rule package. One way to counteract the many incentives that push against the new approach is to lash Next Gen to the rule option benefits. You cannot have one without the other. The Next Gen provisions are what make it possible to rely on the 100 percent compliance

assumption. If you make the rule less compliance friendly or take out the drivers that assure results, it's back to the drawing board. No keeping the standard and dropping the Next Gen. Rule writers today wouldn't dream of requiring pollution control equipment but deleting any costs for operation and maintenance of that equipment in a bid to improve the net benefit bottom line. They couldn't get away with ignoring such obvious facts in a naked attempt to torque the accounting. The same needs to be true for Next Gen. Those provisions are in there to make the rule work, and without them it won't. No line-item veto. These are the costs for achieving these benefits. They go together.

The same goes for adding new provisions that make compliance considerably less likely. Like deciding that existing plants will be exempt, also known as grandfathering. Or adding lots of vague qualifiers to the compliance obligation.[87] Those are not minor changes; they go to the heart of implementation. You didn't analyze the compliance outcomes for a rule with those dramatically compliance-altering features, so there is no basis to rely on the 100 percent compliance assumption. It isn't that you can't have such provisions—although they are often a bad idea—it's that the compliance analysis has to include them. If it doesn't, no 100 percent compliance assumption for you. The confusion and uncertainty that would be created for the benefit-cost analysis with the 100 percent compliance assumption rug pulled out will hopefully inspire rule writers not to go down that road. Late hits to the compliance framework won't be worth it.

3. Don't succumb to the "enforcement will do it" assumption. Enforcement is there to address the outliers, the determined violators, and the unanticipated problems that pop up. Enforcers will have their hands full with that. Counting on enforcement to do the basic job of solid implementation is just punting. No "miracle happens here" in the math equation.

The goal of the group of ideas outlined in this section is to force a basic Next Gen analysis for every rule. You can't get to the end without it. There are graduate level ideas in Next Gen, and any strategy to make rules more effective should encourage those. But everyone has to pass the high school equivalency exam.

[87] See Cary Coglianese, Gabriel Scheffler, and Daniel E. Walters, "Unrules," *Stanford Law Review*, Vol. 73 (2021): 885, for a fascinating analysis of the carveouts often added to regulations.

Step three: issue OMB guidance that requires agencies to account for compliance performance

The new framework, and revised EPA guidance requiring rule writers to use it, will cause real change for some programs and rules. Converts will see the benefits and start developing even better ways to build compliance into rules. But that won't be the norm. The beliefs that compliance is good and that it is up to enforcement to ensure it happens are deeply engrained. Not my lane, not my problem, way too complicated will present often insurmountable barriers.

That's why internal guidelines that can be ignored with little chance of contradiction won't do the trick.[88] We need those, but they alone won't make it happen. OMB has to direct that it be done. That's just applying the lessons of Next Gen to our own work. In rules we know that providing industry with a complicated voluntary tool won't result in widespread adoption, and it won't work for rule writers either. The existing system has created the expectation that assuring compliance in rules requires almost no effort or thought. Any new strategy, no matter how elegant, by definition adds work. It has in its favor that it would increase the effectiveness of agency rules—the reason it is adopting a rule in the first place—but arrayed in opposition to that are overpowering reasons against, including I don't believe that's needed, we never did it that way before, I don't have time, I have no idea how good compliance will be, why are these compliance people sticking their noses into my rule?, why can't enforcement just do their job?, and many more. The batting average will not be high enough to make it into the major leagues.

The target audience for an OMB directive isn't just the regulatory agencies. It's also OMB itself. The staff in the Office of Information and Regulatory Affairs in OMB (OIRA) have a reputation—whether deserved or not—of leaning against regulation and putting costs for industry at the top of the importance list.[89] They expressly favor some kinds of regulatory strategies—like performance standards and markets[90]—although the evidence to support their effectiveness is thin, and

[88] The agency doesn't always follow the requirements of the ADP, as we were reminded by a recent EPA Inspector General report, finding that there was a wide variation in the EPA's adherence to its ADP, ranging from 44% to 100%. EPA OIG, "EPA Does Not Always Adhere to ADP," At a Glance. The EPA OIG didn't include in its analysis the issue of some offices being cut out of the rule-development process, which as I have explained in this chapter is a key failure of the ADP for Next Gen. EPA OIG, "EPA Does Not Always Adhere to ADP," 12.

[89] See Lisa Schultz Bressman, "Flipping the Mission of Regulatory Review," *The Regulatory Review*, Feb. 18, 2021: "OIRA has long had the reputation of housing career economists with outmoded training and an anti-regulatory orientation." See also Lisa Heinzerling, "Inside EPA: A Former Insider's Reflections on the Relationship Between the Obama EPA and the Obama White House," *Pace Environmental Law Review*, Vol. 31, No. 1 (Winter 2014): 325–69 (describing how the actual practice of OMB review of regulations is quite different from what the executive orders say, and that environmental rules take a particular beating in the process).

[90] See OMB, "Circular A-4," 8–9.

despite the fact that they can be among the most compliance-unfriendly ways to proceed.[91]

An OMB directive would remedy this situation by requiring that every regulatory impact analysis reviewed by OMB include an examination of the compliance risks, available compliance evidence, a description of structural drivers inserted in the rule to improve compliance, and why the agency thinks that compliance will be strong.

EPA already knows a lot that can inform compliance-resilient rule design. It has a vast trove of enforcement and compliance data and experience. EPA managers have guidance, checklists, and multiple tools for building compliance into rules and identifying issues that are likely to lead to significant compliance failures.[92] This book describes a multitude of strategies that can be effective.

But EPA has barely scratched the surface because use of these tools is entirely voluntary. Remember that the people who decide to use these tools, or not, are the same people who largely accept the assumption that compliance is already good and believe worrying about that is enforcement's problem. They aren't avid consumers of the evidence proving that's not right, because those facts are an uneasy fit with the standard beliefs. Yes, they care deeply about having effective rules, but the list of things they have to worry about is already long without adding new problems. If rule writers are not required to robustly consider compliance, with a backstop from OMB that means the obligation cannot be ignored, not many will willingly take it on.

The first of no doubt many objections to this proposal is likely to be, "Wait, OMB should do a general directive for all agencies to address EPA's compliance problems?" I completely agree. That would be crazy. Except count me skeptical that this problem is unique to EPA. I see parallel situations in many other regulatory programs. The news is filled with examples in other regulatory arenas of disastrous compliance fails, followed inevitably by calls for stronger enforcement. Sounds pretty familiar. Here are just a few of recent vintage.

The Food and Drug Administration (FDA) is grappling with widespread and serious violations in generic drug manufacturing around the globe, with a "dangerous chasm between what regulations required of drug companies and how some of those companies actually behaved."[93] One FDA inspector uncovered

[91] See discussion in chapter 6.

[92] Most of the training and guidance for building compliance into regulations at EPA was developed by David Hindin, who was the career leader of our Next Gen work at EPA. He lays out the key principles in David A. Hindin and Jon D. Silberman, "Designing More Effective Rules and Permits," *George Washington Journal of Energy & Environmental Law* (Spring 2016). See also the analysis by another person who worked with us at EPA: Roberts, "Integrating Compliance and Regulatory Design," 545.

[93] Jonathan Lambert, "'Bottle of Lies' Exposes the Dark Side of the Generic Drug Boom," *NPR*, May 12, 2019.

unsafe practices and deliberate attempts to fool regulators in over 75 percent of the drug plants he inspected in India and China. Such unlawful actions contributed to generic drugs with toxic impurities, unapproved ingredients, and dangerous particulates.[94] It's not a small problem: 80 percent of the active ingredients used in both generic and brand-name medications in the United States come from abroad, the majority from India and China.[95] Katherine Eban's 2019 book *Bottle of Lies* describes the valiant but ultimately fruitless attempt of innovative FDA inspectors to solve this problem through surprise inspections, rigorous investigations, and enforcement. Eban describes the FDA's oversight as "a system built on wishful thinking and infrequent scrutiny, which yielded disastrous results."[96]

The Internal Revenue Service (IRS) has a well-documented tax-cheating problem. High-income individuals are underreporting their income, contributing to an annual $600 billion loss of taxes that were legally owed but unpaid.[97] Wages and salaries are subject to third-party reporting (your employer tells the IRS how much you were paid) and also withholding (your employer deducts your taxes owed from your pay). As a result, taxpayers only misreport 1 percent of such income.[98] In stark contrast, taxpayers misreport more than half of income for which there is no third-party reporting.[99] Adopting the Next Gen–style structural solution of third-party reporting for higher-income individuals would go a long way toward closing the gigantic tax gap.[100] After decades of persistent nonpayment of owed taxes, changing the law to require some third-party reporting is finally getting some traction.[101] It is notable that clarity about the problem and the solution is only possible because of the IRS's national research

[94] Katherine Eban, "Americans Need Generic Drugs. But Can They Trust Them?," *New York Times*, May 11, 2019.

[95] Eban, "Americans Need Generic Drugs"; David Dobbs, "A New Book Argues That Generic Drugs Are Poisoning Us," *New York Times*, May 13, 2019.

[96] Katherine Eban, *Bottle of Lies* (Harper Collins, 2019), 226.

[97] See Charles O. Rossotti, Natasha Sarin, and Lawrence H. Summers, "Shrinking the Tax Gap: A Comprehensive Approach," *Tax Notes*, December 15, 2020, https://www.taxnotes.com/featured-analysis/shrinking-tax-gap-comprehensive-approach/2020/11/25/2d7ht. Note that there are diverse ways to calculate the tax gap, which is why various reports peg the number at different levels. While they may not be exactly the same, they are all huge.

[98] GAO, "Tax Gap: Multiple Strategies Are Needed to Reduce Noncompliance," Statement of James R. McTigue, Jr., GAO-19-558T (May 9, 2019), 8.

[99] GAO, "Tax Gap," 8. The Treasury Department Inspector General says that where there is neither withholding nor information reporting, the IRS believes tax compliance is as low as 37%. Treasury Inspector General for Tax Administration, "Understanding the Tax Gap and Taxpayer Noncompliance," Testimony of the Honorable J. Russell George (May 9, 2019), 2.

[100] GAO, "Tax Gap," 7–8, 14; Treasury IG, *Understanding the Tax Gap*, 9.

[101] "How to Collect $1.4 Trillion in Unpaid Taxes," editorial, *New York Times*, March 20, 2021, https://www.nytimes.com/2021/03/20/opinion/sunday/unpaid-tax-evasion-IRS.html; US Dept. of Treasury, "The American Families Plan Tax Compliance Agenda" (May, 2021), 1–2, 18–20, https://home.treasury.gov/system/files/136/The-American-Families-Plan-Tax-Compliance-Agenda.pdf.

program, which has rigorously documented both the violations and the reasons for them.[102]

The Federal Aviation Administration (FAA) got snagged by the most-companies-comply mode of thinking when it started allowing airplane manufacturers to certify their own compliance with safety standards. This previously little-known program was brought forcibly to our attention in 2019 when two Boeing Max 737s crashed, killing 346 people. It turned out that FAA safety certification for Boeing was largely delegated to Boeing employees.[103] Letting industry police their own compliance. What could go wrong? As one report put it: "The assumption that a 'for profit' company that is faced with significant financial incentive will always make appropriate compliance findings contradicts human nature and is not supported by experience in other industries."[104] Boeing subsequently paid $2.5 billion to settle criminal charges that it conspired to defraud the FAA in connection with the certification of the 737 Max.[105] What's unusual about this example isn't that the FAA's delegation system is so obviously a huge compliance risk, it's that Congress finally recognized a structural flaw in the delegation program and moved (belatedly) to start to fix it.[106]

The *New York Times* has documented that a good deal of the information submitted by nursing homes to the US Centers for Medicare & Medicaid Services (CMS) in support of its much-touted star rating system is wrong. The CMS star rating system for nursing homes is widely used by consumers. In search of the coveted, and profitable, five-star rating, many nursing homes are inflating staffing levels and underreporting accidents, health problems, and numbers of patients on dangerous antipsychotic medicine.[107] *New York Times* investigators

[102] GAO, "Tax Gap," 6.

[103] See Aaron C. Davis and Marina Lopes, "How the FAA Allows Jet Makers to 'Self-certify' that Planes Meet U.S. Safety Requirements," *Washington Post*, March 15, 2019, https://www.washingtonpost.com/investigations/how-the-faa-allows-jetmakers-to-self-certify-that-planes-meet-us-safety-requirements/2019/03/15/96d24d4a-46e6-11e9-90f0-0ccfeec87a61_story.html; Dominic Gates and Mike Baker, "Engineers Say Boeing Pushed to Limit Safety Testing in Race to Certify Planes, Including 737 Max," *Seattle Times*, May 5, 2019. The Department of Transportation (DOT) Inspector General found that there were instances when a Boeing employee worked on an aircraft design and then changed hats and approved that same design as safe on behalf of the FAA under the delegation program. DOT OIG, "Weaknesses in FAA's Certification and Delegation Processes Hindered Its Oversight of the 737 MAX 8," Report No. AV2021020 (February 23, 2021), 33.

[104] 2017 Report by the National Air Traffic Controllers Association, cited in House Committee on Transportation and Infrastructure, "Final Committee Report on the Design, Development and Certification of the Boeing 737 Max" (September 2020), 64.

[105] See Niraj Chokshi and Michael S. Schmidt, "Boeing Reaches $2.5 Billion Settlement with U.S. over 737 Max," *New York Times*, January 7, 2021.

[106] Congress had pushed the FAA to expand the certification delegation program but reversed course after the fatal crashes. See Dominic Gates, "Congress on the Brink of Major FAA Oversight Reform in Wake of Boeing 737 Max Crashes," *Seattle Times*, December 21, 2020.

[107] Jessica Silver-Greenberg and Robert Gebeloff, "Maggots, Rape and Yet Five Stars: How U.S. Ratings of Nursing Homes Mislead the Public," *New York Times*, March 13, 2021, https://www.nytimes.com/2021/03/13/business/nursing-homes-ratings-medicare-covid.html.

found that in a three-year period, health inspectors wrote up about 5,700 nursing homes—more than one in every three in the country—for misreporting data, including nearly 800 homes with top ratings. The Health and Human Services Inspector General found that nursing homes only reported 16 percent of incidents where residents were hospitalized for potential abuse and neglect.[108] The California Attorney General sued a nursing home chain with more than 68,000 facilities across the nation for submitting false information to CMS in an attempt to manipulate the star rating system.[109]

These are just a few examples from headlines in the last few years. It turns out EPA's compliance challenges aren't that unusual. That's what you would expect, because there's no reason to think that compliance is worse for environmental rules than it is for other regulatory programs. The real world is messy in every field of endeavor. Companies look for ways around, people make mistakes, it is easy to not pay attention, most follow the path of least resistance, and some cheat. If visibility into violations is low, and opportunities to evade or obfuscate or ignore abound, we know what happens. There is little point in railing against these facts of life. The question is, what are we going to do about it?

The time is right. President Biden has called for a fresh look at OMB's standards for regulatory action.[110] OMB is charged to review and modernize regulatory review so that it affirmatively promotes regulations that advance essential values, including environmental stewardship, public health, and racial justice. No agency is more affected by OMB's regulatory review than EPA; energy and environmental regulations represented more than 80 percent of the benefits and about two-thirds of the costs of all significant federal rules.[111]

Conclusion

The twin beliefs that underlie nearly all environmental rules—that companies will comply just because you write a requirement in a rule, and it is up to

[108] Department of Health and Human Services Office of the Inspector General, "Incidents of Potential Abuse and Neglect at Skilled Nursing Facilities Were Not Always Reported and Investigated," A-01-16-00509 (June 2019), 11–12.

[109] Mallory Hackett, "California AG Claims Large Nursing Home Chain Manipulated CMS Star Ratings," *Health Care Finance News*, March 16, 2021, https://www.healthcarefinancenews.com/news/california-ag-claims-large-nursing-home-chain-manipulated-cms-star-ratings.

[110] Presidential Memorandum, "Modernizing Regulatory Review" (January 20, 2021), https://www.whitehouse.gov/briefing-room/presidential-actions/2021/01/20/modernizing-regulatory-review/. The Presidential Memorandum specifically calls for updating OMB's 2003 Circular A-4. Revesz, "New Era for Regulatory Review."

[111] Joseph E. Aldy, "Modernizing Regulatory Review for Energy, Environmental Policy," *Environmental Forum* (Environmental Law Institute, March/April 2021).

enforcement to make compliance happen—have been cemented in place by EPA guidance.

Some have responded to criticism of the 100 percent compliance assumption by suggesting we insert a more realistic compliance assumption in its place. I am opposed to that for this simple reason: our goal is to make the rules better, not more accurately track their demise. Let's not chuck the 100 percent compliance assumption, let's design rules that are worthy of it.

The suggestions in this chapter apply Next Gen learning to EPA's regulatory process. Regulated companies don't reliably do things just because you tell them to, and rule writers don't either. We tried the voluntary approach at EPA, and predictably, it didn't work. You have to create a process where doing the new thing is impossible to avoid. There is static and pushback at the outset, because of course. Asking people to change is a steep uphill climb. Once it gets going though, the initially resistant can be the best advocates.

Regulations are one of the main ways governments protect people. The public is counting on regulators to make sure that happens. We can do a lot better than we are doing now, by being clear-eyed about implementation realities. It is possible to make compliance the path of least resistance. We just have to set aside the blinders that have obscured the path thus far and give ourselves a strong shove in the right direction.

5

Next Gen Strategies

A Playbook

Although every regulation needs a design tailored to its specific problem, some Next Gen strategies are so important and broadly valuable that they deserve a place in just about every program. This chapter is about those widely useful tools, the workhorses of Next Gen.

Some of the strategies in common use already—like monitoring and reporting—are not living up to their potential and could be a great deal more powerful with a Next Gen upgrade. Others—like data-substitution requirements and shifting the burden of proof—are underappreciated and should be added to the consider-every-time list. All of these tools have stand-alone benefits, but their real power is in how they work in combination with other Next Gen ideas. Monitoring is essential, but its impact is greatly amplified if the data is reported to regulators. Government awareness is good, but public disclosure adds the power of public accountability. Each element has a role to play as part of the larger compliance-driving structure.[1]

Next Gen tools are just like any other part of a regulation. Regulators can't assume that companies will do them just because the rule tells them to. Even the best Next Gen ideas aren't immune from the realities that affect all other parts of a regulation; if there are many ways to evade, avoid, or ignore a Next Gen idea, it won't be that effective. So compliance drivers like monitoring or reporting also need to include provisions that make them more likely to be implemented. That's Next Gen 101.

Monitoring

The centerpiece of a strong regulatory outcome is measuring what regulators care most about. Reliable monitoring, especially in close to real time, significantly

[1] For a well-analyzed identification of principles to inform more effective rules, written by two EPA employees who helped lead that effort from the inside, see David A. Hindin and Jon D. Silberman, "Designing More Effective Rules and Permits," *George Washington Journal of Energy & Environmental Law* (Spring 2016).

Next Generation Compliance. Cynthia Giles, Oxford University Press. © Cynthia Giles 2022.
DOI: 10.1093/oso/9780197656747.003.0006

improves the chances that the regulation will accomplish what it set out to achieve. It isn't enough by itself—it works as part of a suite of strategies to drive better performance—but credible measurement can be the foundation of a powerfully effective regulatory structure.

The confidence that strong measurement creates provides more room for innovative regulatory options. Accurate pollution measurement is what supported cost-saving trading of pollution allowances in the Acid Rain Program, for example; precise monitoring of SO_2 at every regulated source allowed creation of a market, which otherwise would have been impossible.

Emerging techniques to measure pollution are the biggest technological change that could dramatically affect outcomes. Pollution monitoring is rapidly getting better, faster, cheaper, smaller, and more mobile. Especially when combined with powerful new information technologies and the ability to crunch previously daunting amounts of data, these technological innovations can be a game changer. This chapter includes discussion of the compliance benefits of requiring sources to monitor their own pollution or other regulated behavior. Use of advanced monitoring strategies by regulators is explored in chapter 10. What are the key principles for effective monitoring strategies?

Direct Measurement Is Best

One of the biggest compliance problems environmental programs face is the reality that when EPA monitors actual emissions, it finds that pollution is much worse than is being reported. An example: actual measurement of flares at two large refineries found that they were emitting up to 25 times the amount of pollution that the companies had reported based on estimates.[2] The net effect: a lot more pollution was going into the surrounding communities than was revealed by the estimated emissions.[3]

Government's information about pollution from regulated sources is frequently off the mark because reported pollution is estimated rather than measured.[4] Many air sources rely on EPA's "emission factors" or other estimates to

[2] See Cynthia Giles, "Next Generation Compliance, Partnership with States and Local Air Agencies," Presentation at the National Association of Clean Air Agencies (May 5, 2014), slide 8, for a graph displaying data on the emissions disparities at these refineries, available online at http://4cleanair.org/Spring2014/Giles.ppt.

[3] EPA, "EPA Enforcement Targets Flaring Efficiency Violations," EPA Enforcement Alert (August 2012), https://www.epa.gov/sites/default/files/documents/flaringviolations.pdf.

[4] Ralph Smith, "Detect Them Before They Get Away: Fenceline Monitoring's Potential to Improve Fugitive Emissions Management," *Tulane Environmental Law Journal*, Vol. 28, No. 2 (Summer 2015): 446–47. See also sources cited in chapter 2, section titled "For Some Important Programs, EPA's Understanding of Noncompliance Is Wrong" (stationary sources of air pollution).

report their pollution levels instead of actually measuring their emissions.[5] EPA's emission factors—over half of which EPA itself describes as "poor" quality—are averages and were never intended to predict what an individual facility emits. EPA itself recently issued an Enforcement Alert warning facilities about using such estimated emissions in compliance reporting, citing the example of health-concerning emissions from heated oil storage tanks where emission factors understated actual emissions by a factor of 100.[6] Time and again actual measurement reveals that actual pollution is much worse—sometimes hundreds of times higher—than estimates claim.[7]

While the evidence piles up that emission factors systematically underestimate emissions, new monitoring technologies are a possible paradigm changer.[8] Advances in measurement technology make more accurate and reliable measurement possible for many pollutants.[9] Some monitoring technologies create a visual image of pollution that can't be seen with the naked eye, a quick and powerful detection strategy.[10] Satellites are increasingly viable as a site-specific pollution tracker.[11] The cost of direct measurement is already reasonable for some pollutants, and prices are rapidly declining.[12]

New monitoring technologies can also help identify pollution from dispersed "fugitive" sources—like valves, pumps, and compressors—at a large facility.[13]

[5] EPA, "AP-42: Compilation of Air Emissions Factors," https://www.epa.gov/air-emissions-factors-and-quantification/ap-42-compilation-air-emissions-factors (list of emission factors); Rachael Leven, "Most of EPA's Pollution Estimates Are Unreliable. So Why Is Everyone Still Using Them?" *Center for Public Integrity*, January 29, 2018, https://publicintegrity.org/environment/most-of-the-epas-pollution-estimates-are-unreliable-so-why-is-everyone-still-using-them/ (explaining the many flaws in use of emissions factors).

[6] EPA, "EPA Reminder About Inappropriate Use of AP-42 Emission Factors," EPA Enforcement Alert, November 2020, https://www.epa.gov/sites/default/files/2021-01/documents/ap42-enforcementalert.pdf.

[7] See chapter 2, "For Some Important Programs, EPA's Understanding of Noncompliance Is Wrong"; EPA OIG, "EPA Can Improve Emissions Factor Development and Management," EPA OIG Report #2006-P-00017, March 22, 2006, at 8, 10–13.

[8] See Emily G. Snyder et al, "The Changing Paradigm of Air Pollution Monitoring," Environmental Science and Technology, Vol. 47, No. 20 (2013): 11369–377.

[9] Smith, "Fenceline Monitoring," 434.

[10] For video showing how infrared cameras reveal otherwise invisible pollution, see Jonah M. Kessel and Hiroko Tabuchi, "It's a Vast, Invisible Climate Menace. We Made It Visible," *New York Times*, December 12, 2019, https://www.nytimes.com/interactive/2019/12/12/climate/texas-methane-super-emitters.html.

[11] Brady Dennis, "How Satellites Could Help Hold Countries to Emissions Promises Made at COP26 Summit," *Washington Post*, November 9, 2021, https://www.washingtonpost.com/climate-environment/2021/11/09/cop26-satellites-emissions/.

[12] Adam Babich, "The Unfulfilled Promise of Effective Air Quality and Emissions Monitoring," *Georgetown Environmental Law Review*, Vol. 30, No. 4 (Summer 2018): 571, 575.

[13] See, e.g., EPA, "Standards of Performance for New, Reconstructed, and Modified Sources and Emissions Guidelines for Existing Sources: Oil and Natural Gas Sector Climate Review," Proposed Rule, *Federal Register*, Vol. 86 (November 15, 2021): 63146, 63175–177 ("EPA 2021 Proposed Methane Rule") (description of a workshop exploring advances in monitoring technology and a proposal for changing EPA's regulatory approach based on those monitoring innovations).

Monitoring at the fence line is also possible, as EPA required refineries to do for benzene in a 2015 rule.[14] Unsurprisingly, fenceline monitoring revealed emissions of benzene at many refineries that were well above what they should have been.[15] Fenceline monitoring can work for other sectors,[16] and for water pollution as well.[17]

New measurement technologies for water pollution can help resolve decades-long disputes that have hampered cleanup. One example is the so-called "lab on a chip" technologies.[18] We are close to the day when monitoring can quickly identify the source of pathogens in surface water, making it easy to figure out if the problem is leaking septic tanks or the upstream chicken farm.[19] That measurement clarity can cut through arguments about who is responsible, making it easier to solve the compliance problem.[20]

It's not just pollution that can be directly measured. A game changer for illegal fishing is the ability to track vessel location and know if a ship is in a no-fish zone.[21] Some creative researchers—using ocean-going birds fitted with radar detectors—discovered that 28 percent to 37 percent of fishing vessels turned off their electronic tracking devices to evade detection, proving both the value of innovative monitoring and the need to remain vigilant.[22]

[14] See Babich, "The Unfulfilled Promise," 599; Gina McCarthy and Janet McCabe, "Foreword," *Harvard Environmental Law Review*, Vol. 41, No. 2 (2017): 322–23.

[15] "Environmental Justice and Refinery Pollution," *Environmental Integrity Project* (April 2021), https://environmentalintegrity.org/reports/environmental-justice-and-refinery-pollution/.

[16] Hedrick Strickland and Bob Fraser, "On the Fence about Fenceline Monitoring?," *Air and Waste Management Association* (August 2018) ("As passive monitoring proves its utility in the refinery sector and becomes mainstream, more opportunities to utilize it for 'next-generation compliance' could be forthcoming for other types of facilities."); Jacob Hollinger, "EPA's Next Generation Compliance Initiative—The Agency's Latest Proposed Rule for Refineries Shows the Initiative in Action and Provides a Glimpse of the Future for Other Industries," *Energy Business Law*, May 27, 2014.

[17] The water equivalent of fenceline monitoring is checking water quality both upstream and downstream of a water discharge point, as Colorado requires some water pollution sources to do. Colorado Regulation 85.6(2) (2012), https://www.colorado.gov/pacific/sites/default/files/WQ_nonpoint_source-Regulation-85.pdf.

[18] See Ning Wang, Ting Dai, and Lei Lei, "Optofluidic Technology for Water Quality Monitoring," *Micromachines*, Vol. 9, No. 4 (April 2018): 158, https://doi.org/10.3390/mi9040158.

[19] EPA, "Advanced Monitoring," EPA Office of Enforcement and Compliance Assurance, Next Generation Compliance, EPA/Association of Clean Water Administrators Annual meeting, August 11, 2016, https://www.acwa-us.org/wp-content/uploads/2017/05/ACWA-OECA-Water-Monitoring-Fact-Sheets-combined-2016-Annual-Meeting-8-11-16.pdf.

[20] See GAO, "Wider Use of Advanced Technologies Can Improve Emissions Monitoring," GAO-01-313, June 2001, at 58 (studies show that animals were the primary source of fecal-related bacteria in certain waters, not leaking sewers or septic systems).

[21] Jane Lubchenko, "People and the Ocean, 3.0: A New Narrative with Transformative Benefits," in Daniel C. Esty ed., *A Better Planet: Big Ideas for a Sustainable Future* (Yale University Press, 2019), 78; Global Fishing Watch, "Revolutionizing Ocean Monitoring and Analysis," https://globalfishingwatch.org/map (interactive map showing active fishing vessels using data from satellites and other sources).

[22] Katherine Kornei, "They're Stealthy at Sea, but They Can't Hide from the Albatross," *New York Times*, January 27, 2020, https://www.nytimes.com/2020/01/27/science/albatross-ocean-radar.html.

Regulations that require actual measurement, instead of allowing estimates or completely skipping any obligation to monitor and report, would reveal that many sources are in violation of environmental rules. That's a major reason industry often opposes mandatory measurement; they fear it will make violations more difficult to get away with.[23] My point exactly. Required direct measurement should always be considered for environmental standards, because it is a first and huge step toward fixing noncompliance and knowing how bad the violations are.

Mandatory direct monitoring has another salutary effect as well: it creates a market for measurement technologies, giving monitoring companies incentive to develop better, cheaper monitors. During the Obama EPA, I asked many authors of new pollution measurement technologies what EPA could do to support further innovation. They uniformly said: "Make monitoring mandatory." Regulatory requirements lead to advances in technology that reduce compliance costs, as long experience has repeatedly demonstrated.

Continuous Monitoring Has the Most Compliance Power

Just as some measurement is better than none, frequent measurement is far better than very intermittent. Frequency that approaches "continuous" monitoring has tremendous power to improve compliance and dramatically cut pollution. Continuous emission monitoring systems (CEMS) are already in use for many air pollutants, and the list is growing quickly. CEMS are widely viewed as the best compliance-monitoring option for large sources of air pollution.[24]

The extremely intermittent methods used to monitor pollution from many air sources today have proven unreliable, both because they happen so infrequently—less than once every five years in some cases—and because they involve so many opportunities to go astray. A recent EPA Inspector General investigation in one state found that most tests for particulate matter—which is one of the most health-damaging pollutants—were unreliable because they failed to follow the quality assurance and calibration obligations, without which the test results are meaningless.[25] The Office of Inspector General (OIG) reports that EPA staff confirmed that the same problems occur across the country.

One of the most obvious benefits of continuous monitoring is rapid detection of pollution problems so violations can be prevented or quickly fixed. An Exxon

[23] Babich, "The Unfulfilled Promise," 571. It's not just industry that worries about accurate monitoring; states also resist upgrading monitors for fear it would boost their pollution readings, requiring them to crack down on polluters. GAO, "Opportunities to Better Sustain and Modernize the National Air Quality Monitoring System," GAO Report GAO-21-38 (November 2020), 29.

[24] EPA, "Inappropriate Use of AP-42 Emission Factors."

[25] EPA OIG, "More Effective Oversight is Needed for Particulate Matter Emissions Compliance Testing," EPA OIG 19-P-0251 (July 2019), 12, 15–16.

Mobile refinery, for example, released large amounts of dangerous benzene when a plug failed at 2 a.m. and the ensuing benzene release was not detected until an inspector arrived seven hours later; that one leak released more benzene than the refinery reported emitting over the two preceding years.[26] Leaks in underground storage tanks can be immediately detected through inside-the-tank monitors that transmit continuously to a central office, a big advance for protecting from leaks over twice-a-year measurement by a person with a dipstick.[27]

Real-time information on pollution also has a lot more operational relevance for plant managers. When grab sample results don't come back from the lab for days or sometimes months, the world has moved on. The opportunity to explore explanations for elevated readings as they are happening will be lost. Putting real-time emissions information into the hands of plant managers elevates its operational visibility and gives mangers a chance to find the reasons for problems, learning that can have significant long-term benefits.

Continuous monitoring can also reveal wide variability that is obscured with sampling that averages daily, weekly, or monthly performance. Armed with that more nuanced understanding, companies discover more effective solutions to improve compliance. After a dairy plant started monitoring every five minutes, it discovered that phosphorus loading was highly variable. Relying on that near-continuous monitoring, the plant shifted from treatment decisions using average pollution loads to an automated system that dispensed treatment chemicals to address pollution measured in real time, improving compliance and saving money.[28] The City of South Bend, Indiana had a similar experience when it installed smart monitors throughout the sewer system; continuous monitoring allowed the city to redirect flow when backups were detected, reducing violating discharges of raw sewage by 70 percent and saving the city hundreds of millions of dollars.[29]

Another big advantage of continuous monitoring is that it allows an automated response that can fix the problem before noncompliance occurs. When readings show that a facility is getting close to a violation, it can automatically change inputs. Violation avoided. That's what the refinery operators of the violating flares mentioned earlier in this chapter did after the violations came to

[26] Smith, "Fenceline Monitoring," 448–49.

[27] EPA "Total Petroleum Puerto Rico Corp. Agrees to Spend $1.6 Million to Improve Leak Detection in at Least 125 Gas Stations Across Puerto Rico and U.S. Virgin Islands," News Release, March 9, 2015, https://archive.epa.gov/epa/newsreleases/total-petroleum-puerto-rico-corp-agrees-spend-16-million-improve-leak-detection-least.html.

[28] Hach, "Improving Compliance Through Real-Time Phosphorus Control," *Treatment Plant Operator Magazine*, May 7, 2018. https://www.tpomag.com/online_exclusives/2018/05/improving_compliance_through_real_time_phosphorus_control_sc_001fa.

[29] Luis Montestruque, "Hi-tech Sewers Can Help Safeguard Public Health, Environment and Economies," *American City and County*, February 17, 2021, https://www.americancityandcounty.com/2021/02/17/hi-tech-sewers-can-help-safeguard-public-health-environment-and-economies/.

light.[30] Even if the response isn't automated, continuous monitoring can be equipped with alarms or automated text notices that draw attention to problems and invite active decision-making to prevent violations.

The huge increase in confidence about accuracy of pollution measurement that comes from continuous monitoring also makes it possible to embrace innovative regulatory approaches. Government isn't going to be willing to try a market-trading strategy for pollution if the data on which it is based is known to be unreliable. With protection of public health in the balance, EPA will be understandably reluctant to give up known-to-be-effective strategies in favor of approaches that can't be reliably measured. Continuous monitors help break that impasse. Certainty about results allows more flexibility as to methods.

Continuous monitoring is also a big deterrent to negligence and fraud. When companies know that increases in pollution can be detected in real time, they are far less likely to engage in risky or prohibited practices, like dumping prohibited chemicals down the drain or increasing emissions at night. By definition, monitoring all the time also eliminates a big problem frequently encountered with intermittent sampling: intentionally avoiding times when pollution is likely highest. It is still possible to cheat even with continuous monitors,[31] but it requires a lot more sophistication and determination and is much easier to catch, all of which make cheating far less likely.

Use of continuous emissions monitors has been associated with emissions reductions and increased rates of compliance. A study conducted by EPA's Midwest regional office involving data from more than 1,100 facilities found that use of CEMS, coupled with an obligation to tell EPA what the company was going to do about excess emissions, reduced pollution, especially from the highest-emitting sources.[32]

For all of these reasons, continuous monitoring is likely a big part of the answer to some of our more vexing compliance problems. Government struggles to assure compliance at over 1 million oil and gas wells in the United States, for example, because emissions are invisible, intermittent, and unpredictable. That makes it challenging to impose accountability for the threats to health and climate that result from widespread violations in this sector. Close to continuous

[30] EPA, "Marathon Petroleum Company—LP and Catlettsburg Refining—LLC Settlement (Flaring)," April 5, 2012, https://www.epa.gov/enforcement/marathon-petroleum-company-lp-and-catlettsburg-refining-llc-settlement-flaring#mitigation.

[31] DOJ, "Former Berkshire Power Manager Sentenced For Conspiring to Tamper with Air Pollution Monitors," Press Release, US Attorney's Office, District of Massachusetts, May 31, 2017, https://www.justice.gov/usao-ma/pr/former-berkshire-power-manager-sentenced-conspiring-tamper-air-pollution-monitors.

[32] GAO, "Wider Use of Advanced Technologies," 27.

monitoring, which may soon be possible through a combination of satellites, aerial monitoring, and on-the-ground monitors, could be the game changer this challenging compliance problem needs.[33] The same is true for the stunning amount of pollution from ocean-going ships. Thus far the regulatory strategy has been to mandate that vessels use low-sulfur fuel, but knowing which on-board fuel ships are burning hundreds of miles from shore makes that a compliance nightmare. Requiring these large ships to continuously monitor emissions through CEMS could be the breakthrough needed to achieve compliance with these essential-for-public-health measures, as Seema Kakade and Matt Haber have so convincingly argued.[34] Continuous monitoring for trucks through on-board diagnostics and real-time communication to a central location through telematics has tremendous potential to address the serious pollution problems caused by unlawful engine alterations.[35]

Continuous monitoring is now technically viable for many pollutants in both air and water. But industry often opposes them. Why? The same reason that they resist any kind of monitoring: fear that continuous monitoring might reveal violations.[36] Even in the early days of CEMS, the EPA CEMS manager opined that the reason facilities don't want their emissions monitored on a full-time basis was that it would then be possible to know if a facility was exceeding emission limits, and for how long and by how much.[37] While there are many advantages for industry in knowing more about their own operations—including the potential to save money—it is indisputable that better information could reveal problems. That's one of the reasons that continuous measurement is not going to happen in more than a small handful of companies unless it is required. Every proposed environmental regulation should include a serious look at continuous monitoring for whatever provisions are most essential because of its tremendous power to drive much better outcomes.

[33] See discussion of methane pollution issues and potential solutions in chapter 9.

[34] Seema Kakade and Matt Haber, "Detecting Corporate Environmental Cheating," *Ecology Law Quarterly*, Vol. 47 No. 3 (2020): 771–822.

[35] See, e.g., Alex Crissey, "Let's Get Visible: Telematics and Visibility Are Fleet Equipment's 2020 Truck Trend of the Year," *Fleet Equipment Magazine*, January 4, 2021, https://www.fleetequipmentmag.com/truck-trend-of-the-year-2020-telematics-visibility/ ; Eric Miller, "CARB Approves Amendments to Onboard Diagnostic System Regulations," *Transport Topics*, July 26, 2021, https://www.ttnews.com/articles/carb-approves-amendments-onboard-diagnostic-system-regulations (changes to rules for onboard diagnostics for trucks to improve real-time diagnostic information); "National Compliance Initiative: Stopping Aftermarket Defeat Devices for Vehicles and Engines," EPA, https://www.epa.gov/enforcement/national-compliance-initiative-stopping-aftermarket-defeat-devices-vehicles-and-engines (extensive aftermarket alterations of vehicles is a significant contributor to air pollution).

[36] See, e.g., GAO, "Wider Use of Advanced Technologies," 6.

[37] GAO, "Improvements Needed in Detecting and Preventing Violations," RCED-90-155, September 1990, at 24.

Citizen Monitoring Can Be an Additional Motivator

Citizen science can apply additional pressure for compliance. Knowing that neighbors can check what's happening and blow the whistle can inspire better performance. You never know when a nearby resident might be watching.[38] A simple cell phone can be enough to record obvious lead paint violations and notify government.[39] Nonprofit organizations can do sophisticated measurement and publish the results, putting pressure on violating companies.[40] Aggregating large numbers of low-tech monitoring stations is an intriguing method for screening pollution problems.[41]

What do these possibilities mean for rule design? Number one: don't get in the way; at minimum don't make it any harder for outside groups to do their own work to press for better outcomes. Second, make as much facility-compliance data available to the public as possible, to facilitate community engagement and scrutiny. Finally, consider building citizen science into the rule. When remote monitoring is feasible—as it is for oil and gas emissions, for example, through aerial surveys and increasingly with satellites—government might create a pathway for credible citizen science to trigger regulatory obligations. In the right situations, crowdsourcing compliance monitoring can be a powerful compliance motivator.

Remember the Lessons of Next Gen When Designing Monitoring Requirements

Monitoring obligations are just like any other requirement in a rule. They aren't going to reliably happen just because you write them down. Assuming compliance won't cut the mustard. Nor can regulators count on enforcement to remedy monitoring violations; they are often harder to catch and far more numerous than end-of-pipe-type violations. And more insidious: widespread monitoring violations make government blind to the seriousness of the noncompliance.

Here are some of the ways that monitoring can run off the rails:

[38] That's what the Louisiana Bucket Brigade teaches fenceline communities to do. "The Bucket," https://labucketbrigade.org/pollution-tools-resources/the-bucket/#.

[39] "EPA Cites First RRP Violator, with an Assist from YouTube," *The Journal of Light Construction*, July 1, 2011, https://www.jlconline.com/how-to/jlc-report-epa-cites-first-rrp-violator-with-an-assist-from-youtube_o (EPA enforcement action based on anonymous video tip via YouTube).

[40] EDF, "Methane Research Series: 16 Studies," *Environmental Defense Fund*, https://www.edf.org/climate/methane-research-series-16-studies.

[41] For example, Purple Air automatically uploads from privately installed monitors and displays the data on an interactive map in real time, https://www2.purpleair.com/.

Strategic shut down. Monitors expected to reveal some unpleasant news can just be shut down. For alleged maintenance, for example. That's what's happening with air quality monitors in some areas of the United States at risk of exceeding pollution thresholds that would trigger big consequences.[42] Significant percentages of fishing vessels turn off electronic tracking devices that would reveal fishing in illegal areas.[43]

Don't monitor or monitor incorrectly. Some sources, intentionally or not, don't conduct the required monitoring. Especially when there are multiple locations in one facility that are supposed to be checked, it is exceptionally difficult for regulators to know if the company is skipping some required monitoring or doing it improperly.[44]

Dodge violations by oversampling in cleaner areas. That's what many drinking water systems do to avoid going over the percentage threshold that would obligate corrective action: add some additional known-to-be-clean samples to the mix, so the bad samples are no longer above the triggering percentage.[45]

Monitor where you expect pollution will be less. That's happening with national air quality monitors, recent research shows.[46] Despite requirements to sample where lead contamination is most likely, many drinking water suppliers don't do that.[47] One consulting firm advertised strategic placement of air monitors: "EPA offers a great deal of flexibility in determining sampler placement. By carefully considering the facility configuration, the location of known sources, proximity of neighboring facilities, and

[42] Yingfei Mu, Edward A. Rubin, and Eric Zou, "What's Missing in Environmental (Self-) Monitoring: Evidence from Strategic Shutdowns of Pollution Monitors," *National Bureau of Economic Research* (April 2021, revised October 2021), https://doi.org/10.3386/w28735.

[43] Uki Goni, "Hundreds of Fishing Fleets That Go 'Dark' Suspected of Illegal Hunting, Study Finds," *The Guardian*, June 2, 2021, https://www.theguardian.com/environment/2021/jun/02/fishing-fleets-go-dark-suspected-illegal-hunting-study.

[44] For an example, see EPA, "Reduction of Hazardous Waste Air Emissions," EPA Compliance Advisory, April 2018, https://www.epa.gov/sites/default/files/2018-05/documents/rcraaircompliance advisory.pdf. See also Littice Bacon-Blood, "Water Employees Convicted in Brain-Eating Amoeba Investigation," *NOLA.com*, May 13, 2015, updated July 18, 2019 (two employees of a drinking water provider convicted for skipping required monitoring in a drinking water system where a potentially deadly brain-eating amoeba was found).

[45] See discussion of this strategy—called "sampling out"—in chapter 1, section titled "Programs with Pervasive Violations: Four Examples," sections on drinking water; chapter 2, section titled "For Some Important Programs, EPA's Understanding of Noncompliance Is Wrong," section on drinking water.

[46] Corbett Grainger, Andrew Schreiber, and Wonjun Chang, " Do Regulators Strategically Avoid Pollution Hotspots when Siting Monitors? Evidence from Remote Sensing of Air Pollution," Working paper (2017).

[47] Brenda Goodman, Andy Miller, Erica Hensley, and Elizabeth Fite, "Lax Oversight Weakens Lead Testing of Water," *WebMD*, June 12, 2017, https://www.webmd.com/special-reports/lead-dangers/20170612/lead-water-testing.

seasonal weather patterns, monitoring networks can be designed to minimize the chances of exceeding the action level."[48]

Mess with the monitors or the data. Straight-ahead cheating is an ever-popular option. Water dischargers have submitted false monitoring data.[49] Utilities have tampered with air monitors.[50] Drinking water suppliers have lied about water quality sampling.[51] Lead paint inspectors have falsely cleared properties as containing no lead paint.[52]

Monitors that don't deliver. The promised immediate detection of leaks from oil pipelines, for example, isn't proving so reliable; federal data show that leak detection technology only identified about 7 percent of pipeline spills since 2010.[53]

All of this is unsurprising. Companies will not be at their most diligent when it comes to ferreting out information that government will use to hold them accountable. As with any regulatory obligation, when the incentives line up against compliance, and it is next to impossible for regulators to find out, violations will be widespread. Of course.

The good news is that a lot of these issues are fixable. Regulators just need to ask themselves: Is it possible for companies to avoid, ignore, or game the monitoring requirements, and if so, how would they do that? Then tee up a Next Gen strategy to make that hard or impossible. There will be less strategic down time of monitors when blank spaces in the self-monitoring report result in worse-than-likely assumptions, as data-substitution provisions require. Rules can also provide less discretion on where and how monitoring is done; every time impossible-to-observe judgment is applied, the opportunity to fudge goes way up. Don't set up a situation where gaming is allowed, even invited, such as obligations triggered by a percentage of samples taken. Make sure the rule communicates that monitoring and reporting violations are very important, through penalties, for example.[54] And definitely don't buy into the canard that

[48] Strickland, "On the Fence about Fenceline Monitoring," 2.

[49] Ken Ward Jr., "Lab Official Admits Faking Coal Water Quality Reports," *Charleston Gazette-Mail*, October 9, 2014, updated October 27, 2017.

[50] DOJ Press Release, "Berkshire Power."

[51] DOJ, "Former Town of Cary Employee Pleads Guilty to Falsifying Drinking Water Sampling Results," Press Release, US Attorney, Eastern District of North Carolina, September 26, 2016, https://www.justice.gov/usao-ednc/pr/former-town-cary-employee-pleads-guilty-falsifying-drinking-water-sampling-results.

[52] "Former Detroit Lead Inspector Sentenced for Fraud," *Environmental Protection*, February 10, 2011, https://eponline.com/articles/2011/02/10/former-detroit-lead-inspector-sentenced-for-fraud.aspx.

[53] Mike Soraghan, "Giant N.C. Spill Shows Gaps in Pipeline Safety," *E&E News*, February 25, 2021.

[54] That's what EPA's recent climate rule for phasedown of HFCs does, by explicitly stating that failure to file the required reports can be a basis for EPA to withhold allowances that the companies need to operate. EPA Final Rule, "Phasedown of Hydrofluorocarbons: Establishing the Allowance

failures to monitor or report are minor "paperwork" problems that deserve less scrutiny when violations are being tallied. Because monitoring is so central to the entire compliance enterprise, the monitoring provisions have to be just as compliance resilient as the standard itself.

Reporting

Reporting—regulated entities' obligations to provide information to government—has been a foundational element of environmental laws since the beginning of EPA. How can reporting design help drive better compliance?

Self-reporting

Reporting by regulated facilities to government is the backbone of regulators' knowledge about compliance. It makes sense that facilities are required to self-report: they are in the best position to know what's happening in their own companies. That's not the only reason to require self-reporting though. Mandating reports to regulators serves three important functions. First is informing the company about its own operations. If the company has to file a report that includes data, the company has to collect the data. Now the company is aware of its obligations and how it is doing, which is an essential first step toward compliance. Second is keeping government informed so regulators can take corrective action if needed. The third value of self-reporting is that it puts more compliance pressure on the regulated. Once the report is submitted, the company knows that regulators can see the violations. Government might enforce. And it might make the information available to the public, expanding the circle of parties who can press for action. Those risks induce some companies to fix problems on their own rather than wait to get hammered.

Self-reporting obligations can be sophisticated and powerful, as the Greenhouse Gas Reporting Rule has proved.[55] But every reporting program can be strengthened by attention to a few basics.

Allocation and Trading Program Under the American Innovation and Manufacturing Act," *Federal Register*, Vol. 86, No. 190 (October 5, 2021): 55171 ("Final Rule for HFC Phasedown").

[55] The Greenhouse Gas Reporting Program is the gold standard of a reporting program that makes the most of Next Gen–type features, including common metric and format electronic reporting, automated pre-submittal checks that identify potential errors, mandatory certification, and robust post-submission data analytics to identify anomalies for follow-up. See a description of the Greenhouse Gas Reporting Program Report Verification at https://www.epa.gov/sites/production/files/2015-07/documents/ghgrp_verification_factsheet.pdf.

Required reporting of facts—not just conclusions—packs the most compliance punch.

Facts are harder to fudge than conclusory statements about compliance. The measured parts per million of pollutant X, the date you installed the required equipment, and your total sales of chemical Y this year are examples of facts that might be required in reporting. They are true or they aren't: no shades of gray. Plus, hard data gives regulators a lot more information about what's happening on the ground. It often matters if the pollutant is 7 parts per million or 700. After Colorado mandated a reporting checklist of facts to which hazardous waste companies had to respond yes or no, the government-verified percentage of fully compliant facilities went from 31 percent to 84 percent.[56]

Reports should be designed to make it impossible to avoid admitting violations

Reporting isn't just a fact-gathering exercise, it is also a method to draw the companies' and the governments' attention to what's most essential. The most powerful design requires an unambiguous statement of the relevant fact, immediately followed by a compliance determination: Is this a violation, yes or no? The report cannot be submitted without an answer. This simple design choice makes violations salient and helps achieve all three purposes of reporting: the company has admitted a violation so certainly knows it has a problem, government can easily spot admitted violations, making its compliance-monitoring job much easier, and the public knows where to look to find out who is violating and by how much. On the other hand, if the report allows a narrative description of facts or compliance status or allows generic and broad statements of compliance decoupled from particular facts, opportunities for avoidance abound. Massachusetts' air permit reporting form is an example of impossible-to-dodge compliance reporting: it requires companies to check the box yes or no for each compliance requirement—not just for emissions limits but also for monitoring, testing, recordkeeping, and reporting.[57]

Certifications by a senior official under penalty of perjury provide compliance a boost

These certifications elevate the profile of compliance within the company; the senior official will have to be informed what the report says and may question

[56] See the description of this program in Colorado Department of Public Health and Environment, "2014 Annual Report to the Colorado General Assembly," February 1, 2015, at 10, https://spl.cde.state.co.us/artemis/heserials/he171318internet/he17131820l4internet.pdf.

[57] See "Operating Permit Annual Compliance Certification and Report," MassDEP Operating Permit & Compliance Program, Massachusetts Department of Environmental Protection, available at https://www.mass.gov/guides/massdep-operating-permit-compliance-program.

why the company has so many violations. The fact that false reports expose the official to personal, not just corporate, accountability helps to increase accuracy. And of course, they are admissions in a potential enforcement action.

Regulators can test the resilience of their reporting design by asking themselves three questions. First, will a company that reports have looked at everything that it should to understand how its own operation is doing? Report design should ensure that the necessary monitoring is done and that the most essential facts jump off the page. Second, once government gets the report will it be manifestly clear at a glance—for individual facilities and across the entire sector—who is complying and who isn't? The report should be set up so that the worst problems are quickly flagged, and all the reports considered together present a complete picture for the entire program. And third, but by no means last, if the self-reports are made public, is it possible for nonexperts to make sense of them? If the violating company knows that the reports will be publicly available, they should be worried.

If regulators struggle to create a reporting system that will achieve these objectives, maybe the problem is the underlying rule doesn't have a good compliance design. If the compliance obligation isn't clear in the rule, that confusion cannot be fixed through the reporting form. Reporting clarity also requires a definitive approach to measurement and monitoring; rules with badly designed monitoring obligations, like allowing estimates rather than actual measurement, also mean that self-reports won't really answer the compliance question. If regulators are having a tough time translating the rule into a yes/no format, maybe that's because the rule itself doesn't identify exactly what the firm is supposed to do and what facts prove it did or didn't. This is why rule writers shouldn't defer design of reporting obligations to others; viewing the rule through the reporting lens can help to identify important holes that need to be plugged before the rule is promulgated.[58]

The reporting obligations for the nation's water pollution discharge system are widely viewed as a model of clarity. They require a statement of the permit limit, the monitored actual level of pollution, followed by a check-the-box yes/no admission of violation. This self-reporting structure clearly identifies both the fact of a violation and its extent. In 1990, Congress attempted to import the effective

[58] The Court of Appeals for the Third Circuit recently trashed Pennsylvania's air rules for coal-fired power plants for this kind of reporting flaw. Pennsylvania's regulations imposed less stringent air emissions limits when firms were operating at lower temperatures but permitted plants to use lower operating temperatures as an excuse for higher emissions without reporting actual operating temperatures to substantiate their claims. The Court found that it was "fanciful" to base a regulatory enforcement regime on such an "honor code" approach. "It is a strange regulatory system indeed," the Court noted, "that is based on the good faith of the regulated entity to keep records which may be prejudicial to its operation and profitability." On the basis of this "gaping loophole," among other deficiencies, the Court found that the Pennsylvania regulation violated the Clean Air Act. Sierra Club v. EPA, 972 F.3d 290, 308, 309 (3d Cir. 2020).

water-reporting system into the air program.[59] It hasn't turned out that way. Congress's direction to beef up monitoring and reporting as an essential step toward reducing air pollution violations has been undercut by political maneuvering, court challenges, and state implementation lapses.[60]

Next Gen applies to all aspects of a rule, including reporting. Companies don't reliably do what's required just because the obligation is written in black and white. If regulators are counting on compliance with reporting obligations, but don't take steps to build in compliance drivers for those elements of the rule, that's a major weakness. Reliable self-reporting is so important to effective implementation that it requires its own Next Gen strategies—regulatory designs intended to make reporting hard to avoid and much more accurate.

At minimum, rules shouldn't provide incentives to skip required reports. Almost all rules have less severe consequences for failing to report than they do for admitting a standards violation. That opens the door for deciding to pass on reporting instead of admitting a violation of standards. And it allows a low-consequence way out for companies that would rather focus on other things.

Consider requiring the strongest compliance-forcing type of reporting: real-time data with no retroactive changes. Direct reports of data from continuous emission monitors, for example, are much harder to avoid or change than end-of-quarter summary reports. That same is true for real-time recording of each transaction in the commercial chain, as EPA recently did in a rule to control the strong climate forcing chemicals referred to as HFCs.[61] It takes a lot of sophistication and foresight to mess with that data without getting caught, so most companies won't try.

Adopt Next Gen ideas like automatic consequences to encourage compliance with monitoring and reporting obligations. If companies know that a missed report or a report with a blank space means automatic substitution with data that is likely worse than what accurate monitoring would have shown, companies are motivated to file reports and keep monitors in working order. Other automatic consequences are possible too, like penalties. The goal of such automatic outcomes is to change the dynamic so that avoiding or ignoring the monitoring or reporting obligation is more hassle and expense than just doing what's required.

[59] David P. Novello, "The New Clean Air Act Operating Permit Program: EPA's Final Rules," *Environmental Law Reporter*, Vol. 23, No. 2 (February 1993): 10081.

[60] EPA OIG, "Substantial Changes Needed in Implementation and Oversight of Title V Permits If Program Goals Are to Be Fully Realized," Report No. 2005-P-00010, March 9, 2005, at 1, 2. Nor has the situation improved much since the IG's 2005 report. Claudia Copeland, "Clean Air Permitting: Implementation and Issues," Congressional Research Service Report 7-5700, RL33632, September 1, 2016.

[61] "EPA Final Rule for HFC Phasedown," 55183–186. See discussion of the HFC phasedown rule in chapter 4.

The lessons of Next Gen for self-reporting are fairly straight forward, but there are two important complications. The first is that many programs are administered through individualized permits. The rule might be terrific, but if the permit isn't, it's the permit that controls. Any regulation that will be implemented through permits therefore needs sufficiently clear direction about how permits must be written so that Next Gen strategies in rules make it through to permits unscathed.

The second—and related—challenge is federalism. For the great majority of environmental programs, federal rules are just the start. States will write their own regulations to implement the federal program and will also write the permits based on state rules. An excellent federal rule can founder on the rocks of state indifference, resource constraints, or sometimes open hostility to accountability for regulated companies. Chapter 10 describes our 50-year-old model of federalism and proposes updates to bring environmental regulation into the modern era. Monitoring and reporting, no less than standard setting, need to be part of our revised federalism strategy. The foundation provided by monitoring and reporting can collapse—taking effective implementation with it—when 50 states head in different directions. That's what has happened in the Title V air-permitting program, where wide disparities in state monitoring and reporting implementation have significantly undercut the purpose of the law.[62]

Third-Party Verification and Auditing

Self-reporting isn't reliably accurate. Raising your hand to admit a violation in a report to regulators can bring unwanted attention, hassle, and expense. Doing the monitoring and recordkeeping necessary to report can be inconvenient or a low priority. Employees might not know how to do it right or may not want to tell their employer the truth. If companies or their employees would rather not follow the rules, and they know it is almost impossible for government to find out, many take the path of least resistance. It happens in every program.[63] In addition to the Next Gen monitoring and reporting provisions already discussed in this chapter, how can regulators ensure that self-reports, on which program integrity so heavily depends, are as accurate and reliable as possible?

One option is to require that self-reporters hire an outside party to examine their work. Third-party verification, certification, or auditing—there are

[62] EPA OIG, "Substantial Changes Needed in Implementation"; Copeland, "Clean Air Permitting." Greenhouse gas permitting suffers from similar problems. Matt Haber and Seema Kakade, "Revitalizing Greenhouse Gas Permitting Inside a Biden EPA," *Environmental Law Reporter*, Vol. 51, No. 5 (May 2021): 10, 384.

[63] See the many examples cited in chapter 2.

multiple names and flavors of this approach—can bring additional pressure and scrutiny to bear on self-reports, thereby improving accuracy.[64] If the companies required to self-report know that someone with expertise will be checking, they understand that it is harder to get away with shoddy or fraudulent work. When it works well, third-party auditing can greatly increase the chances that companies decide it is in their interest to do it right.

For all their possible value, there is also a significant caution about third-party auditors.[65] Time and again EPA and researchers have discovered that third-party auditors can be unreliable. It is common for third-party auditors to spin, or sometimes just lie, in favor of the company paying their fee.[66] They want to be hired for the same job again, and maybe even want other work from the audited firm. That isn't going to happen if the outside reviewers hold the audited firm's feet to the fire. An all-clear audit will be much more favorably received. No express agreement is necessary for that to happen; it is self-evident to everyone.

That is why any third-party certification strategy should include independence-supporting features. First, certification should generally focus on facts rather than judgments, so the audit answers questions with verifiable answers. Second, the audit report should go directly to government, instead of being submitted through the audited company.[67] And third, the rules should require that certifiers maintain independence from the audited company. Alleged auditors who are employees of the audited firm aren't third parties of course, never mind "independent." Nor are auditors who do other work for the company they are supposedly checking. At minimum, third-party verifiers need

[64] For one illustration, see EPA's rule requiring all producers of composite wood products to have their products tested by an EPA-recognized third-party certifier to ensure their products are compliant with the formaldehyde emissions requirements. See EPA, "Formaldehyde Emission Standards for Composite Wood Products," https://www.epa.gov/formaldehyde/formaldehyde-emission-standards-composite-wood-products.

[65] See Lesley K. McAllister, "Regulation by Third-Party Verification," *Boston College Law Review*, Vol. 53, No. 1 (January 2012): 22–23.

[66] Two examples: "EPA Notice of Intent to Revoke the Ability of Genscape to Verify RINs as a Third Party Auditor," January 4, 2017, https://www.epa.gov/fuels-registration-reporting-and-compliance-help/notice-intent-revoke-ability-genscape-verify-rins (third-party verifiers for Renewable Identification Numbers (RINs) under the Renewable Fuel Standard ignored obvious evidence of fraud and certified millions of RINs that were found to be fraudulent); Justin Marion and Jeremy West, "Dirty Business: Principal-Agent Problems in Hazardous Waste Remediation," *Semantic Scholar* (2019) (private third parties hired to assess severity of issues at cleanup sites manipulate scores in favor of clients to avoid triggering higher regulatory oversight). For a survey of the evidence about compromised third-party auditors, see Jodi L. Short and Michael W. Toffel, "The Integrity of Private Third-Party Compliance Monitoring," Harvard Kennedy School Regulatory Policy Program Working Paper, No. RPP-2015-20 (November 2015, revised December 2015), https://hbswk.hbs.edu/item/the-integrity-of-private-third-party-compliance-monitoring.

[67] EPA acknowledged that independent third-party auditors have a "well-documented record of fostering compliance" in its recent rule phasing down climate-damaging HFCs. That rule found that self-audits don't have the proven benefits that third-party audits provide and explicitly required third-party auditors to submit the results of their audit to EPA before sending them to the auditee. "EPA Final Rule for HFC Phasedown," 55179–181.

rigorous independence criteria—like a prohibition on being hired for other work from the audited firm for a number of years—or else they are little different from consultants. And it is helpful to require that auditors meet professional licensure requirements, placing some oversight responsibility on an independent board. Adding other indicia of independence or consequences for cheating—like a ban on any certifier that falsifies data or looks the other way when it uncovers evidence of error or fraud—can help to make third-party oversight more meaningful.

But when a lot rides on the accuracy of the third-party certification—such as when those audits are the only meaningful check on compliance and accuracy—hard-to-oversee independence criteria are probably not enough to ensure reliable results.[68] For such cases, the gold standard for audit reliability is random assignment of auditors. With random assignment, the company being audited can't count on the cooperation of its auditor, because it doesn't know in advance who that auditor will be. Sure, collusion can still happen, but it's a risky bet. If the assigned auditor ends up not playing along, it's too late to fix it, and now it's close to a sure thing that the violations will be revealed, with potentially significant consequences. Not playing by the rules becomes perilous and straight-out fraud is tough to get away with. Exactly what you want to have happen.[69]

A robust study of the random assignment approach was tested in India, under a regulatory regime similar to that in the United States. Instead of having the regulated parties select the verifier, the audited firms were assigned an auditor randomly from an approved list. Audited firms paid their fees into a central pool from which the third-party firms were compensated. State inspectors confirmed that this strategy really delivered for environmental protection: randomly assigned firms reported pollution levels that were 50 percent to 70 percent higher than for control group firms hired and paid by the audited company (i.e., were much more accurate), and independent auditors were 80 percent less likely to falsely report the company as complying. Pollution also declined significantly, mainly the result of big reductions from the worst polluters who now had

[68] That's the case for nearly every carbon offset program, for example, and for the Renewable Fuel Standard, where fraud has been so widespread. See discussion in chapter 8.

[69] How hard companies fight for the ability to select their own auditors is one indicator of the importance that auditor independence has in promoting compliance. One illustration is the legal fight by the makers of wood heaters—notoriously bad polluters—against an EPA 2015 rule that allowed EPA to randomly select heaters for pollution audit testing using any EPA-approved laboratory. The manufacturers of wood heaters wanted EPA to be forced to use the auditor chosen by the company. Though their flimsy legal challenge was ultimately unsuccessful, their attempt to eliminate independent review shows they appreciate its power for holding companies accountable. Hearth, Patio & Barbeque Association v. EPA, 11 F.4th 791 (D.C. Cir. 2021).

nowhere to hide.[70] That's hard proof of how robust reporting design can deliver greater reporting accuracy and reductions in pollution.

Third-Party Information Reporting

Unlike third-party auditors, who check the reports prepared by the regulated party and verify their accuracy, third-party information reporting provides data to the government that the government can use to check what the regulated party tells them. We all have personal experience with third-party information reporting: that's what's happening when our employers tell the IRS how much we were paid. When we file our income tax returns, we are aware that the IRS already knows how much we made. There is little point in telling the IRS that your income was something different. With today's electronic data-matching capability, taxpayers know for a fact that the IRS will catch inaccuracies. That's the major reason the IRS has such impressive compliance rates—99 percent, according to a recent IRS statement—for individual wage income.[71] In notable contrast, taxpayers misreport more than half of income for which there is no third-party reporting, like capital gains and partnership earnings.[72]

Third-party information can help government identify violators, which is incredibly useful. But its far more significant power is to deter companies from violating in the first place, because they know they won't get away with it.[73] The good results derive both from government having the information and the regulated party knowing the government has it.

When does third-party information reporting work best? The factors that have been shown to be important for third-party reporting in the tax context also make sense for environmental regulations. It is most effective if the information comes from an entity with an arm's-length relationship with the regulated party, so the possibility of collusion is small and there is no easy way to get around having to provide the third-party information. And it helps that the number of

[70] Esther Duflo, Michael Greenstone, and Nicholas Ryan, "Truth-Telling by Third-Party Auditors and the Response of Polluting Firms: Experimental Evidence from India," *Quarterly Journal of Economics*, Vol. 128, No. 4 (November 2013): 1499–1545, https://doi.org/10.1093/qje/qjt024.

[71] GAO, "Tax Gap: Multiple Strategies Are Needed to Reduce Noncompliance," Statement of James R. McTigue, Jr., GAO-19-558T (May 9, 2019), 8. Compliance is also aided by withholding; because you have already paid your taxes owed, government doesn't have to chase you down for the money.

[72] GAO, "Tax Gap," 8. The Treasury Department Inspector General says that where there is neither withholding nor information reporting, the IRS believes tax compliance is as low as 37%. Treasury Inspector General for Tax Administration, "Understanding the Tax Gap and Taxpayer Noncompliance," Testimony of the Honorable J. Russell George, May 9, 2019, at 2.

[73] See the discussion of third-party reporting in the tax context in Leandra Lederman, "Reducing Information Gaps to Reduce the Tax Gap: When Is Information Reporting Warranted?," *Fordham Law Review*, Vol. 78 (2010): 1735.

information reporters is small, compared to the regulated parties, and that the information reported provides all of the necessary information to quickly match it to the regulated parties' obligations.

Can we tap the power of third-party information reporting to improve compliance with environmental rules? It isn't a strategy that has been widely used in the environmental context, but given its demonstrated effectiveness in the right circumstances, it is on the should-always-be-considered list.

EPA's recent rule for controlling the powerful climate-forcing chemicals known as HFCs contains a version of third-party information reporting.[74] EPA's rule requires each party to a transaction involving HFCs to enter it in real time into an ongoing electronic log that records HFCs from creation or import to ultimate end user sale. Subsequent purchasers along the chain will have a tough time going off the compliance rails, because each container, its contents, and its transaction history will have already been reported to regulators by other companies, and that data can't be retroactively altered.

Real-time reporting can make companies into their own "third-party" information reporters. If a pollution monitor is reporting directly to regulators in real time, for example, it is much harder to change the data after the fact, or claim the monitor wasn't working. The company can't know exactly how the 24-hour averaging is going to turn out before it happens; the fact that all the data is already submitted makes manipulation much harder to do.[75] Similar approaches might be possible for chemical testing studies; if you can't use a study in your application for government approval unless you notified government in real time that the study was starting, it is harder to deep-six studies that don't turn out the way the company hoped. It won't guarantee compliance, but it at least gives regulators a fighting chance of knowing what to ask about. Any strategy that makes it tough for a company to rewrite history—even when not technically "third-party" reporting—adopts some of the same incentives that make third-party information reporting so effective as a compliance driver.

Electronic Reporting

Nearly every Next Gen strategy discussed in this chapter either depends on or is strengthened by electronic reporting. E-reporting is an absolutely must-have element of any effective Next Gen plan. Recognizing the many advantages of

[74] EPA, "Phasedown of Hydrofluorocarbons," 55183–186.

[75] See, e.g., Michael Greenstone, Guojun He, Ruixue Jia, and Tong Liu, "Can Technology Solve the Principal-Agent Problem? Evidence from China's War on Air Pollution," National Bureau of Economic Research Working Paper 27502 (July 2020), https://www.nber.org/papers/w27502 (automatic monitoring and real-time reporting significantly reduced manipulation of air pollution data).

electronic reporting, EPA in 2013 established a presumption that reporting required in new rules be electronic.[76]

Some of the benefits are obvious: electronic reporting is faster, more accurate, and lower cost. Time isn't wasted entering paper-reported data into electronic systems or dealing with the errors that transfer introduces. Government gets the information immediately and does not have to decipher handwriting or guess where the decimal point goes. The cost savings to states for shifting from paper to electronic submission of water-discharge monitoring reports was estimated at $24 million a year.[77] The switch to electronic for hazardous waste manifests, which track movement of hazardous waste from generator to disposal, is expected to save states and industry users $50 million annually.[78]

Converting paper reports to electronic is only a baby step; the real compliance power of e-reporting is its potential for leap-ahead strategies. What are e-reporting's Next Gen advantages?

Built in compliance checks

Electronic submissions can improve accuracy and completeness of reports. Just like you can't put your zip code into the phone number space when you shop online, electronic forms can reduce errors and omissions. They can refuse submissions with obviously impossible data or essential fields left blank. Online pre-submission quality checks can be built in, as they were for the Acid Rain Program's e-reporting program.[79] EPA's greenhouse gas reporting is another example: the system won't accept reports until errors are fixed, and it flags problematic data as requiring resolution after submission.[80] The more complex the information being submitted, the more valuable these pre-submission tools are at reducing erroneous and incomplete reports.

Opportunity for more powerful compliance data

E-reporting opens up the possibility of better reporting methods. Date- and geospatial-stamped photographs might be both more accurate and more

[76] EPA "E-Reporting Policy Statement for EPA Regulations," September 30, 2013, https://www.epa.gov/sites/production/files/2016-03/documents/epa-ereporting-policy-statement-2013-09-30.pdf.

[77] EPA, "National Pollutant Discharge Elimination System (NPDES) Electronic Reporting Rule," *Federal Register*, Vol. 80 (October 22, 2015): 64065.

[78] EPA, "Learn about the Hazardous Waste Electronic Manifest System (e-Manifest)," https://www.epa.gov/e-manifest/learn-about-hazardous-waste-electronic-manifest-system-e-manifest.

[79] John Schackenbach, Robert Vollaro, and Reynaldo Forte, "Fundamentals of Successful Monitoring, Reporting, and Verification under a Cap-and-Trade Program, *Journal of the Air & Waste Management Association*, Vol. 56, No. 11 (February 2012): 1578.

[80] EPA, "Greenhouse Gas Reporting Program: Report Verification," https://www.epa.gov/sites/production/files/2017-12/documents/ghgrp_verification_factsheet.pdf.

efficient evidence of the installation of equipment, for example.[81] Electronic submissions could also allow for higher volumes and complexity of data than can be conveyed on paper, like videos or streaming of monitoring data in real time. How about location- and date-stamped infrared camera footage proving that tanks were not leaking at the time of an oil and gas site self-inspection? We have barely begun to tap the compliance potential of new modes of electronic reporting.

Real-time checking to prevent violations

Many violations are allowed to occur unimpeded now, because it hasn't been practical to have checkpoints that can operate in real time. E-reporting changes that. For example, it used to be tough to know if a product stopped at the border was legally registered and that each particular shipment was cleared. Today that's possible. EPA is requiring that real-time automated checking as part of its 2021 HFC regulation.[82] In Europe, illegal HFC imports waltz across the border because customs officials have no practical way to know what's legal and what isn't. EPA's border check system will eliminate that giant compliance hole by requiring all imports to be cleared through a real-time data system before they can enter the country.

Designs that reduce violations

The recently launched shift to e-reporting for hazardous waste shipments is an example of a reporting system that could be used to dramatically improve compliance. Instead of a handwritten form using the equivalent of carbon paper, an electronic form can refuse unlawful actions, like initiating a hazardous waste shipment to a facility not licensed for that kind of waste.[83] Not just better tracking, better compliance.

Real-time reporting to decrease opportunities for fraud

One of the reasons that plainly unlawful actions escape notice now is that it very tough to spot the falsehood in the sea of unconnected data. For example, plenty of self-proclaimed "small quantity" hazardous waste generators could lie about their quantity status and thus avoid more stringent regulations because regulators with mounds of paper records aren't able to put 2 and 2 together. With

[81] Digital picture reporting was allowed as an alternative for well completions in EPA's 2016 Oil and Gas rule, *Federal Register*, Vol. 81 (June 3, 2016): 35870 (preamble section titled "Final Standards Reflecting Next Generation Compliance and Rule Effectiveness").

[82] EPA, "Phasedown of Hydrofluorocarbons," 55186–187.

[83] E-manifest could do this, but as yet it does not. This idea is presented as an illustration of the potential of e-reporting, unfortunately not yet realized.

e-reporting, they can. Allegedly small generators that actually ship large amounts of hazardous waste can readily be flagged by an electronic reporting system. Yes, that helps identify violators for enforcers, but even more to the point, it deters companies from taking the illegal action in the first place. Whenever you make fraud much harder to do or get away with, you reduce the number who try.

E-communication as a two-way street

There is no reason information flows only in one direction. When firms report electronically, they can also receive responses from government that can aid compliance. Automated assistance that is focused on the company-specific problem revealed in the self-report reaches its audience at a time when it is most likely to be useful. Private companies can also develop e-reporting tools that build in advice and suggestions, as has happened for tax filing.

E-reporting as the foundation of public accountability

When government receives reported data electronically, it is a simple matter to share that data with the public. Most transparency measures—to inform the public and also to ratchet up compliance pressure—depend on e-reporting, as is discussed further later in this chapter.

Common formats, common data: a new era for federalism

As the discussion thus far makes plain, electronic reporting using common reporting formats and shared data systems has the greatest compliance power. And it is by far the easiest and cheapest way to go. It has another important virtue too: it creates the opening to re-envision federalism. The relationship between federal and state governments has been stuck in an outdated paper-oriented model from the 1970s. Instead of a tussle over who controls the information, which dominates the interaction between state and federal governments today, we can get back to the original vision of states as laboratories for innovation, as I discuss in chapter 10.

Transparency

Public access to government information—aka transparency—is a potentially formidable strategy for better compliance. The drinking water program provides a powerful example. A 1996 "right to know" law required drinking water suppliers to tell their customers about the companies' violations of safe drinking water obligations. That's health-related information customers are entitled to, but it turned out to also motivate water suppliers to do a better job. Drinking water companies that were required to mail the compliance reports to customers reduced health-based violations by 40 percent to 57 percent.[84] That's an impressive public health gain from the simple obligation to mail a report.

Transparency promotes better behavior. Some companies might want to avoid the risks of citizen enforcement, investor blowback, or reputational damage if their poor performance becomes known. Industries can also fear that widespread violations will lead to calls for tougher regulations or more significant penalties. And sometimes negative news hitting the press is how management finds out about their own company's bad record. There's a positive effect too: high performers win praise and can provide benchmarks for other companies to aim toward. Because it is effective, low-cost, and serves multiple values at the same time, transparency is a must-consider Next Gen idea for every rule. What are the elements of a strong transparency strategy?

For transparency to work as a compliance driver, the information has to be readily accessible to the public. Allegedly public information that is only available on written request or has to be examined on-site in person is not readily accessible.[85] For transparency to affect companies' behavior, they have to think that there is real potential for violations to be noticed and acted on. If as a practical matter very few people will actually see the information, it loses its deterrent punch. Today that means being available on the web to anyone at any time. Regulations should make it crystal clear that's where the information will reside.

Making compliance information available online was the motivating force behind a 2015 EPA regulation for water pollution dischargers. The NPDES e-reporting rule mandated that water polluters had to report electronically into a data system shared by EPA and states and would be provided to the public online.[86] In theory, the public was already entitled to see that information, but the rules varied by state and sometimes were restricted to in-person and on paper, that is, it wasn't actually available, to the public or to EPA. Now it will be; companies who previously thought their violations were hard to spot know that they are now visible to neighbors, advocacy groups, investors, and competitors. The risk

[84] Lori S. Bennear and Sheila M. Olmstead, "The Impacts of the 'Right to Know': Information Disclosure and the Violation of Drinking Water Standards," *Journal of Environmental Economics and Management*, Vol. 56, No. 2 (September 2008): 117–18. The authors suggest that the most likely explanation for this big improvement was political: suppliers feared a political backlash to disclosures of serious violations and took action to prevent that.

[85] Nor is information only required to be kept on site by the regulated entity. See, e.g., Aman Azhar, "Pollution from N.C.'s Commercial Poultry Farms Disproportionately Harms Communities of Color," *Inside Climate News*, October 13, 2021, https://insideclimatenews.org/news/13102021/north-carolina-commercial-poultry-farms-justice-communities-of-color/ (poultry farmers in North Carolina keep records of waste pollution on-site and don't provide them to regulators unless there is a (rare) investigation, making records about important pollution inaccessible to the public).

[86] EPA, "NPDES E-reporting Rule," 64063, 64065.

of unfavorable attention for violations goes way up, inspiring some companies to do a better job.[87]

Rule writers should consider what information the public wants and what will motivate companies to act and make sure it is included in the rule's monitoring and reporting obligations. That's what EPA did in the 2015 rule regulating coal ash, the waste material produced by the burning of coal to make power. The regulation required companies to monitor for contamination in the groundwater below coal ash storage ponds and to post that data on the web.[88] Four organizations used that public data to identify places at high risk and also to set up a website to translate the complicated monitoring data into a user-friendly format, helping other communities use the information to address concerns in their states.[89]

Experience has shown that using standardized metrics and formats, which allow aggregation of data and also comparisons over time and across facilities, greatly increases accountability.[90] Along with mandatory electronic reporting, it should go without saying. The Greenhouse Gas Reporting Rule is a model of strong reporting design that enables a vast array of public transparency and accountability tools.[91]

Standardized information is important for another reason too: it allows government to make sense of the data for the public. Regulators can transform large volumes of data to useful information through user-focused search tools that zoom out for the big picture nationwide or home in on one sector, state, or facility. One of the most popular pages on EPA's website, with millions of queries every year, is a transparency tool writ large: Enforcement and Compliance History Online (ECHO).[92] ECHO is one-stop shopping for data on compliance

[87] EPA, "NPDES E-reporting Rule," 64065, 64069–070. Note that implementation of this rule was repeatedly delayed during the Trump administration, so has not yet reached its full potential.

[88] EPA, "Hazardous and Solid Waste Management System; Disposal of Coal Combustion Residuals from Electric Utilities," *Federal Register*, Vol. 80 (April 17, 2015): 21301.

[89] Kari Lydersen, "Toxic Coal Ash Pollution in Illinois Raises Drinking Water Concerns," *Energy News Network*, November 29, 2018, https://energynews.us/2018/11/29/midwest/toxic-coal-ash-pollution-in-illinois-raises-drinking-water-concerns/; "Ashtracker: Tracking Groundwater Contamination at Coal Ash Dumps," Ashtracker.org, https://ashtracker.org/. For background on the tortured history of amendments and legal challenges to this rule under the Trump administration, see the Harvard Environmental and Energy Law Program's Regulatory Tracker, https://eelp.law.harvard.edu/portfolios/environmental-governance/regulatory-tracker/ (Coal Ash Rule).

[90] Bradley Karkkainen, "Information as Environmental Regulation: TRI and Performance Benchmarking, Precursor to a New Paradigm?," *Georgetown Law Journal*, Vol. 89, No. 2 (January 2001): 257, 289.

[91] See Lavender Yang, Nicholas Z. Muller, and Pierre Jinghong Liang, "The Real Effects of Mandatory CSR Disclosure on Emissions: Evidence from the Greenhouse Gas Reporting Program," National Bureau of Economic Research, Working Paper 28984 (July 2021), https://www.nber.org/papers/w28984.

[92] https://echo.epa.gov/. Showing that the public uses these transparency tools, there are about 3.5 million queries to ECHO a year as of 2020. EPA OIG, "Total National Reported Clean Air Act Compliance-Monitoring Activities Decreased Slightly During Coronavirus Pandemic, but State Activities Varied Widely," Report No. 22-E-0008, November 17, 2021, at 6.

with major environmental laws by more than 800,000 regulated facilities. Search tools allow users to tailor their search, with data presented as maps, charts, or tables. It also has analytic tools that can identify the top violators by chemical, industry category, or geography. EPA's Greenhouse Gas Reporting page is another versatile data access system that flags violating facilities, profiles key industries, provides graphs and other visual presentations of the data, and allows user-directed searches.[93] Visualizations and constant updating with new data and tools means these sites are used. Because they are used, the companies whose data is reported there better up their game.

Government-created tools to summarize and highlight key information are valuable, but government should also make the underlying data available. Most people won't be able to follow all those details, but some will. Experts can pull what is most useful and turn that into pressure for improvements. That kind of data-driven advocacy is one of transparency's most powerful pathways for change. One example: after EPA required all refineries to do fenceline monitoring for benzene and made that data available online, the Environmental Integrity Project used the data to spotlight the refineries that had repeatedly high benzene emissions.[94] Expert interpretation of public data focuses public and government attention where it is likely to matter most.

In designing a transparency program, regulators should not limit the disclosures to the uses they can anticipate. Creative outside researchers, advocates, and reporters can use the data in unexpected ways and uncover information that significantly informs the policy debate. Like the recent deep dive into methane releases from the oil and gas sector, which revealed the wide variation among companies in pollution control effectiveness.[95] Or the follow-on report by the *New York Times*, discovering that some of the large oil and gas companies claiming to be moving away from fossil fuels are actually selling their most polluting assets, allowing companies to claim they are slashing emissions while the same pollution continues under a different, less visible, owner.[96] Or the research team that discovered required-to-report utilities shifted production, and thus emissions, to plants not required to report.[97]

The features of monitoring and reporting programs that make them resilient to compliance pressure—actual measurement, real-time electronic reporting,

[93] https://www.epa.gov/ghgreporting.
[94] Environmental Integrity Project, "Environmental Justice and Refinery Pollution," April 28, 2021, https://environmentalintegrity.org/reports/environmental-justice-and-refinery-pollution/.
[95] M.J. Bradley, "Benchmarking Methane and other GHG Emissions of Oil and Natural Gas Production in the United States," Clean Air Task Force, June 2021, https://www.catf.us/wp-content/uploads/2021/06/OilandGas_BenchmarkingReport_FINAL.pdf.
[96] Hiroko Tabuchi, "Here Are America's Top Methane Emitters. Some Will Surprise You," *New York Times*, June 2, 2021, https://www.nytimes.com/2021/06/02/climate/biggest-methane-emitters.html.
[97] Yang, "The Real Effects."

standardized metrics and formats—are central for transparency as well. That's why these essential qualities for rule effectiveness need to be considered from the start of rule design, not postponed to the end as they often are today. If the rule doesn't ensure accurate and timely monitoring, clear and unambiguous reporting, and standardized electronic formats easily translated to internet use, transparency options will be greatly constrained. If you don't think about that until the rule is nearly done, it's too late.

The flip side of the best-practices coin is the danger of transparency when the underlying data is highly dubious. If the information is shaky or known to be incomplete—as is true for a lot of EPA's compliance-related data as discussed throughout this book—sharing it far and wide is just as likely to mislead as inform. No regulator wants to be between a rock (the obligation to share the data they have) and a hard place (knowing that data is incomplete or unreliable). But if regulators find themselves there, transparency has to be closely accompanied by high-visibility disclosures about data limitations. EPA knows, for example, that its publicly disclosed data about drinking water noncompliance wildly understate actual violations.[98] That major data flaw used to be front and center in EPA's public summaries of the data, but not anymore. GAO recently criticized EPA for the same failing in water pollution discharge data.[99] Here's the Next Gen take-away: avoid that uncomfortable spot through rules that use Next Gen strategies to get far more accurate and robust information.

While transparency can be powerful as part of a multifaceted regulatory approach, it isn't the magical elixir that some have claimed. The most touted transparency program—the Toxics Release Inventory (TRI)—is a case in point. Researchers have discovered that much if not most of the pollution reductions usually attributed to TRI were the result of changes in calculation methods and spillovers from conventional command-and-control-style regulations.[100] The vaunted hygiene-grades-in-restaurant-windows program has faced similar skepticism in subsequent analysis.[101] Leakage—transferring pollution to firms

[98] See discussion in chapter 1, section titled "Programs with Pervasive Violations: Four Examples," sections on drinking water; chapter 2, section titled "For Some Important Programs, EPA's Understanding of Noncompliance Is Wrong," section on drinking water.

[99] GAO, "Clean Water Act: EPA Needs to Better Assess and Disclose Quality of Compliance and Enforcement Data," GAO-21-290 (July 2021).

[100] Karkkainen, "Information as Environmental Regulation," 288, n.136 (citing evidence that as much as one-third of reported TRI reductions in some years may be attributable to changes in estimation methods or reporting requirements); Linda Bui, "Public Disclosure of Private Information as a Tool for Regulating Environmental Emissions: Firm-Level Responses by Petroleum Refineries to the Toxics Release Inventory," Census Bureau, Center for Economic Studies, Working Paper CES-05-13 (October 2005), 22, https://www.census.gov/library/working-papers/2005/adrm/ces-wp-05-13.html (reviewing evidence that TRI was likely responsible for less of the reductions in toxic releases than "devotees are inclined to believe").

[101] Daniel Ho, "Fudging the Nudge: Information Disclosure and Restaurant Grading," *Yale Law Journal*, Vol. 122, No. 3 (December 2012): 522.

not required to be so transparent—can also undercut the hoped-for gains.[102] And in no sense of the word are transparency programs voluntary, despite popular nomenclature; they are government mandates, subject to enforcement and penalties. Public accountability through transparency isn't an alternative to regulation, it's part of the regulatory toolbox to help ensure that regulatory mandates deliver.

Data Analytics and Machine Learning

The recent gigantic expansion in data analytic capacity opens new vistas for detecting and solving environmental problems. These tools can be a game changer in some programs.

Even simple analysis of information submitted in self-reports can make a powerful difference. For example, water pollution dischargers report both the concentration and the volume of their discharges; simple multiplication allows EPAs computers to immediately determine not just who is in violation, but the total amount of unlawful pollution they discharged—the so-called "load over limit." EPA makes that calculation available to the public in a searchable online database so it is easy to see which facilities contribute the most unlawful pollution of concern in each water body.[103] Similarly, the shift to electronic tracking for hazardous waste shipments will make it simple to immediately spot which self-proclaimed small-quantity generators are in reality shipping large quantities and should therefore be meeting tighter safety standards.[104] Hopefully, we are well past the days when EPA was legitimately criticized for not noticing that some heavily industrialized states were reporting zero serious air violators.[105]

The era of big data makes it possible to up regulators' game by bringing in additional information that can hone models to identify the places regulators need to focus attention. A machine learning model using inspection data could double—or more—the effectiveness of inspectors in finding violating water dischargers.[106] Statistical analysis blending housing data and information about

102 Tabuchi, "Top Methane Emitters"; Yang, "The Real Effects."
103 EPA, "Water Pollutant Loading Tool Frequently Asked Questions," https://epa.gov/trends/loading-tool/resources/faq.
104 See Bracewell, "EPA's 'Next Generation' Enforcement Hitting Region 6 Facilities Now," June 15, 2012, https://bracewell.com/insights/epas-next-generation-enforcement-hitting-region-6-facilities-now, describing how EPA Region 6 was able to find hazardous waste generators not meeting legal requirements through an examination of manifests, information that will be available electronically nationwide once e-manifest is fully implemented.
105 EPA OIG, "Consolidated Report on OECA's Oversight of Regional and State Air Enforcement Programs," E1GAE7-03-0045-8100244 (September 1998), 9.
106 M. Hino, Elinor Benami, and N. Brooks, "Machine Learning for Environmental Monitoring," *Nature Sustainability*, Vol. 1 (October 1, 2018): 583-88).

location of lead service lines can help identify neighborhoods at high risk of exposure to drinking water containing lead.[107] Algorithms can use satellite imagery to map concentrated animal-feeding operations—the first step toward flagging the unpermitted sites that contribute to significant water quality problems.[108]

More in-depth analytic tools can help spot program weaknesses that require attention. For example, Lori Bennear's careful statistical analysis of drinking water data showed that many drinking water systems were skirting their obligations by taking more known-to-be-clean samples to keep their sampling results under the action threshold—a practice so common it has a name: "sampling out."[109] Another study made use of the nonintuitive Benford's Law to figure out that air polluters in North Carolina may be underreporting emissions to squeak under the threshold for higher fees and tighter emissions controls.[110] Predictive analytics holds promise for preventing the worst catastrophes by spotting the likely serious violators in advance.

Analytic tools are also central to confronting environmental injustice; EJSCREEN and other sophisticated data aggregation and prioritization systems help to focus government's attention on communities disproportionately affected by pollution and environmental threats.[111] And we also need to remain mindful of the ways in which seemingly neutral analytic tools can unwittingly exacerbate negative distributive effects.[112]

Why include these data analytics possibilities in a book about rule design? Because analytics are only possible if the type, quality, and speed of the available data supports them. If monitoring is too infrequent to provide confidence in the results, or failure to report is so widespread that data is significantly incomplete, no amount of sophisticated analytics will be able to usefully inform policy. If the design hasn't done all it can to ensure reporting accuracy and completeness, data errors and omissions will overwhelm the targeting tools that analytics could produce. Rule writers should have one eye on the analytics that might be possible

[107] GAO, "Drinking Water: EPA Could Use Available Data to Better Identify Neighborhoods at Risk of Lead Exposure," GAO-21-78 (December 2020).

[108] Cassandra Handan-Nader and Daniel Ho, "Deep Learning to Map Concentrated Animal Feeding Operations," *Nature Sustainability*, Vol. 2 (April 2019).

[109] Lori S. Bennear, Katrina K. Jessoe, and Sheila M. Olmstead, "Sampling Out: Regulatory Avoidance and the Total Coliform Rule," *Environmental Science & Technology*, Vol. 43, No. 14 (June 2009): 5176.

[110] Christopher F. Dumas and John H. Devine, "Detecting Evidence of Non-Compliance in Self-Reported Pollution Emissions Data: An Application of Benford's Law," 2000 Annual Meeting, American Agricultural Economics Association, available at https://ideas.repec.org/p/ags/aaea00/21740.html.

[111] EPA, "EJSCREEN: Environmental Justice Screening and Mapping Tool," https://www.epa.gov/ejscreen.

[112] Elinor Benami et al., "The Distributive Effects of Risk Prediction in Environmental Compliance: Algorithmic Design, Environmental Justice, and Public Policy," Proceedings of the 2021 ACM Conference on Fairness, Accountability, and Transparency, March 3, 2021, https://dl.acm.org/doi/proceedings/10.1145/3442188.

when they are designing a rule's monitoring and reporting requirements, to maximize the analytic strategies that will ultimately be possible.

Other Widely Applicable Next Gen Ideas

Simplicity

There are many factors that push in the direction of complexity in environmental rules. One is the reality of the underlying complexity of the real world, where there are always a range of facility types and circumstances, and a future that is difficult to predict. The more the rule attempts to reflect the real world's diversity of firms, locations, experience, and costs, the more complicated it becomes. Then there are the economists who seek maximum efficiency in results, which almost always means more differentiation among the firms affected by a rule and greater flexibility either in what the requirements are or how to meet them, again leading to more complexity.[113] And of course there are the many stakeholders affected by the rule, whose principal interest is in how the rule affects them and whose acquiescence can sometimes be obtained by carving out an exception or providing more flexibility that accounts for their individual situation. As the rule takes on more and more complicated provisions, there is often no good place for the conversation that starts with "Wait a minute, is this going to work?"

Complexity is where violators hide. If sophisticated expertise, special equipment, and extended examination are required to figure out if an entity is complying, then, as night follows day, there will be a lot of violations. There is a mountain of evidence proving that. Sometimes complexity-induced confusion causes mistakes. It definitely makes it easier for regulated companies to make compliance a lower priority. Intentional violators have reason to think they will never be noticed amidst the general chaos of compliance uncertainty, and if by chance they are, regulatory complexity arms them with a lot of throw-dust-in-the-air-type excuses. The New Source Review Program for emissions from large sources of air pollution is the poster child for a rule that creates opportunities to dodge compliance obligations by claiming to be covered by a complicated array of exceptions, exclusions, and conditions.[114]

[113] See Schackenbach, "Fundamentals of Successful Monitoring," 1580: "EPA has consistently followed the principle that a high degree of flexibility in the regulations is desirable provided that environmental goals are not sacrificed. However, it should be noted that added regulatory flexibility is often accompanied by great rule complexity and length. Therefore, before adding new compliance options to a regulation this should be taken into account." See also David L. Markell and Robert L. Glicksman, "Dynamic Governance in Theory and Application, Part I," *Arizona Law Review*, Vol. 58, No. 3 (2016): 606.

[114] See discussion in chapter 1, section titled "Programs with Pervasive Violations: Four Examples," section on New Source Review.

Compliance simplicity is possible even in complex situations. But it requires good design. In the Acid Rain Program, for example, the complicated issue of controlling SO_2 and installing and conducting quality assurance for pollution monitors required lots of detailed provisions. But the rule ultimately boiled all that detail down to a very simple question: Do you have enough allowances to cover your emissions, yes or no? A multitude of great Next Gen–style provisions went on behind the scenes to make the answer to that question easy to figure out and impossible to duck. In that rule, as in many others, actual monitoring, electronic reporting, and many other Next Gen tools provide the foundation for eventual simplicity in compliance design.

Impossibility

Many of the Next Gen ideas discussed in this book are about ways to encourage compliance or discourage firms from violating. The goal is getting closer to compliance as the default setting. Another way to achieve much better compliance is to make violations impossible, or close to it.

Leaded gasoline provides two examples of the impossibility approach. In an effort to reduce pollution from motor vehicles, EPA prohibited the sale of leaded gasoline for most passenger cars, although it continued to be available for trucks. How to prevent someone at the gas station from using leaded fuel in a passenger car? EPA set up a physical restriction: the filler inlet to the car (where you put the pump nozzle) had a specified size, and the required size of the leaded gasoline nozzle was too big to fit. Only the nozzle for the unleaded fuel would work. It was literally impossible to make a mistake.[115]

The outright ban on lead in gasoline that was ultimately adopted in 1995 is another kind of impossibility strategy. You can't use leaded gasoline in your car because you can't buy it anywhere. When the health impacts are severe, and safe implementation is near impossible—as is sometimes true, for example, for dangerous pesticides and toxic chemicals—a straightforward ban can be the best, and sometimes only, strategy.

Today's sophisticated monitoring and information technologies provide another way to use an impossibility approach. If real-time monitoring shows that a violation is about to happen, a rule could require automated feedback loops that change the company's process to avoid the violation. Such automated changes also create additional incentives for better compliance: companies will seek to

[115] 40 C.F.R. § 80.22; EPA, "History of Fuel Tank Filler Restrictor" in the Preamble to the 2002 rule, "Prohibition on Gasoline Containing Lead or Lead Additives for Highway Use: Fuel Inlet Restrictor Exemption for Motorcycles," *Federal Register*, Vol. 67 (May 24, 2002): 36765.

avoid the business consequences of such unexpected changes to production by finding and fixing problems before they become serious.

EPA adopted this approach in the settlement of a clean water case. The agreement required that a company's boats used for waste disposal in underwater areas employ global position system technology to create a virtual fence around the disposal area.[116] The "Geo Fence" prevents dumping until the scow is inside the permitted dump site coordinates, thus eliminating the element of human error, a common cause of misdumping.[117]

The same idea could work for pesticide application: sprayers can be equipped with technology that maps agricultural boundaries and measures wind speed and direction, to ensure that application is only taking place in the approved areas and avoids drift.[118]

Automation is the first cousin to impossibility. If the complying action is automatic, a violation can only happen if businesses deliberately and intentionally violate, not most companies' first choice. In the oil and gas industry, for example, hatches left open on tanks is one of the ways that significant amounts of pollution are released to the air. Sure, you can have endless ways to remind employees to close the hatches. Or you can require that hatches be equipped to close automatically, as New Mexico is proposing to do in an oil and gas regulation.[119]

Nothing is ever truly impossible. Determined and skilled violators can find a way around a seemingly insurmountable barrier. But a system that makes violations close to impossible will eliminate most noncompliance. That will free up enforcement resources to focus on the dedicated offenders, and significantly increase the chances that they will be caught and suffer the consequences. Regulatory designs that make violations impossible, rather than just discouraging them, have the additional advantage of applying equally to all, creating a more level playing field. Seeing that everyone is being held to the same standard helps to reinforce a norm of strong compliance.

[116] EPA "Companies Fined and Take Action to Comply with Ocean Dumping Requirements," News Release, EPA Region 1, November 6, 2015, https://archive.epa.gov/epa/newsreleases/companies-fined-and-take-action-comply-ocean-dumping-requirements.html. Such automated systems don't always prevent violations though. EPA "Quincy, Mass. Dredging Company to Pay Penalty for Violations of the Ocean Dumping Act," News Release, EPA Region 1, March 15, 2021, https://www.epa.gov/newsreleases/quincy-mass-dredging-company-pay-penalty-violations-ocean-dumping-act.

[117] See Norman Bourque, Frank Belesimo, and Tim Mannering, "Scow Geofence System (SGS) and Its Application in Dredged Material Disposal," Dredging Summit and Expo 2018 Proceedings, June 25, 2018, https://www.westerndredging.org/phocadownload/2018_Norfolk/Proceedings/5a-1.pdf.

[118] Association of Equipment Manufacturers, "Modern Sprayer Technology," *Emerging Technologies Overview,* May 12, 2021, slides 8, 14, https://www.epa.gov/sites/default/files/2021-05/documents/aem-emerging-technologies-ppdc-may2021.pdf.

[119] New Mexico Environment Department, "Oil and Gas Sector—Ozone Precursor Pollutants," Proposed Rule, May 6, 2021, § 20.2.50.123(B), https://www-archive.env.nm.gov/air-quality/wp-content/uploads/sites/2/2021/03/Proposed-Part-20.2.50-May-6-2021-Version.pdf.

Shifting the Burden of Proof

Regulated companies know (or are in a position to find out) about their own operations and pollution. Regulators mainly rely on the companies' information to find out what's going on. This information asymmetry is a challenge for compliance monitoring. One way to deal with it is by shifting the burden of proof.

The norm today is that government has to prove a company is in violation to prevail in an enforcement action. Sometimes government can do that through information the company itself is required to provide, but still, the burden of proof, as it is called in legal proceedings, rests with the government. That creates a situation where companies have incentive to be evasive, or to deliberately remain in the dark, so if asked they can truthfully claim not to be aware. Delay and obfuscation can serve the company's interests in postponing the day of reckoning.

One option for changing this unhelpful dynamic is to shift the burden to the company. Once credible evidence of a serious problem is provided—like satellite evidence of a significant pollution event, for example—it could then be up to the company to prove that it is not in violation. That makes sense because the company is best positioned to know or figure out what happened.

Shifting the burden shifts the company's attitude too. Instead of finding it advantageous to look the other way, companies that have to carry the burden of proof are best served by being in the know. They want those monitors working and they want solid recordkeeping so they can prove what happened. They want to avoid getting that credible evidence notice, and they want to know what happened, because not knowing means they lose. If companies collect more information about their own performance as a defensive measure, they might find out that yes, in fact, they do have a problem. Knowledge is the first step toward fixing it.

Shifting the burden also makes the threat of enforcement more credible, which enhances deterrence. Instead of the common situation now, when prying the necessary evidence from the company is like pulling teeth, it is in the company's interest to bring that evidence forward itself.

If a regulation allows citizen science to provide the credible evidence that shifts the burden—which in the right situation makes a lot of sense—that changes the dynamic even more. The risk of a credible evidence notice won't be constrained by the tight resources and political headwinds hindering regulators. Citizens with access to reliable monitoring equipment or other evidence made available by government transparency programs can blow the whistle in a way that companies cannot ignore. EPA has announced that it is considering this approach for methane super-emitters in regulations for oil and gas.[120] It is almost

[120] "EPA 2021 Proposed Methane Rule", 63115, 63177 (see *supra* note 13).

always more hassle and more expense for companies to have to engage in unplanned and rushed investigation and response. Rather than repeatedly facing that problem in response to credible evidence notices, companies might decide to stay on top of possible violations and fix them as they go. Perfect. Exactly what should happen.

Data Substitution

The problem of missing or inaccurate data plagues all environmental programs. Failure to monitor or to report required information is the most common violation for many rules.[121] Accurate and complete self-reports are the foundation of effective implementation oversight. Without them, regulators are blind to potentially serious problems. For some regulated companies, that's the point; it is easier and less costly to skip monitoring and reporting than it is to disclose significant violations. One in-depth study found that monitoring and reporting violations were strong and statistically significant predictors of health-based violations.[122]

Most environmental rules do not contain strong incentives to ensure that monitoring and reporting are done and done right. For many, just the opposite. They are remarkably casual about monitoring and reporting failures. Monitoring and reporting misses are both less noticed and treated as less critical than violations of other provisions. Putting our Next Gen hats on, it is easy to predict what happens: the stretched-too-thin facilities correctly perceive that monitoring and reporting are lower priority violations, so don't feel pressure to do what's necessary to comply. The entities who know that monitoring or reporting accurately is likely to reveal big problems take the path of least resistance and bypass the reporting obligation rather than fess up. This happens everywhere, from drinking water to air monitors.[123]

Data-substitution provisions are a powerful tool to flip these perverse incentives. The concept is simple: if you miss a report, or you fail to do what's necessary for your monitoring to be reliable, regulators will assume the worst. If a rule adopts a data-substitution provision, no one will have missing data: they will either have self-reported data or they will have data that is automatically inserted.

[121] See, e.g., GAO, "EPA Needs to Better Assess," 33. It isn't just regulated companies that are failing to report; states also have significant problems with completeness and accuracy in reporting. GAO, "EPA Needs to Better Assess," 24. See discussion of the many state-reporting failures in chapter 2.

[122] GAO, "Drinking Water: Unreliable State Data Limit EPA's Ability to Target Enforcement Priorities and Communicate Water Systems' Performance," GAO-11-381 (June 2011), 16.

[123] See EPA, "Economic Analysis for the Final Revised Total Coliform (Drinking Water) Rule," September 2021, at 4–5 ("Low compliance with monitoring and reporting may occur if systems would rather incur a Monitoring/Reporting violation rather [*sic*] than risk an MCL violation by sampling."); Mu, Rubin, and Zou, "What's Missing in Environmental (Self-)Monitoring" (some local governments skip air pollution monitoring when they expect air quality to deteriorate).

The key is to make the consequences of that substituted data be worse for the reporting entity than just doing it right. That's what the Acid Rain Program did; if air monitors weren't working or failed quality control checks, EPA automatically inserted worst-case assumption data. Required substitute data got increasingly conservative (i.e., worse for the company) as the extent of unreported or invalid data went up.[124] Because all emissions had to be covered by purchased allowances, these higher-than-likely emissions assumptions imposed a direct financial cost on the company. This simple data-substitution provision inspired companies to maintain monitors and do the required quality control checks because that was cheaper than the alternative. If a similar approach were taken for quality control fails in air pollution stack tests, we would not see the widespread problems that are evident today.[125]

Data substitution is a straightforward way to align private interests with the public good. The regulated entities are the ones with the ability and authority to comply. They can order that necessary replacement part or apply an accurate label to the product or prioritize collecting the required data, or not. The rule's design shouldn't make it easy to skip those things, or force government to identify the problems and try to sanction bad choices after the fact. Data substitution inspires the regulated to do it right the first time. That's the idea behind a labeling provision in EPA's recently finalized rule to phasedown climate damaging HFCs: if the label for an HFC container doesn't clearly state what HFCs are inside, EPA will assume the container contains the most damaging HFC, which will cost the company a lot more money.[126] Don't like that outcome? It's on you to fix it. Data substitution is an elegant Next Gen solution: the sensible choice for the company is the one that best protects the public.

Automatic Consequences

Enforcement cases take time. Government has to collect data to figure out that there is a violation, then needs more information to decide what the consequences should be. Negotiation with the company can be laborious, followed by appeals and delays associated with legal process. All of that is appropriate and worth it for serious problems that deserve that level of regulators' attention. But it also dilutes the deterrent impact of enforcement because regulated entities know that they may never be found out, and even if they are, the consequences for violations will be delayed and are negotiable.

[124] EPA, "Part 75 Emissions Monitoring Policy Manual—2013, at 15-2, https://www.epa.gov/sites/default/files/2019-10/documents/part_75_emissions_monitoring_policy_manual_10-18-2019.pdf.
[125] EPA OIG, "More Effective Oversight" (see *supra* note 25).
[126] "EPA Final Rule for HFC Phasedown," 55178.

One option for changing that dynamic is to make negative outcomes more likely, much faster, and more predictable through automatic consequences. That's what the Acid Rain Program did by setting a fixed penalty—higher than the cost of buying an allowance—for failure to have the required number of pollution allowances.[127] Violating was certain to cost more than complying.

Penalties are not the only option for automatic consequences. Data substitution, for example, is a form of automatic consequences for monitoring and reporting violations. What works best will depend on the circumstances of the individual rule. All that's required are three things: (1) rule design that forces companies to self-disclose the violation (there's no need to prove it—the company has admitted it); (2) consequences for violating that are less attractive than doing what's required and that happen automatically, without any additional engagement by regulators; and (3) backbone when the inevitable complaints in particular circumstances arise; if regulators cave and agree to reduce consequences for individual violators, pretty soon the deterrence advantages are lost and the agency is dragged into the individual case negotiation process again. Yes, automatic consequences are rigid. That's the point.

Contingent Regulation

Usually, when environmental rules are adopted, that's the end of the story. If it turns out to have been a bad regulatory design, or compliance is horrible, too bad. The only option is to go back and start the rule-writing process over. Even when changes are clearly needed, starting over is extremely unlikely because of the expense, time, and political risks of reopening the entire rule to new political maneuvering. Instead of this all-or-nothing framework, in which the dynamics almost always favor doing nothing, there is another option: contingent regulation.[128]

Contingent regulation builds adaptation into the rule: if X happens, then the standard will change to Y. That's what EPA did in the Renewable Fuel Standard—if net lands in farming went up, contrary to expectation, more stringent recordkeeping obligations would take effect.[129] EPA wouldn't have to reopen the rule or figure out what to do; that was already decided in the rule as

[127] See the description of the Acid Rain Program in chapter 1.

[128] This section draws on the insights and naming conventions in Justin R. Pidot, "Governance and Uncertainty," *Cardozo Law Review*, Vol. 37, No. 1 (October 2015): 113–84. See also Lori S. Bennear and Jonathan B. Wiener, "Built to Learn," in Esty, *A Better Planet*, 357.

[129] This "aggregate compliance" approach adopted by EPA for the Renewable Fuel Standard didn't work to support the greenhouse gas reduction purpose of RFS, but it illustrates how EPA has already used the concept of contingent regulation. The Renewable Fuel Standard and the aggregate compliance approach are discussed at length in chapter 8.

promulgated. Once the regulator publishes a notice that the factual predicate has been met, the contingent rules—already adopted as part of the final regulation—kick in. Contingent regulation could be used to allow a quick and preset response if compliance turns out to be bad, allowing stronger provisions to take effect automatically.

For example, regulated companies might object to a data-substitution provision, claiming it is unduly harsh for what they expect to be rare monitoring failures. A contingent rule could say, OK we hope you are right, but if it turns out that monitors are not working or companies fail to report more than 1 percent of the time, data substitution will be required. Another illustration: if regulated parties argue for less stringent standards for companies that are claimed to have lower risk or smaller impact—this happens all the time—EPA could say, OK for now, but if EPA finds more than five companies falsely claiming to qualify for lower standards, that category ends, and everyone has to meet the stricter standards.

Contingent regulation isn't an excuse for regulators to dodge making the tough call now. Lots of times the tougher standard is the obviously right choice from the get-go and there is little basis for the pushback. However, sometimes the likely outcome isn't that clear or is genuinely uncertain. Contingent regulation might fit the bill in that case: it sidesteps dueling predictions and bases a decision about additional requirements on ascertainable facts at a defined point in the future.

Automatic triggering of Next Gen–style compliance measures has many benefits beyond forcing better compliance. It motivates collection of data and attention to the factors necessary to determine the contingency, which will by itself be valuable. It makes for better rules by calling the companies' bluff about likely future outcomes; if companies resist an automated reporting system, for example, because they claim compliance will be strong without it, why oppose a rule contingency mandating automated reporting if more than 1 percent of companies fail to report? Perhaps the companies are not as confident in their prediction as they claim. Good to know. It also motivates compromise; competing predictions about the necessarily uncertain future will eventually have a factual outcome, so unrealistic predictions will get less traction. And, best of all, it inspires industry action to try to make sure that their optimistic forecast is what actually happens. Next Gen at its finest.

6

The Ideologues

Performance Standards and Market Strategies

There is near-universal admiration in environmental policy circles for performance standards and market strategies in environmental rules.[1] Virtually everyone unites in trashing "command-and-control" regulation.[2] Is one of these approaches best for assuring compliance?

[1] *See, e.g.*, Cary Coglianese, "The Limits of Performance-Based Regulation," *University of Michigan Journal of Law Reform*, Vol. 50 (2017): 525, 553 (noting the "seemingly unbridled enthusiasm for performance based regulations by regulatory commentators and officials around the world"); Laura Montgomery, Patrick McLaughlin, Tyler Richards, and Mark Febrizio, "Performance Standards vs. Design Standards: Facilitating a Shift Toward Best Practices," *Mercatus Center George Mason University*, Working Paper, June 26, 2019, at 33, https://www.mercatus.org/publications/regulation/performance-standards-vs-design-standards-facilitating-shift-toward-best ("performance standards have been touted as best practice in regulatory rulemaking since at least 1980"); Timothy F. Malloy, "The Social Construction of Regulation: Lessons from the War Against Command and Control," *Buffalo Law Review*, Vol. 58, No. 2 (2010): 267, 343 (majority of legal scholars are advocates for market-based regulation); Jody Freeman and Charles D. Kolstad, "Prescriptive Environmental Regulations versus Market-Based Incentives," in Jody Freeman & Charles D. Kolstad eds., *Moving to Markets in Environmental Regulation* (Oxford University Press, 2007), 3, 4 ("the superiority of market-based instruments has developed into a near orthodoxy"); Frank Ackerman and Kevin Gallagher, "Getting the Prices Wrong: The Limits of Market-based Environmental Policy," Global Development and Environment Institute, Working Paper 00-05, 2000, at 1 ("Market based policies are fast becoming the recommended policy panacea for all the world's environmental problems"); National Academy of Sciences, Engineering, and Medicine Transportation Research Board, *Designing Safety Regulations for High-Hazard Industries* (The National Academies Press, Special Report 324, 2018), at 18. https://www.nap.edu/catalog/24907/designing-safety-regulations-for-high-hazard-industries (many rule types seek to call themselves performance-based because of the "political legitimacy" ascribed to performance as a tool of governing); Daniel H. Cole and Peter Z. Grossman, "Beyond Compliance Costs: Comparing the Total Costs of Alternative Regulatory Instruments," in Kenneth R. Richards and Josephine van Zeben eds., *Policy Instruments in Environmental Law* (Edward Elgar Publishing, 2020), 32, 39 (noting a "consensus in the literature" favoring economic instruments for environmental protection); Shi-Ling Hsu, "Prices Versus Quantities," in Richards and Van Zeben, *Policy Instruments*, 183, 186 (market mechanisms are a "presumptively favored" means of regulating); Jason Scott Johnston, "Tradable Pollution Permits and the Regulatory Game," in Freeman and Kolstad, *Moving to Markets*, 353 ("Indeed, so powerful is the standard economic argument for tradable pollution permit regimes that their relative scarcity in American environmental regulation now stands as something of an unexplained paradox").

[2] See, e.g., Malloy, "The Social Construction of Regulation," 268–69 (arguing that there is a "war" against command and control and that "bashing traditional regulation has become something of a national pastime among legal scholars"); Wendy Wagner, "The Triumph of Technology-Based Standards," *University of Illinois Law Review* (December, 2000): 85 n.6, 107 (noting that virtually all of the literature is critical and that "law scholars who have publicly applauded the use of technology-based standards can be counted on one hand"); Daniel H. Cole, "Explaining the Persistence of 'Command-and-Control' in US Environmental Law," in Richards and Van Zeben, *Policy Instruments*, 157,159 ("command and control is often used as a term of derogation"); Ackerman and Gallagher,

Next Generation Compliance. Cynthia Giles, Oxford University Press. © Cynthia Giles 2022.
DOI: 10.1093/oso/9780197656747.003.0007

The policy discussion suffers from continued nomenclature confusion.[3] Loosely speaking, most people think a regulation is a performance standard if it tells the regulated what to do but not how to do it.[4] Many of today's EPA pollution regulations meet this definition but are nevertheless disdained by performance standard purists.[5] Some people see market mechanisms as a subset of performance standard approaches, while market devotees think market mechanisms are a category unto themselves and everything that isn't a market approach is the dreaded command and control.[6] Some policy scholars, observing that the labels have become so politically freighted that they are losing all meaning, have suggested abandoning these terms altogether and creating a new lexicon.[7]

All this ideological fervor is misplaced. Performance standards, market strategies, and even the much-maligned command and control are all approaches that can succeed, or dramatically fail. The key to widespread compliance is having a well-crafted rule that picks a strategy that matches the problem. Every regulation, including performance standards or market approaches, must be well designed for the rule to realize the intended objectives and achieve widespread compliance. When the necessary regulatory safeguards are not built in, every kind of rule can struggle.

The rhetorical positions in this theory debate—unbridled enthusiasm for performance standards and market strategies and condemnation of command and control—do not resonate with most practitioners. EPA's current rule-writing practice looks nothing like the rigid one-size-fits-all, Soviet-style

"Getting the Prices Wrong," 2 (command and control "frequently stigmatized"); Daniel C. Esty, "Red Lights to Green Lights: From 20th Century Environmental Regulation to 215t Century Sustainability," *Environmental Law*, Vol. 47, No. 1 (Winter 2017): 10, 15, 46 (argues for shift from "government mandates" to a regulatory regime of "price signals," where government can "get out of" the "old command and control regime").

[3] Scholars and policy advocates use a wide variety of labels to mean close to the same thing. See, e.g., National Academy of Sciences, *Designing Safety Regulations*, 16 (explaining that the terms "prescriptive," "technical," "design-specific," "technology-based," "command-and-control," and "one-size-fits-all" are often used interchangeably). The terms are also used inconsistently. See, e.g., National Academy of Sciences, *Designing Safety Regulations*, 16–18.

[4] Coglianese, "The Limits of Performance-Based Regulation," 532; Montgomery, "Performance Standards vs. Design Standards," 3, 5; National Academy of Sciences, *Designing Safety Regulations*, 16.

[5] See, e.g., Malloy, "The Social Construction of Regulation," 313–18; Coglianese, "The Limits of Performance-Based Regulation," 534 n.32 (what some people call technology-based standards are actually performance standards); Wagner, "The Triumph of Technology-Based Standards," 90 (EPA's air toxic regulations contain a quantitative pollution limit that is derived from what the top performing sources can achieve).

[6] See, e.g., Coglianese, "The Limits of Performance-Based Regulation," 535 (market strategies are a type of performance standard); Robert Stavins, "Market-Based Environmental Policies: What Can We Learn from U.S. Experience (and Related Research)?," in Freeman and Kolstad, *Moving to Markets*, 19 (everything not markets is command and control); Cole, "Explaining the Persistence," 159 (performance standards are classified as command and control).

[7] See National Academy of Sciences, *Designing Safety Regulations*, 2, 19, 30.

characterization attributed to it by the market proponents.[8] Nor is the innovative rational actor of economic theory frequently encountered by inspectors in the field; it is common to find firms that have failed to adopt better and cheaper pollution-reduction technologies. Most pollution standards adopted by EPA are performance-based, but you would never know that by reading the blistering critiques.[9]

A few brave souls have pointed out the lack of evidence for these soaring claims of universal policy superiority, noting that scholars have uncritically adopted these positions despite their "astonishing lack of empirical support."[10] The absence of evidence does not slow them down. One recent paper describes what it characterizes as the five known empirical studies on performance-based regulations, noting that all five found that the studied regulation did not achieve the desired objective. The paper nevertheless concludes with a rousing call for a "more adamant devotion to adopting performance-based standards."[11] There are not that many market examples for empirical study, but most of the successful ones are limited to air pollution and fisheries.[12] Professors Daniel Cole

[8] See Malloy, "The Social Construction of Regulation," 331 (referencing Professor Stewart's "oft-repeated comparison of command and control to 'Soviet-style central planning' ").

[9] See Malloy, "The Social Construction of Regulation," 315 (most EPA air rules set an emission limit and don't mandate use of any particular control technology); U.S. Congress, Office of Technology Assessment, *Environmental Policy Tools: A User's Guide*, OTA-ENV-634 (Washington, DC: U.S. Government Printing Office, September 1995), 11, 14–16 (finding that most EPA programs set emission limits derived from what high performing controls can achieve [which it calls "design standards"], and that explicit technology specifications are rarely used), https://www.princeton.edu/~ota/disk1/1995/9517/9517.PDF; David M. Driesen, "Design, Trading, and Innovation," in Freeman and Kolstad, *Moving to Markets*, 436, 448 (noting that environmental statutes usually encourage performance standards).

[10] Malloy, "The Social Construction of Regulation," 345. See also Cary Coglianese and Jennifer Nash, "The Law of the Test: Performance-Based Regulation and Diesel Emissions Control," *Yale Journal on Regulation*, Vol. 34, No. 1 (2017): 80 (describing the striking absence of empirical studies on performance standards despite the widespread belief in their superiority, noting that "conventional wisdom's unbridled enthusiasm for these standards has rested almost exclusively on theory and intuition"); National Academy of Sciences, *Designing Safety Regulations*, 4 (noting that claims about advantages and disadvantages of regulatory types are too often anecdotal and that systematic empirical research is lacking); Montgomery, "Performance Standards vs. Design Standards," 20 (noting that the lack of broader analyses of the effectiveness of performance standards remains a "gap in the literature"); Driesen, "Design, Trading, and Innovation," 450 (empirical evidence of emission trading's superiority in stimulating innovation is "surprisingly thin").

[11] Montgomery, "Performance Standards vs. Design Standards," 34. The authors of this article include a tiny nod of the head to the contradiction between the evidence and their conclusion by adding the qualifier "where feasible and appropriate," but this is light ballast for the article's enthusiastic push for more use of performance standards. Note that a preference for performance standards is enshrined in federal guidelines for writing regulations. See Executive Order No. 12,866, *Federal Register*, Vol. 58 (October 4, 1993): 51735; Office of Management and Budget, "Circular A-4" (September 17, 2003), https://obamawhitehouse.archives.gov/omb/circulars_a004_a-4/. See also National Academy of Sciences, *Designing Safety Regulations*, 117.

[12] Stavins, "Market-Based Environmental Policies," 35 (noting that the three successes with tradable permits—acid rain, leaded gasoline, and CFCs—involved air pollution and stating that there is almost no evidence in other areas); Tom Tietenberg, "Tradable Permits in Principle and Practice," in Freeman and Kolstad, *Moving to Markets*, 63, 86.

and Peter Grossman have pointed out that the broad consensus favoring market approaches is based on studies that ignore the full range of costs, and thus provide an insufficient basis to conclude that market approaches are superior.[13]

Flexibility: A Strength and a Weakness

The principal theoretical benefit of both performance standards and market strategies is their flexibility; they allow firms to make choices about how best to comply, which can reduce firms' compliance costs, especially when there is a lot of variation among the regulated firms.[14] The intuitive appeal of this perspective has contributed to its widespread adoption. But the same flexibility that holds promise for reducing compliance costs creates additional compliance challenges.

The flexibilities that make these approaches economically attractive can undermine the objective that was the reason for the rule in the first place. One study of the impact of a market strategy for reformulated gasoline illustrates the tradeoffs.[15] EPA adopted regulations about the volatile organic compound (VOC) content of fuel in an effort to tackle ozone pollution. The rule set a limit on VOC content but allowed companies to choose how to comply. Companies, not surprisingly, chose their least costly option. Unfortunately, the least costly option also meant than there was no discernable impact on ozone because the VOC that companies elected to reduce to meet the standard was not a principal contributor to ozone formation. California, by contrast, adopted a standard that specified which VOCs had to be reduced. That more rigid approach increased the costs of compliance, but it also had a significant benefit in improved air quality.[16] In this case, the more flexible performance approach might have been lower cost,

[13] Cole and Grossman, "Beyond Compliance Costs," 39.

[14] Coglianese, "The Limits of Performance-Based Regulation," 545; Malloy, "The Social Construction of Regulation," 289 (economic efficiency is the most widely used justification for the recommended shift toward alternative regulatory schemes based on market principles). The second most frequently cited rationale for performance standards and market approaches is their theoretical strength at encouraging innovation. This rationale likewise lacks empirical support. See Coglianese, "The Limits of Performance-Based Regulation," 541–42 (the common understanding that performance standards encourage innovation is not correct); Driesen, "Design, Trading, and Innovation" (claims that market approaches do a better job than traditional regulations of encouraging innovation lack both theoretical and empirical support); Malloy, "The Social Construction of Regulation," 272, 308 (lack of empirical support for claims of economic efficiency and technological innovation).

[15] Maximilian Auffhammer and Ryan Kellogg, "Clearing the Air? The Effects of Gasoline Content Regulation on Air Quality," *The American Economic Review* (October 2011): 2687. See also the useful description of this rule, and other fuels requirements, in Joseph E. Aldy, "Promoting Environmental Quality Through Fuels Regulations," in Ann Carlson and Dallas Burtraw eds., *Lessons from the Clean Air Act: Building Durability and Adaptability into U.S. Climate and Energy Policy* (Cambridge University Press, 2019), 159, 161; Coglianese, "The Limits of Performance-Based Regulation," 561–62.

[16] Auffhammer and Kellogg, "Clearing the Air?," 2719–20.

but it did not achieve the desired benefit. The more inflexible California standard had higher costs but got the job done. In the zeal to reduce compliance costs we should not lose sight of the reason we adopt rules: to achieve an environmental benefit and protect the public.[17] A lower cost but ineffective regulation is not a better deal for the public than a strategy that may cost more but produces the necessary results.

One of the main difficulties for both performance and market approaches is that they only work if they build in a way to reliably measure performance. That is a problem for all types of regulation, but for rules that specify only ends and not means, measurement of ends is even more important.[18] A rule requiring a specified type of pollution control need only determine if that method is in fact deployed. A rule that sets a pollution standard and leaves it to the regulated to decide on a compliance method requires a way to measure the pollution. Regulations creating tradable pollution credits won't have a functioning market or achieve the pollution-reduction goal unless everyone can count on the fact that the tradable unit reflects an actual reduction in pollution, and you can't know that without measurement.[19]

Reliable pollution measurement is more complicated than many assume. Many companies currently report using emission estimates instead of actual measurement, and often those estimates prove to be wildly inaccurate. For example, EPA found that measured emissions from two refineries' industrial flares were over 20 times higher than the estimate.[20] The Acid Rain Program—the most touted example of an effective pollution trading program—would not have achieved its 99 percent compliance rate without continuous emission monitors and the regulatory provisions that forced companies to use them.[21] Where,

[17] *See* Coglianese, "The Limits of Performance-Based Regulation," 561 (the federal gasoline standards failed because they gave firms too much flexibility); Kenneth Richards and Josephine van Zeben, introduction to *Policy Instruments*, 1,7 (agreeing that a policy instrument's measure of success should be primarily the extent to which it achieves the desired environmental objectives but noting that most of the policy literature focuses on minimizing the cost of compliance).

[18] See Coglianese and Nash, "The Law of the Test," 86 n.328; Montgomery, "Performance Standards vs. Design Standards," 15–16 (also noting that "Measurement may be one area where performance standards suffer by comparison with prescriptive standards"). See also National Academy of Sciences, *Designing Safety Regulations*, 105 (noting that for many problems a measure may be difficult to find), and 108 (cautioning about the problem of manipulation of performance metrics).

[19] See Cole and Grossman, "Beyond Compliance Costs," 36 (noting that the absence of reliable and cost-effective monitoring can be disabling for a market strategy); James Salzman and J.B. Ruhl, "'No Net Loss': Instrument Choice in Wetlands Protection," in Freeman and Kolstad, *Moving to Markets*, 323, 342 (developing a measure that captures the value of the credit being traded is the "critical first step" in any trading-based mechanism).

[20] See the discussion of the use and abuse of emissions estimates and how far they often vary from real life in chapter 2. One example, cited in chapter 2, is field investigations at refineries finding that actual emissions were between 4 and 448 times higher than the estimated emissions. Similar violations at multiple refineries led EPA to issue an enforcement alert about the problem. See "EPA Enforcement Targets Flaring Efficiency Violations," EPA Enforcement Alert, EPA 325-F-012-002 (2012), https://www.epa.gov/sites/production/files/documents/flaringviolations.pdf.

[21] See the description of the Acid Rain Program in chapter 1, and later in this chapter.

when, and how measurement is done matters too; intermittent measurement or sampling done at locations entirely at the discretion of the regulated will likely not present an accurate picture of the facts. If a company only measures pollution occasionally, or does it incorrectly, it isn't possible to know what is going on.[22] Measurement regimes also have to include ways to reduce operator error and gaming. What economists politely call "strategic behavior" occurs unfortunately too frequently. As just one example, drinking water operators can, and do, game the monitoring system by taking additional samples to artificially lower the percentage exceeding standards, or by sampling where the water is expected to be clean, as a way to avoid triggering the obligation to do more to protect drinking water safety.[23]

These monitoring complexities are usually ignored or casually brushed aside by advocates of performance-based and market strategies.[24] The leading scholar on performance standards in environmental rules puts it this way: "It may seem almost a truism to note that performance standards depend on the ability of government agencies to specify, measure, and monitor performance. But it is often not acknowledged how difficult, if not impossible, it sometimes can be to

[22] See the discussion of measurement challenges in chapter 2, section titled "For Some Important Programs, EPA's Understanding of Noncompliance Is Wrong" (discussion of monitoring flaws in the drinking water and air stationary source programs), and chapter 5, section titled "Monitoring."

[23] See discussion of monitoring loopholes for drinking water lead and pathogen rules in chapter 1, section titled "Programs with Pervasive Violations: Four Examples." Examples of gaming and outright fraud in monitoring are legion in environmental rules. Seema M. Kakade and Matt Haber, "Detecting Corporate Environmental Cheating," *Ecology Law Quarterly*, Vol. 47, No. 3 (2020): 812–14 (cheating on air pollution–related records and monitoring in the shipping industry is widespread); Yingfei Mu, Edward A. Rubin, and Eric Zou, "What's Missing in Environmental (Self-)Monitoring: Evidence from Strategic Shutdowns of Pollution Monitors," *National Bureau of Economic Research*, Working Paper 28735 (April 2021), https://doi.org/10.3386/w28735 (statistical evidence that local governments skip pollution monitoring when air quality is expected to be poor); Daniel Nicholas Stuart, "Strategic Non-Reporting Under the Clean Water Act," chapter in "Essays in Energy and Environmental Economics," PhD diss., Harvard University 2021, https://nrs.harvard.edu/URN-3:HUL.INSTREPOS:37368502 (nonreporting increases when water pollution discharge levels are expected to exceed permit limits); Department of Justice U.S. Attorney's Office District of Massachusetts, "Western Massachusetts Power Plant Owner and Management Companies Sentenced for Tampering and False Reporting," Press Release, March 23, 2017, https://www.justice.gov/usao-ma/pr/western-massachusetts-power-plant-owner-and-management-companies-sentenced-tampering-and (criminal prosecution for tampering with air pollution monitoring equipment). Despite this reality, it is common to encounter vague and unsupported assertions that gaming and fraud are rare (see, e.g., Esty, "Red Lights to Green Lights," 19) or can readily be solved through higher penalties (see, e.g., Montgomery, "Performance Standards vs. Design Standards," 27).

[24] Mark A. Cohen and Jay P. Shimshack, "Monitoring, Enforcement, and the Choice of Environmental Policy Instruments," in Richards and Van Zeben, *Policy Instruments*, 76, 78 n.14 (scholars "regularly ignore or assume away monitoring and enforcement issues" when considering the choice of policy instruments). See, e.g., Johnston, "Tradable Pollution Permits, 371 (noting only in passing that the effectiveness of market strategies "hinges on" accurate monitoring and effective enforcement, but then quickly moves on as though those precursors can safely be assumed); Stavins, "Market-Based Environmental Policies," 26 (very briefly noting that market approaches do not eliminate the need for monitoring and enforcement, implying that these activities are outside the scope of instrument choice).

obtain reliable and appropriate information on performance."[25] Environmental economists advocating for market approaches typically assume "perfect (and incidentally, costless) monitoring."[26] Some authors dispatch these challenges by making unrealistic claims that monitoring is easy and cheap.[27]

Many of our existing environmental pollution rules, nearly all of which are performance standards, do not clear the measurement hurdle. They rely on estimates or guesses about pollution. They require only very occasional monitoring or allow the regulated to select a time or place for that monitoring that is most likely to produce a favorable outcome. And they turn a blind eye to evidence that the monitoring data that is submitted doesn't reflect reality, due to confusion, incompetence, gaming, or flouting of monitoring and reporting requirements.[28]

The good news is that advances in monitoring and information technology hold promise for expanding our monitoring reach. Measurement technologies are becoming more mobile, smaller, cheaper, and more accurate, making continuous monitoring a possible game changer for some problems.[29] But not all.[30] There are many situations in which reliable, affordable measurement is not possible.[31] The attempt to allow offsets in wetlands protection, for

[25] Coglianese, "The Limits of Performance-Based Regulation," 558–59.

[26] Cole and Grossman, "Beyond Compliance Costs," 33 (citing C.S. Russell, Winston Harrington, and William J. Vaughn, Enforcing Pollution Control Laws, at 3 (Resources for the Future 1986)). Cole and Grossman also note the "dearth of empirical information on the costs of monitoring under various environmental protection regimes." Cole and Grossman, 39. A notable exception is Salzman and Ruhl's excellent analysis of wetlands mitigation banking and how the impossibility of measuring the outcomes we care about doomed that trading program. Salzman and Ruhl, "No Net Loss."

[27] See, e.g., Esty, "Red Lights to Green Lights," 46.

[28] See chapter 2, section titled "For Some Important Programs, EPA's Understanding of Noncompliance Is Wrong" (discussion of monitoring flaws in the drinking water and air stationary source programs), and chapter 5, section titled "Monitoring."

[29] Cynthia Giles, "Next Generation Compliance," *Environmental Law Reporter*, Vol. 45, No. 3 (2015): 10206–207. CEMS on ships, for example, hold promise for addressing the gigantic but underappreciated impact of air pollution from ships that is only recently regulated. See Kakade and Haber, "Detecting Corporate Environmental Cheating," 805–20. Lab-on-a-chip technologies could help to solve the difficult problem of rapidly identifying which animal species are contributing pathogens to surface water. See Ning Wang, Ting Dai, and Lei Lei, "Optofluidic Technology for Water Quality Monitoring," *Micromachines*, Vol. 9, No. 4 (April 2018): 158, https://www.mdpi.com/2072-666X/9/4/158/htm. Tracking of emissions from notoriously difficult to measure oil and gas wells might become cost-effectively possible through satellites. See Mike Lee, "The Key for EPA Rules? Inside the Methane Tech Revolution," *E&E News*, October 25, 2021, https://www.eenews.net/articles/the-key-for-epa-rules-inside-the-methane-tech-revolution/ ;Brady Dennis, "How Satellites Could Help Hold Countries to Emissions Promises Made at COP26 Summit," *Washington Post*, November 9, 2021, https://www.washingtonpost.com/climate-environment/2021/11/09/cop26-satellites-emissions/.

[30] We are nowhere near the monitoring nirvana that some enthusiasts claim. See, e.g., Esty, "Red Lights to Green Lights," 46 (asserting that every pollution source no matter the size can be equipped with pollution monitoring devices, so that "market mechanisms are now feasible in almost all pollution contexts"). In reality, monitoring isn't available or feasible for many environmental problems. Cole and Grossman, "Beyond Compliance Costs," 37; Coglianese, "The Limits of Performance-Based Regulation," 558–59.

[31] See, e.g., Coglianese, "The Limits of Performance-Based Regulation," 558–59; Cole and Grossman, "Beyond Compliance Costs," 37.

example—permitting destruction of wetlands on the desired site in exchange for development of wetlands elsewhere—is doomed by the impossibility of reliably measuring whether the "new" wetlands actually replace the functions of the wetlands destroyed.[32] If we cannot be sure that the things being swapped are equal in value, treating them as equivalent will inevitably lead to compliance shortfalls that undermine the regulation's purpose.[33]

Increased flexibility for companies from performance standards and market strategies has another effect too: increased costs for government. More variability in companies' compliance strategies makes it harder to have a uniform and simple monitoring and reporting structure. Inspectors will have a more complicated job. And flexibility can introduce a degree of uncertainty and discretion in figuring out compliance, adding confusion, opportunity for strategic evasion, and administrative burden for government.[34] These hidden costs of regulatory structure choices can add up.[35]

The additional burden on government from more flexible standards is usually ignored in the literature that promotes performance standards and market strategies. Only the costs of compliance for the regulated firms count.[36] This

[32] Due to widespread acknowledgment that the wetlands mitigation banking program had failed, based in significant part on the impossibility of measurement (see Salzman and Ruhl, "No Net Loss"), new regulations were adopted in 2008. A 2019 assessment of this newer approach makes clear that although wetlands mitigation banking has made some progress from the early days of near universal disastrous compliance fails, the serious foundational problems remain. Palmer Hough and Rachel Harrington, "Ten Years of the Compensatory Mitigation Rule: Reflections on Progress and Opportunities," *Environmental Law Reporter*, Vol. 49, No. 1 (2019): 10018–27. The assessment identifies two studies of the new wetlands mitigation banking program: one concluding that performance standards set out in site-specific plans were too vague to be meaningful, and the other finding it is doubtful that there will be adequate long-term funding to ensure the replacement sites are maintained. Hough and Harrington, 10026. These continued issues are not surprising given this program's insurmountable structural hurdles, including the impossibility of reliable measurement of the extremely complex issue of wetlands function and its dependence on high-intensity and high-quality government involvement for every project in perpetuity. Despite (because of?) the structural flaws that make it impossible to assure the underlying protection goal, use of wetlands mitigation banking is increasing. Hough & Harrington, 10025.

[33] Salzman and Ruhl, "No Net Loss" 342; Coglianese and Nash, "The Law of the Test," 86; Cole and Grossman, "Beyond Compliance Costs," 36.

[34] See Coglianese, "The Limits of Performance-Based Regulation," 548–51; Cohen and Shimshack, "Monitoring, Enforcement," 80 (noting that monitoring and enforcement may be easier and cheaper with a command-and-control strategy because determining compliance is quicker and easier); Cole and Grossman, "Beyond Compliance Costs," 33, 34 (pointing out the sizable differences in measuring or monitoring costs from one environmental protection instrument to another, and noting that it will generally be cheaper for the government to administer uniform standards than economic instruments).

[35] See, e.g., National Academy of Sciences, *Designing Safety Regulations*, 100 (describing the challenges of new more flexible rules for offshore drilling that required both additional staff and a change in the type of expertise needed; either the agency revamps its capacity in response to the rule, or the rule will be ineffective). See also Coglianese, "The Limits of Performance-Based Regulation," 448–553 (describing how more flexible tools can increase government costs).

[36] See Cole and Grossman, "Beyond Compliance Costs," 33 (noting that discussions of the choice of instrument for environmental protection have typically focused on which instrument will create the lowest costs of compliance, "as if that were the sole concern").

myopia is an outgrowth of the widespread but unfounded belief that most companies comply.[37] The assumption—usually unstated—that compliance with regulations just happens, or that the costs to government of ensuring compliance are the same for every type of rule, creates a powerful bias in favor of performance standards and market strategies.[38] If your desired approach theoretically reduces firms' cost of compliance, and you assume that firm compliance costs are the only way in which regulatory costs vary, why wouldn't you believe that your strategy is always preferable?

There are a few encouraging signs that practical considerations, like the feasibility of monitoring and the challenges of ensuring compliance, are starting to elbow their way to the policy table. Some scholars acknowledge that in selecting a regulatory approach the full range of costs must be considered, including the reality that more flexibility for the regulated can dramatically increase costs for the regulator. And they are discovering that this more complete analysis can upend traditional wisdom; the theoretically preferable performance standard or market approach can turn out to be both less effective and more costly than the oft-derided command and control.[39]

However, even the few policy scholars who acknowledge that infeasibility and inefficiency can make performance standards or market strategies unworkable go astray by adopting the universally assumed and nearly always wrong premise that compliance issues are solely the responsibility of enforcement.[40]

For all the reasons discussed at length in chapter 1, enforcement will never be able to assure widespread compliance for rules that create many ways around compliance; without strong rule design that makes compliance the path of least resistance, the compliance effort is doomed no matter what enforcement does.

[37] See discussion of this near-universal assumption in chapters 1 and 2; Cohen and Shimshack, "Monitoring, Enforcement," 78 n.14 (noting that "the great bulk of the literature on the economics of environmental regulation simply assumes that polluters comply with existing directives").

[38] Cole and Grossman, "Beyond Compliance Costs," 35 (noting that many economists employ "simplifying assumptions" about administrative costs that create a strong bias in favor of economic instruments and lead to a presumption that market approaches are always preferable overall, a bias that persists to the present day); Cohen and Shimshack, "Monitoring, Enforcement," 78 (noting that ignoring monitoring and enforcement when considering alternative instruments might lead policymakers to choose a policy that "in theory" looks better but in practice has worse environmental or economic outcomes). See also Stavins, "Market-Based Environmental Policies," 26.

[39] Cole and Grossman, "Beyond Compliance Costs" (market instruments can turn out to be less efficient than command-and-control alternatives when the limits of monitoring and the cost of ensuring compliance are included); Coglianese, "The Limits of Performance-Based Regulation," 547–52; Cohen and Shimshack, "Monitoring, Enforcement," 78.

[40] See, e.g., Freeman and Kolstad, "Prescriptive Environmental Regulations," 7; Coglianese, "The Limits of Performance-Based Regulation"; Cole and Grossman, "Beyond Compliance Costs," 33, 36; Cohen and Shimshack, "Monitoring, Enforcement," 79; National Academy of Sciences, *Designing Safety Regulations*, 97. Although these scholars are ahead of the pack because they at least grapple with the often-ignored reality that poor compliance will undermine the goals of the regulation, they still look to enforcement to solve compliance problems. The belief that compliance is the job of enforcers is ubiquitous in the environmental policy literature.

Because these scholars start from the enforcement-is-responsible assumption, the compliance costs they consider are monitoring and enforcement costs. That is a significant improvement over the vast majority, who just pretend there are no government costs. But it falls short of the insight that enforcement alone can't do it; compliance drivers need to be built into the rules, not stapled on at the back end.

Nor will the suggested solution solve the problem. Some of these scholars argue that rules should consider total costs—not just costs for regulated firms—and therefore advocate that government costs like monitoring and enforcement be added to the cost-benefit analysis. In this telling, the additional expense for government of more complicated monitoring and more difficult enforcement should be added to the tally sheet before deciding which approach is most efficient.[41] Again, this is a notable advance over paying no attention to implementation costs, but still ignores hard reality: government isn't going to significantly increase expenditures to implement a complicated performance or market rule. In the theoretical world of cost-benefit analysis, policy advocates may think they solve the problem by adding the additional government costs to the hypothetical balance sheet. But back in the real world, budgets don't depend on cost-benefit analysis. The agency has its allocated budget, and that's it. Hundreds of rules compete for implementation attention. If the new rule is by far the most important thing happening in the agency, you have a chance. Otherwise, no way.

What really happens is government doesn't have the resources to take on these more complex tasks and so it just doesn't do the additional work to assure that the standards are met, and the public health objectives achieved. In that situation—unfortunately too common—government doesn't know if the regulation has achieved its purpose.[42] This reality needs to be part of regulatory design. An approach that would work great if only government had a 200 percent increase in resources doesn't make practical sense.

This book argues that the response to this costs dilemma is not to throw our hands in the air and give up. On the contrary. When we accept that it is rule design—not primarily enforcement—that determines compliance outcome, we are freed from the paralyzing expense of depending on enforcement to force fit compliance on millions of regulated sources. Where we stand today is between a rock (widespread violations) and a hard place (fixing noncompliance primarily through enforcement is ludicrously unaffordable).[43] Fortunately, Next Gen

[41] Coglianese, "The Limits of Performance-Based Regulation," 449–50; Cole and Grossman, "Beyond Compliance Costs," 33, 35, 39.

[42] See discussion of this problem across many programs in chapter 2. See also Tietenberg, "Tradable Permits in Principle and Practice," 71 n.10 (noncompliance not only makes it more difficult to reach stated goals, it sometimes makes it more difficult to know whether the goals are being met).

[43] This isn't the result of recent budget cuts. Those cuts have hurt, but enforcement resources have never been, and will never be, large enough to be the principal means of ensuring compliance. Nor

says there is another way. Everyone would like to use performance standards and market strategies—when they are the best fit for the problem—despite their additional complexity. The answer isn't to use the often-favored approach of pretending that the additional complexity doesn't exist. Instead, we should apply the principles of Next Gen to see if the complexity and compliance problems are solvable, at a reasonable cost, by building compliance drivers into the rule.

Market Strategies Face Additional Challenges

Market strategies face all of the implementation hurdles that other regulations do. But market mechanisms also have additional challenges.[44] The need for certainty about performance is more acute. If the market can't be sure that a ton equals a ton, the market won't serve its function and can't be counted on to produce the desired pollution outcome.[45] It is very hard to parse this in a market once it is launched, so spending time and money to get verification correct up front is even more important for market mechanisms than it is for other approaches.

Getting markets right takes more effort than traditional rules, not less.[46] Markets that push toward, and not against, the environmental or health objective aren't formed by setting a price; they are crafted through conscientious and thoughtful rule design. An effective system for trading pollution credits has to pay careful attention to defining what is traded and by whom; how that will be monitored; what quality assurance obligations the parties have and how those will be verified; where the trades will occur and who will authenticate and administer the trades; what price collars are needed, if any; how firms will report; how the necessary information will be made publicly available; how gaming, mistakes, incompetence, and fraud will be prevented and dangerous hot spots avoided; how violations will be detected; and what the consequences of violations will be. These are just some of the elements of a successful trading program.

For most environmental problems, setting up a market will be considerably more complicated than it was in the Acid Rain Program, which benefited from the small number, homogeneity, and sophistication of the regulated coal-fired

is it desirable to aspire to that. We can't, and shouldn't strive to, achieve widespread compliance by millions of regulated sources using exclusively our most expensive tool.

[44] See also Ackerman and Gallagher, "Getting the Prices Wrong."

[45] See, e.g., Driesen, "Design, Trading, and Innovation," 449 (noting that trading relies on good monitoring and that "when good measurement proves impossible, trading will not succeed"). The Renewable Fuel Standard market trading program, for example, has been plagued by fraud, undermining confidence in the market and provoking persistent political turmoil. See discussion in chapter 8, section titled "Renewable Fuel Standard Fraud."

[46] See Freeman and Kolstad, "Prescriptive Environmental Regulations," 14.

power plants.[47] Fees and taxes are similarly complex; take a look at the tax code if you think taxes are simple to define and administer. And that doesn't even begin to cover the ongoing oversight that is an essential component of any market approach. The idea that government just sets a price and then its work is done is way off the mark.[48]

A market strategy also requires political backbone. The whole concept of a market approach is letting the market shake out the best and cheapest way to get the desired outcome. Markets need certainty and predictability to do that. Having set up the design and the structure of the market, government needs to get out of the way and let the market function. If government intervenes to protect individual market participants in response to their pleas for special treatment, or changes the targets midstream, it works against market principles. When government loses its nerve in this way, it does more damage to program integrity than occurs when these choices are made in a more conventional permit situation.[49]

And markets are often bad at addressing fairness and distributional effects. Markets don't care about those things, but government should.[50] Efficiency is good, but a market-based regulation has to address equity as well. Transferring pollution or risk from one place to another through market trading can end up shifting health threats as well. That's what happened in the Acid Rain Program,

[47] Even the comparatively straightforward monitoring and reporting system set up by the Acid Rain Program was complicated; it required hundreds of pages of guidance, as one indicator of complexity.

[48] Hsu, "Prices Versus Quantities," 184 (economic theory would de-emphasize the traditional mode of regulation: what would be left for governmental mandate would be the level of the tax, or the quantity of allowable pollution that could be traded); Esty, "Red Lights to Green Lights," 46 (government would have to do the analysis to set a price but then would be able to "get out of" the time-intensive and expensive regulatory requirements of the old command-and-control regime). For a comprehensive assessment of the factors that should be considered in constructing a cap-and-trade market, based on experience with the Acid Rain Program, see John Schackenbach et al., "Fundamentals of Successful Monitoring, Reporting, and Verification under a Cap-and-Trade Program," *Journal of the Air and Waste Management Association*, Vol. 56 (2006): 1576.

[49] As one example, the renewable fuels program attempted to place the burden of verifying credit integrity on the refineries that purchased credits. It included a "buyer beware" fail-safe mechanism; if refiners decided to reduce costs by not checking on the integrity of the credits they purchased, they would bear the financial consequences should the credits turn out to be invalid. That was a great market-embracing idea, but when push came to shove it proved to be politically untenable. Purchasers failed to police the market as the rule envisioned but didn't end up paying the full price as the market strategy had intended. This outcome will make future rule writers understandably more cautious about fully embracing financial drivers as a compliance mechanism. See discussion in chapter 8, section titled "Renewable Fuel Standard Fraud." See also Bradley C. Karkkainen, "Information as Environmental Regulation: TRI and Performance Benchmarking, Precursor to a New Paradigm?," *Georgetown Law Journal*, Vol. 89, No. 2 (2001): 278 (noting that post hoc adjustments may be destabilizing to markets).

[50] See Suryapratim Roy, "Distributional Concerns in Environmental Policy Instruments," in Richards and Van Zeben, *Policy Instruments*, 56, 61, which underscores the essential and often overlooked point that environmental justice includes the equal distribution of both costs and benefits.

for example, where emissions trading caused huge public health damages by moving pollution from low- to high-density population centers.[51] The current pandemic is underscoring just how deadly these disparities are, as communities of color suffer far worse COVID-19 outcomes stemming in part from the historic inequity of disproportionate exposure to air pollution.[52] Allowing firms to pay to take big risks with people's health is not an acceptable outcome. If a market can't be designed to address environmental justice issues, that is telling you that a market isn't the right approach.

Cheerleading for performance standards and market strategies suffers from another blind spot too: a near exclusive focus on permitted air and water pollution discharges. Reading the literature, you might get the impression those are the only kinds of environmental regulations there are.[53] Rarely does one see these theories applied to other important public health programs, like limiting exposure to lead paint and asbestos, ensuring safe disposal of hazardous waste, reducing harm from pesticide applications, preventing accidental chemical releases, requiring chemical manufacturers to disclose adverse health studies, avoiding leaks from underground storage tanks, notifying citizens about drinking water contamination, or preventing the use of dangerous chemicals, to name just some examples. There are scores of important environmental and health protection programs that present design challenges vastly different from those faced in regulation of point sources of air and water pollution. It isn't possible to claim universal superiority of regulatory strategies without grappling with the breadth and diversity of public health programs that require regulation.

The Resurrection of Command and Control

Just as it doesn't make sense to tout performance standards and market strategies as the solution to all problems, it is equally mindless to broadly condemn command and control. All regulations of every stripe are command (the regulation mandates something) and control (regulators will use their authority to make you). Regulations using a market approach also require command and control. If

[51] H. Ron Chan, B. Andrew Chupp, Maureen L. Cropper, and Nicholas Z. Muller, "The Impact of Trading on the Costs and Benefits of the Acid Rain Program," Resources for the Future Discussion Paper, RFF DP 15-25-REV (2017), 4, https://www.rff.org/publications/working-papers/the-impact-of-trading-on-the-costs-and-benefits-of-the-acid-rain-program/ (the trading mechanism caused public health damages of $2.4 billion more than would have occurred had the same program been implemented without trading).

[52] Lisa Friedman and Zoë Schlanger, "Race, Pollution and the Coronavirus," *New York Times*, April 8, 2020, https://www.nytimes.com/2020/04/08/climate/coronavirus-pollution-race.html.

[53] A notable and refreshing exception to the rule is the insightful report from the National Academy of Sciences, which explores the challenges of instrument choice in the context of pipeline and offshore oil and gas safety. National Academy of Sciences, *Designing Safety Regulations*.

they don't, why is a regulation needed? The differences among regulations are in how the regulatory mandates are deployed.

The phrase command and control has ceased to convey any substantive meaning; it is used more like an all-purpose curse to deride anything the author does not like.[54] Professors Ackerman and Stewart piled on by comparing most EPA regulatory strategies to "Soviet-style central planning."[55] There is even a theory of environmental governance, with its own acronym and everything, called—I kid you not—"The Pathology of Command and Control (TPCC)."[56] All this name-calling is to the detriment of thoughtful discussion. Throwing the term around with abandon relieves people of having to say what they mean. Is the particular problem under discussion not suited to a uniform requirement for all regulated firms? Then say that and explain why. If we refuse to accept a disparaging label as though it were evidence, we will force people to articulate their actual objections and not allow them to hide behind what amounts to no more than saying something is bad.[57] That level of vague generality shouldn't pass muster in serious debate.[58]

The command-and-control label also makes it harder to build Next Gen ideas into rules. Rules that work use command and control creatively, to smooth the path to compliance and block the violation exits. Instead of scoffing at the very idea of command and control, we need to focus on using it better.

Many learned the wrong lesson from the Acid Rain Program. That misunderstanding has had an outsized influence because of the Acid Rain Program's central role in the markets-are-the-answer narrative.[59] It is true that the Acid

[54] See National Academy of Sciences, *Designing Safety Regulations*, 16 (noting that "command-and-control" and related terms almost always have negative connotations); Coglianese and Nash, "The Law of the Test," 39 n.14 (noting that command and control is rarely used approvingly and is almost always used to distinguish the writer's own preferred approach from disparaged alternatives); Esty, "Red Lights to Green Lights," 4 (stating that command and control is an outdated regulatory model that no longer fits our current requirements); Cole, "Explaining the Persistence," 159 (command and control used as a term of derogation).

[55] Bruce A. Ackerman and Richard B. Stewart, "Reforming Environmental Law," *Stanford Law Review*, Vol. 37, No. 5 (May 1985): 1334.

[56] See Michael Cox, "The Pathology of Command and Control: A Formal Synthesis," *Ecology and Society*, Vol. 21, No. 3 (September 2016): 33.

[57] Succeeding in getting the derogatory term "command and control" so widely accepted has been described as a "semantic triumph" for the advocates of market mechanisms. Freeman and Kolstad, "Prescriptive Environmental Regulations," 4.

[58] See Driesen, "Design, Trading, and Innovation," 447 ("Most analysts employ a simplistic C&C/economic incentive dichotomy as a substitute for cogent analysis"), and 456 ("The literature's preoccupation with a simplistic and misleading command-and-control/economic incentive dichotomy has led to a failure to adequately address crucial design issues").

[59] Two scholars described the general overreading of the Acid Rain Program by saying it led to "the presumption that if cap-and-trade can work for sulfur dioxide emissions from power plants in the United States, and for fisheries in many locations, then the mechanism can work equally well anywhere in the world to reduce any kind of pollution, from any kind of sources." Cole and Grossman, "Beyond Compliance Costs," 38. See also Nils Axel Braathen, "Flexibility Mechanisms in Environmental Regulations: Their Use and Impacts," OECD Environment, Working Paper No. 151, August 2019, at 18, https://doi.org/10.1787/a6d3ef45-en.

Rain Program had remarkably high compliance rates and therefore achieved its pollution-reduction goals. But the market provisions had nothing to do with that. Take away cap and trade, and the compliance outcome would have been the same.

The reason is command and control. The Acid Rain Program did a masterful job at creating an interlocking set of mandates that made it unlikely regulated plants would violate. The actual amount of pollution was measured in real time through required continuous emission monitoring systems (CEMS). Plants were forced to maintain the CEMS to exacting and very detailed quality standards, because if they didn't, the mandated data substitution provisions would cost them a lot of money, and those increased costs would happen automatically without the need for any government intervention. Every regulated facility had to report frequently, electronically, and in a mandated format to a centralized data system. The data were made available to the public, so there was nowhere to hide. The mandatory centralized electronic reporting in a required format made it comparatively easy for EPA to employ data analytics to spot any anomalies and then challenge companies to explain themselves. All the monitoring and reporting complexity was simplified in the compliance determination: Do you have the permitted authority to emit the tons you reported, yes or no? If not, you automatically owed penalties that were more expensive than just complying.[60] These interconnected provisions created a resilient structure that made complying cheaper and less hassle than violating; in other words, a rule with compliance built in.[61]

These are all classic command-and-control/one-size-fits-all/prescriptive/(insert your favorite term here) requirements. The Acid Rain Program didn't get terrific compliance as a result of the mythical properties of markets, it accomplished that impressive outcome because tough, prescriptive rule design gave the regulated utilities no way out. It was a triumph of command and control. These command-and-control elements don't make the rule bad; they make it effective. They are what was necessary to get the emissions reductions and create

[60] Some have attributed the high compliance rates in the Acid Rain Program primarily to high penalties. See, e.g., Stavins, "Market-Based Environmental Policies," 26; Tietenberg, "Tradable Permits in Principle and Practice," 72. That is the enforcement-sanctions-are-the-reason-for-compliance belief rearing its head again. The penalties in the Acid Rain Program helped—both because they were high and because they were automatic (no waiting to get caught and litigating for years)—but high penalties alone would not have achieved widespread compliance without all of the other compliance-forcing mandates.

[61] See the discussion of the compliance-driving provisions of the Acid Rain Program in chapter 1, section titled "Air Pollution: Acid Rain Program." In some programs, small businesses prefer prescriptive regulations for their comparative compliance certainty. GAO, "Federal Regulations: Key Considerations for Agency Design and Enforcement Decisions," GAO-18-22, October 2017, at 13.

a functioning market. The market intended to reduce costs would never have gotten off the ground without them.[62]

What we should learn from the Acid Rain Program is that careful program design that uses the power of command and control to make compliance the default—that is, Next Gen—works. And that interlocking commands can build a strong foundation for a market that helps to reduce costs. There is no intellectual coherence in praising environmental markets and bashing command and control. As the Acid Rain Program so powerfully demonstrates, the success of markets depends on skillful use of command and control.

There is another reason we should be cautious about using the Acid Rain Program as an all-purpose illustration of the universal utility of markets in environmental rules: the coal-fired power sector was unusually small, homogeneous, well-financed, and sophisticated. Those features, which most other environmental programs do not share, made the tightly designed command-and-control structure possible. A single purpose monitoring technology was available and would work for every company. The companies had the money and the technical sophistication to run the monitoring, install and operate pollution controls, and report extensive data electronically. The data was uniform and easily analyzed. Very few of the programs EPA runs have these advantages. Compare the less than 4,000 similar coal-fired units covered by the Acid Rain Program to the hundreds of thousands of varied industrial and construction stormwater facilities that contribute to serious water pollution, or the over 3 million facilities in diverse industries regulated under the laws that govern the manufacture, use, and distribution of chemicals and you begin to appreciate the entirely different scale and complexity that most programs confront.[63]

That scale and complexity drive a need for more creative compliance strategies, but also mean that some approaches will not get us there. When performance measurement is impossible or unaffordable, or the health imperative can't

[62] Cutting costs of compliance was the central rationale for the cap-and-trade program, and it did help reduce firms' compliance costs, although not nearly as much as was predicted. Chan, "The Impact of Trading," 4; Nathaniel O. Keohane, "Cost Savings from Allowance Trading in the 1990 Clean Air Act: Estimates from a Choice-Based Model," in Freeman and Kolstad, *Moving to Markets*, 194, 224. Cap and trade may have played an important political role too. See Braathen, "Flexibility Mechanisms," 18 n.13 (noting that Congress might not have agreed to the large emission reductions in the Acid Rain Program without the cost reductions that were envisioned by the trading system); Johnston, "Tradable Pollution Permits," 373.

[63] See "Stormwater Discharges from Industrial Activities," EPA, https://www.epa.gov/npdes/stormwater-discharges-industrial-activities (listing 11 categories, including over 25 disparate industrial classifications that are covered by industrial stormwater obligations); "NPDES E-reporting Rule," EPA Final Rule, *Federal Register*, Vol. 80 (October 22, 2015): 64063, 64068, 64081 (noting that there are about 350,000 facilities a year regulated under stormwater regulations); EPA OIG, "Limited Knowledge of the Universe of Regulations Entities Impedes EPA's Ability to Demonstrate Changes in Regulatory Compliance," Report No. 2005-P-00024, September 19, 2005, at 24, https://www.epa.gov/sites/production/files/2015-11/documents/20050919-2005-p-00024.pdf (size of TSCA regulated universe).

be squared with the potential for creating hotspots, or the increased flexibility for thousands of different types of facilities creates compliance loopholes that are technically or politically impossible to close, performance standards and market strategies won't work. Sometimes a straight-ahead ban is the only way to reliably protect the public. Sometimes a one-size-fits-all mandate is the most effective way to get the job done. That was the case for controls on sewage discharges; the flexible outcome-driven strategy favored by many economists as more efficient completely failed in the face of political opposition and technical overload. It took a uniform and inflexible directive to accomplish the goal of cutting sewage pollution.[64]

Blanket criticism of command and control and uncritical promotion of performance standards and market strategies get in the way of creativity and innovation in governance. Yes, we need regulators to get out of a rut that generally ignores how well a rule will function in the world. But the idée fixe that performance standards or markets are the solution to all problems is no better. We should be expanding our understanding of the available tools, not narrowing our focus to a small number of presumptively favored approaches.

Part of tearing down the ideological barriers to Next Gen in rule design is avoiding the tendency to want to cram every rule into a single category. Those classifications lead to sometimes profound misunderstanding. The Acid Rain Program employed both command and control and a cap-and-trade market. Labeling this a "market" rule obscures the essential role of creative command and control and creates the dangerous illusion that the Acid Rain Program stands for the proposition that markets by themselves achieve pollution-reduction goals. Without the foundation of skillful command and control, you won't have a functioning market. Almost no regulation uses just one strategy. If a firm is required to maintain financial assurance of a particular amount to protect against future cleanup costs, and can select among five different financial instruments to satisfy that obligation, is the rule prescriptive or performance-based? If a rule mandates the installation of a specified monitoring technology and requires every company to calculate missing data using a predetermined formula and to report using the identical form, but allows trading of credits, is that rule market-based or one-size-fits-all? To which I say: Who cares? Our goal isn't fighting over label primacy; it is the more exacting practice of building a strong and resilient structure by creatively using all the tools.

[64] See discussion of the uniform mandate requiring secondary treatment for sewage treatment plants in chapter 1, section titled "Programs with Strong Compliance Outcomes: Four Examples." See also William L. Andreen, "Water Quality Today—Has the Clean Water Act Been a Success?," *Alabama Law Review*, Vol. 55, No. 3 (2004): 539 nn.13 and 14 (noting that the widely touted and supposedly more efficient regulatory approach of starting from water quality standards has never worked due to technical gaps and lack of political will).

What matters is designing a rule that fits the problem. We need to select strategies that will address the issue and build a structure that makes those strategies effective. Every rule includes a wide variety of mandates, including who it applies to, what they are supposed to do, how they are supposed to determine compliance and document what they do, how they report, and provisions to address the different circumstances and exceptions that arise in the real world. Every rule. This structure of mandates is the foundation for rule success. Regulations will all include commands by any definition of the word. The question is whether those commands are deftly deployed to ensure that the rule is effective in actual life, and not just in theory. Command and control can be used to impose uniform standards, create markets, establish information-reporting obligations, deploy transparency systems, and scores of other strategies. The key issue isn't how the rule is labeled, it's whether it uses the many available tools of every type to achieve the goal: widespread compliance, at reasonable cost.

Performance standards and market strategies have promise to help tackle environmental issues. But they don't have magical powers, and they are not the right fit for every problem. When performance standards or market strategies make sense, good compliance design is still essential, which will necessarily include—prepare to be shocked—command and control.

7
Ensuring Zero-Carbon Electricity

Climate regulations cannot repeat the regulatory mistakes that have hobbled so many environmental rules in the past. Time's up. What we do next to tackle climate change must work. We know now that the dual assumptions at the foundation of nearly all environmental regulations—that most companies comply, and that it is up to enforcement to take care of the rest—are wrong. In fact, serious violations are widespread. And the principal driver of outcomes isn't enforcement, it's whether the regulations are tightly structured to make compliance the path of least resistance, so compliance is good even if enforcement never comes knocking. These essential truths are the difference between a rule that is great in theory and one that delivers emission reductions in real life.

This book has shown how some rules deliver terrific compliance results but many more don't. The difference isn't a nice-to-have; at the margins we could do better. In program after program, serious widespread violations have undermined the purpose for which the rules were written, achieving only a fraction of the intended gains, or in some cases creating the possibility that we are actually headed in the wrong direction. This is untenable for climate rules. Climate regulations with serious violation rates of 25 percent to 50 percent or higher—all too common in many environmental programs—are the difference in climate between we have a chance, or we don't.

This book's three climate chapters apply the lessons of Next Gen, hard won from decades of experience, to the most pressing issue of our time. They focus on three key areas for urgent immediate action—electricity generation, transportation and biofuels, and methane from oil and gas production—and outline both how government is struggling to implement rules effectively and what regulatory design choices could greatly improve the odds. It is a compliance analysis of both regulatory and legislative options, understanding that some of the suggested strategies will be difficult or impossible without legislative changes.

There has been a shift in the policy discussion about government's approach to climate change. Setting a price on carbon is no longer the one ring to rule them all.[1] It has been replaced with the realization that we have to start by setting the

[1] Benjamin Storrow and Adam Aton, "Burned by Carbon Pricing, Dems Chart New Course on Climate," *E&E News*, February 2, 2021, https://www.eenews.net/stories/1063723981; Matto Mildenberger and Leah C. Stokes, "The Trouble with Carbon Pricing," *Boston Review*, September 24, 2020, http://bostonreview.net/science-nature-politics/matto-mildenberger-leah-c-stokes-trou

Next Generation Compliance. Cynthia Giles, Oxford University Press. © Cynthia Giles 2022.
DOI: 10.1093/oso/9780197656747.003.0008

goal for emissions cuts where science tells us it has to be: 50 percent reduction by 2030, and a 100 percent clean energy economy by 2050. Something remarkably close to a consensus now exists that the first order of business is to focus directly on the biggest carbon-emitting sectors where the solutions are known, clear, and affordable, and set aggressive regulatory standards. Those are electric generation, transportation, and oil and gas production.

That's why this book digs into the compliance problems for these top emitters. It isn't an exploration of all the many policy implications of the choices for these sectors. It asks just one question: Will the regulation actually reduce emissions? In the real world, which is where implementation happens, can we achieve something close to the desired outcome? Sometimes the answer is a qualified yes, it probably can, if the necessary guardrails are built into the rule. But for some, the hoped-for policy strategy can never achieve a good enough result. There is no way to assure broad compliance, or sometimes to even know how close we are. There is too much at stake to put all the chips on strategies where our best knowledge predicts compliance collapse. It is far too late in the game to figure on taking a shot, seeing what happens, and hope to adapt if it doesn't work. Any program with a real chance of catastrophic breakdown isn't a viable option.

Next Gen isn't the band-aid that can make any policy successful. Regulators can't just slap Next Gen ideas on top of an already designed program and call it done. Sometimes Next Gen approaches can plug into policy and considerably strengthen outcomes without unraveling basic design. But the unfortunate reality is that there will be times when a Next Gen analysis concludes that it isn't possible to achieve the desired result with this policy strategy. You just can't get there from here. This is when it is good to remember that we shouldn't get too enamored with one policy approach. Sometimes one has to gaze fondly at the desired policy design and then kiss it goodbye, because the goal is reliable emission reductions, it isn't a particular policy scheme.

Both situations occur in the climate regulations discussed here. Some of the currently used strategies can achieve compliance with strengthened emissions standards, as is true for vehicle emissions. Some cannot, such as renewable fuels. It's better to know that now, and face the facts, before the failure to achieve the emissions reductions becomes all too apparent.

Being realistic about compliance is also essential for environmental justice. Widespread violations of environmental rules have fallen much more heavily on minority and low-income communities. It is almost never feasible to remedy a bad regulatory compliance design through enforcement, and as we have recently

ble-carbon-pricing; Frederick Hewett, "Putting a Price on Carbon: It Was Hot, Now It's Not," *WBUR*, August 3, 2020, https://www.wbur.org/cognoscenti/2020/08/03/carbon-pricing-tax-climate-change-policy-frederick-hewett.

seen, some governments aren't interested in enforcement anyway. Incorporating Next Gen into rules is one of the most important things we can do to protect overburdened communities, because it shields them from high rates of violation and is less dependent on the unreliable commitment of regulators to protect the most vulnerable. While government tackles the existential crisis of climate change, it also has to ensure that there are no disproportionate burdens or benefits.

I am not arguing that compliance considerations are the only marker of effective rules. Nor am I suggesting that a rule should aim for no violations. That's not realistic or achievable. But what we can't have is a climate program where violations overwhelm regulators' ability to accomplish the mission, or a rule that makes it impossible to know how close we are to the necessary emission reductions. Some popular climate policies teeter on the edge of those chasms. Some have already fallen in.

These climate chapters acknowledge that the rule-writing process is messy and compromises are often needed to get something done. Sometimes political reality requires the suboptimal choice. If regulators go in with eyes open to the risk, and constrain it as best they can, that may produce an acceptable, if not preferred, result. What government cannot do is adopt regulations that will, due to predictable, even inevitable, compliance failures, fall far short of the emissions goal. There are enormous compliance deficits in many programs now, with rules that are nowhere close to achieving the intended objectives. This will get dramatically worse with tightened standards and a lot more money in the game, unless the regulations create deliberate and thoughtfully designed strategies to prevent that.

These climate chapters look through a Next Gen compliance lens at three top areas for regulatory action to tackle climate change. Chapter 7 concludes that zero-carbon electricity generation can be a compliance winner if we resist the temptation to allow an offset for energy efficiency. Chapter 8 explains that strong compliance with vehicle emission standards is possible, but conventional transportation biofuels face insurmountable compliance barriers. Finally, chapter 9 details the daunting compliance challenge facing rules to control methane from oil and gas and offers Next Gen strategies that can significantly alter the compliance odds.

It is worth remembering that electricity generation, transportation, and oil and gas production are comparatively easy ones in the array of dramatic actions necessary to address the threat of climate disaster. Government needs to save its resources for the problems that don't have such obvious answers. That means counterproductive but politically convenient loopholes, ambiguities, and complexities are out. Those only burden government with impossible and useless regulatory tasks. Strategies that depend on stringent government oversight are doomed too; government won't have the resources, ability, or in many places the political will, to make that happen.

In writing climate rules, government needs to accept the reality that heavy regulatory pressure leads to widespread serious violations. Wishful thinking about companies' compliance or the ability of enforcement to plug gaping holes will run climate regulations into a ditch. But even as we acknowledge that there are programs with discouraging performance and large obstacles, we can see that there are pathways to yes. Next Gen is about finding those solutions. We can do this. But it requires jettisoning the old assumptions and rejecting the ideas that look great on paper but don't stand a chance in the real world.

Electricity: Clean Energy Can Work, but Only If We Keep It Simple

Electric power generation is one of the two largest sources of carbon dioxide in the United States.[2] Any plan to achieve our climate goals has to include regulatory standards that will get a close as possible to zero-carbon electricity by 2035.[3] The electric generation and transportation goals are tightly linked and work together: as transportation is increasingly powered by electricity, that power must rapidly become much cleaner.

State leadership in cleaning up the energy supply has shown how regulatory standards can effectively push for cleaner energy. There is a spirited debate about the details, and some of the options have serious compliance issues, as is discussed later in this chapter. But here's one thing nearly all of the recently proposed climate policies agree on: the federal legislative strategy should include something very much like the renewable portfolio standards (RPSs) that many states have already adopted.[4]

[2] "Overview of Greenhouse Gases," EPA, https://www.epa.gov/ghgemissions/overview-greenhouse-gases.

[3] "The Long-Term Strategy of the United States: Pathways to Net-Zero Greenhouse Gas Emissions by 2050," US Department of State and US Executive Office of the President, November 2021, at 5, https://www.whitehouse.gov/wp-content/uploads/2021/10/US-Long-Term-Strategy.pdf.

[4] Majority Staff of House Select Committee on the Climate Crisis, 116th Congress, "Solving the Climate Crisis: The Congressional Action Plan for a Clean Energy Economy and a Healthy, Resilient, and Just America," at 37 (2020), https://climatecrisis.house.gov/sites/climatecrisis.house.gov/files/Climate%20Crisis%20Action%20Plan.pdf; "Clean Energy Standards: State and Federal Policy Options and Considerations," Center for Climate and Energy Solutions (C2ES), November 2019, at ix–x, https://www.c2es.org/site/assets/uploads/2019/11/clean-energy-standards-state-and-federal-policy-options-and-considerations.pdf; David Roberts, "At Last, a Policy Platform That Can Unite the Left," *Vox*, May 27, 2020, updated July 9, 2020, https://www.vox.com/energy-and-environment/21252892/climate-change-democrats-joe-biden-renewable-energy-unions-environmental-justice.

RPSs are named for the mix of energy sources required in utilities' "portfolios," which require an increasing percentage of electricity sales to come from renewable sources.[5] Thirty states have some version of such standards.[6] Here's what we have learned: they work. RPSs are credited with driving about half of the increase in renewable generation and capacity since 2000.[7] The developing consensus suggests that the best way to accelerate these trends, driving electricity generation toward clean renewable energy as quickly as possible, is to create a national standard that provides both certainty and cost-cutting opportunities for power generators. Such a national standard can be the backbone of the plan to get to 100 percent clean electricity by 2035.

The great news from a compliance perspective is that it is not that hard to develop a national RPS that will have very high compliance rates. The amount of power that any source generates is easily—and already—measured. The amount of power sold by a utility, likewise. Because the compliance obligation is measured by the amount of clean power over the amount sold, determining compliance is straightforward. There are a lot of details to iron out of course, but the experience of states can guide the way. There are many compliance strengths of this approach: there are a limited number of regulated parties (the electric utilities), a common metric (units of electricity), and an established and reliable measurement system. Ensuring widespread compliance under these circumstances is Next Gen 101. It takes careful and thoughtful design, of course, and there are lots of difficult choices to make. But if regulators resist the temptation to insert too many loopholes and escape hatches, setting a renewable energy standard that achieves high rates of compliance using Next Gen principles is something we know how to do.

The further good news is that a renewable portfolio standard can be both ambitious and allow for growth. Every climate policy requires increased electricity generation; as we move more polluting activities to electric power, the need for electricity will increase. At the same time, we will be cutting the amount of carbon and other pollutants emitted from electricity generation because it will be increasingly clean. Renewable portfolio standards are structured to build in necessary growth in demand while also cutting carbon.

The much harder part from a compliance perspective is what else, in addition to a renewable portfolio standard, is included. The desire to add more ways of

[5] Robert Freedman, Monica Lamb, and Claire Melvin, "Financing Large Scale Projects," in Michael B. Gerrard and John C. Dernbach eds., *Legal Pathways to Deep Decarbonization in the United States* (Environmental Law Institute 2019), 129, 143.

[6] House Committee, "Solving the Climate Crisis," 37. For a summary of state renewable portfolio standards, updated annually, see Galen Barbose, "U.S. Renewables Portfolio Standards 2021 Status Update: Early Release," *Lawrence Berkeley National Laboratory*, February 2021, https://emp.lbl.gov/publications/us-renewables-portfolio-standards-3.

[7] Barbose, "Renewables Portfolio Standards," 5; C2ES, "Clean Energy Standards," 23.

complying and to reduce costs through trading has resulted in a new name for this approach: a clean energy standard. The difference, implied by the name, is what is included. A clean energy standard isn't limited to renewable energy. It potentially includes other zero-carbon sources. States that already have clean energy standards vary in what they allow in the definition of clean: some include nuclear, or fossil fuel with carbon capture, or energy efficiency, for example, while others don't.[8]

This chapter does not address the policy merits of including, or omitting, these additional types of energy in a clean energy standard. It focuses only on compliance and the implications of an expanded definition of clean energy for assuring that we actually achieve the zero-carbon objective. Next Gen operates not in the rarified world of ideology but in the gritty and messy on-the-ground reality of implementation. It asks just one question: Will it work?

Viewed through this lens, there is one candidate for inclusion in a clean energy definition that stands out: energy efficiency. Energy efficiency—using less energy to accomplish the same task—is a pillar of every climate strategy. It is a must-have, can't-live-without component for getting our emissions down to where they have to be. It includes both reducing the total amount of energy needed to do something and changing the times when that energy is needed, to reduce dependence on peak power sources. We know energy efficiency reduces demand, and there is no question that we have to ramp up at a breakneck pace.

But—you knew there was a but coming here—including energy efficiency in a clean energy standard is fraught with peril. There is a world of difference between committing to as much energy efficiency as possible and including it in standard for achieving zero carbon from electric generation. Here's why.

The idea of a clean energy standard is to allow electric generation utilities to make their own decisions about how to achieve the continuously tightening standard. The right mix of renewable and other clean power sources will be up to them. That allows differently situated utilities to select the power portfolio that works for their circumstances and achieves compliance at as a low a cost as possible. Regulators define what the acceptable sources are—what is really zero carbon—and then the utilities buy as much of those sources as they need to comply. The market for clean energy favors the cheapest zero-carbon sources, allowing us to achieve the zero-carbon goal at reasonable cost.[9]

Nearly all power generation can be measured by reliable and well-established methods, and we know how clean they are. Utilities can buy one unit of solar

[8] C2ES, "Clean Energy Standards," 23–27.

[9] Such market-type approaches are all about efficiency and reducing costs. They are less good at addressing equity, which is why the climate justice movement questions use of such strategies. That's an important concern and is touched on later in this discussion of energy efficiency and clean energy standards.

power or one unit of wind power, and we can be certain that both are actually zero carbon. How the utility elects to design their portfolio does not affect the climate outcome. Nearly perfect compliance can be built in; we can have confidence that we are getting the necessary emissions outcome.

But energy efficiency isn't like that. It's about changing how something is done or built with a goal of reducing the amount of energy it consumes. It is complex, extremely hard to measure, impossible to monitor closely, and expensive to evaluate robustly. The built-in incentives encourage participants at every level to overstate the benefits in ways that are hard to detect. And compliance needs to happen at millions of facilities of widely varying types distributed everywhere around the country. Next Gen predicts that in these circumstances, compliance will be poor. The available data tell us that's what's actually happening in energy efficiency; it isn't achieving as much energy savings, and thus carbon reduction, as everyone hoped.

So what? Why worry about the measurement difficulties? We know energy efficiency works, and we know we need more. Any program that achieves that is good, right? Actually, no. The problem isn't the merits of efforts to drive investment in energy efficiency. It's the impact of including such a difficult to measure source of power savings in a market for clean energy.

Why Is Measuring the Impact of Energy Efficiency So Difficult?

The idea of energy efficiency is at base pretty simple: do the same thing but use less energy doing it.[10] When we use energy efficient appliances, swap incandescent for LED bulbs, or upgrade insulation so that less energy is needed to heat a building, we are being more energy efficient. We still get the desired end point—refrigerated food, light, heating in the winter—but we use less energy to get there. But how much energy do we save by deploying those energy efficiency measures? There's the rub.

Conceptually, the energy savings is what energy use is after deploying energy efficiency measures, compared to what it would have been without those actions. You can see the squishiness already creeping in. The savings depend on what you project would have happened in the alternate universe where the efficiency project didn't occur. It isn't something observable in the world; it requires assumptions and creation of a so-called "counterfactual," from which the actual energy use after energy efficiency measures is subtracted. That difference

[10] Of course, it is also possible to reimagine the thing itself, so you get where you are going but in an entirely different way, also reducing energy use. My more colloquial use of the term "energy efficiency" here is not meant to exclude these more creative approaches to cutting energy demand.

is the savings from energy efficiency.[11] It's nowhere near as simple as comparing energy use before and after the project. To throw in some of the complications that happen in real life: What if after I install more energy efficiency, I have another child and also decide to purchase an electric car? Maybe a warming climate means I need less energy in the winter, but because I know I am doing a good thing by improving my energy efficiency, I decide it's OK to buy two big energy-draining TVs.[12] How clear is it now how much energy I saved through more efficient appliances, more insulation, and a switch to LED lights? This is just a small sampling of the mess and complications that enter into figuring out how well energy efficiency projects work. Even experts say they are extremely difficult to measure.[13]

An entire industry has grown up around trying to figure this out, called Evaluation, Measurement and Verification (EM&V).[14] To give you some idea of how complicated this is, California's Evaluation Framework for appraisals of California's energy efficiency programs is 500 pages long.[15] And that's just the framework. Even in the perfect world, figuring out energy savings from efficiency investments is inherently complex.

And, as you may have noticed, we don't live in a perfect world. We cannot devote infinite hours and dollars to building reliable measurement and verification systems. Simplifying assumptions are necessary. So, informed by research, engineers have estimated the level of expected energy savings from certain activities.

[11] See "Evaluation, Measurement, & Verification," American Council for an Energy Efficient Economy (ACEEE), September 25, 2019, https://www.aceee.org/toolkit/2020/02/evaluation-measurement-verification; Noah Kaufman and Karen L. Palmer, "Energy Efficiency Program Evaluations: Opportunities for Learning and Inputs to Incentive Mechanisms," *Energy Efficiency*, Vol. 5 (June 2011): 243, 244.

[12] This really happens. It is called the rebound effect and has been measured and documented. See, e.g., Kenneth Gillingham, Amelia Keyes, and Karen Palmer, "Advances in Evaluating Energy Efficiency Policies and Programs," *Annual Review of Resource Economics*, Vol. 10 (October 2018): 515, https://doi.org/10.1146/annurev-resource-100517-023028.

[13] Kaufman and Palmer, "Energy Efficiency Program Evaluations," 243. See also "Energy Efficiency Program Impact Evaluation Guide: Evaluation, Measurement, and Verification Working Group," Department of Energy State & Local Energy Efficiency Action Network, DOE/EE-0829 (December, 2012), Chap. 3, at 5, https://www4.eere.energy.gov/seeaction/system/files/documents/emv_ee_program_impact_guide_0.pdf (diagram of conceptual framework for measurement of energy efficiency savings).

[14] For a relatively concise summary of EM&V, see ACEEE, "Evaluation, Measurement & Verification." See also "What Is EM&V?," Department of Energy, https://www.energy.gov/sites/prod/files/2014/05/f16/what_is_emv.pdf; Martin Kushler, Seth Nowak, and Patti Witte, "A National Survey of State Policies and Practices for the Evaluation of Ratepayer-Funded Energy Efficiency Programs," American Council for an Energy-Efficient Economy, Report No. U122 (2012), https://www.aceee.org/sites/default/files/publications/researchreports/u122.pdf.

[15] "The California Evaluation Framework," Prepared for the California Public Utilities Commission and the Project Advisory Group, Project Number: K2033910, June 2004, rev. 2006, https://www.cpuc.ca.gov/-/media/cpuc-website/files/uploadedfiles/cpuc_public_website/content/utilities_and_industries/energy/energy_programs/demand_side_management/ee_and_energy_savings_assist/caevaluationframework.pdf.

For ease of administration, these assumptions are the most commonly used way to estimate energy savings from energy efficiency investments, especially in non-industrial settings. They are commonly referred to as "deemed savings," which is the amount of energy you can assume is saved by undertaking a specific type of energy efficiency work.[16] Deemed savings might tell you how much energy savings are deemed to occur from installing an additional two inches of insulation in your attic, just to pick one of hundreds of examples.[17] Deemed savings are universally used to calculate the amount of energy savings claimed for particular energy efficiency measures; 95 percent of states use deemed savings to determine energy savings.[18]

How do those deemed savings stack up against real-world measurement? The data aren't encouraging. A 2017 review of the economics literature concluded that energy savings are often smaller than implied by utility-reported results and that deemed savings in particular tend to overestimate energy savings.[19] The good news is that many studies have found that there was some energy savings from the energy efficiency measures.[20] But some of the few rigorously designed studies find that the actual savings fell far short of what the engineering estimate predicted, possibly delivering less than 40 percent of the promised savings.[21] For lighting upgrades, which account for the majority of energy savings from utilities' residential efficiency programs, there is a "glaring lack of empirical evidence."[22]

[16] ACEEE, "Evaluation, Measurement & Verification" (deemed savings).

[17] See, e.g., "Arkansas Deemed Savings, Installation & Efficiency Standards," Arkansas Public Service Commission, TRM Version 4.0 Volume 2: Deemed Savings (August 2014), 64, http://www.apscservices.info/pdf/10/10-100-R_118_3.pdf. Note that the Arkansas Deemed Savings Standards are 451 pages long.

[18] "National Survey of State Policies and Practices for Energy Efficiency Program Evaluation," American Council for an Energy Efficient Economy (ACEEE), October 15, 2020, at 26, https://www.aceee.org/research-report/u2009.

[19] Gillingham, Keyes, and Palmer, "Advances in Evaluating Energy Efficiency," 517.

[20] Gillingham, Keyes, and Palmer, 527.

[21] See, e.g., Meredith Fowlie, Michael Greenstone, and Catherine Wolfram, "Do Energy Efficiency Investments Deliver? Evidence from the Weatherization Assistance Program," *The Quarterly Journal of Economics*, Vol. 133, No. 3, (January 2018), https://doi.org/10.1093/qje/qjy005 (randomized controlled trial finding that retrofits on average only achieved about 38% of the expected savings predicted by a widely used efficiency audit tool). See also Fiona Burlig, "Making Energy Efficiency Work," in U.S. Energy and Climate Roadmap, Energy Policy Institute at the University of Chicago, 2021, https://epic.uchicago.edu/area-of-focus/making-energy-efficiency-work/ (summary of the studies showing a large difference between expected and actual savings from energy efficiency projects); Cathryn Courtin, "Sacred Cow Gets Controversial, Closer Look: Energy Efficiency," Women's Council on Energy and the Environment (undated), https://www.wcee.org/page/SacredCowGets. See also Lauren Giandomenico, Maya Papineau, and Nicholas Rivers, "A Systematic Review of Energy Efficiency Home Retrofit Evaluation Studies," Carleton Economic Papers 20-19, at 13, 16, Carleton University, Department of Economics (2020), https://ideas.repec.org/p/car/carecp/20-19.html (finding that actual savings for residential retrofit programs ranged from 25% to 85% of predicted savings, with the more rigorous studies finding less energy savings).

[22] Gillingham, Keyes, and Palmer, "Advances in Evaluating Energy Efficiency," 525.

On top of this many-factorial engineering and measurement problem that makes the topic inherently complex, there are a wide variety of self-serving motivations and conflicting incentives that further muddy the waters. Efficiency installers don't always do a good job, because they are poorly trained, rushed, or trying to cut costs.[23] The unscrupulous simply cheat.[24] Just because energy efficiency is good doesn't mean that all the mess and confusion of real life don't apply. They do.

Government has created additional incentives for overclaiming energy savings from energy efficiency. In some places, electric utilities are given financial incentives to reduce energy demand through energy efficiency programs. If you want utilities to commit to more energy efficiency, financial incentives are one obvious way to do it. But anyone who has been paying attention so far knows what will happen when financial incentives exist for companies to do something, but it is difficult to measure if they actually did it: many will claim better performance than they actually had.

A 2011 study of energy efficiency in California by Noah Kaufman and Karen Palmer reveals that the problem of overestimating and overreporting energy savings is systemic.[25] The Kaufman and Palmer study compared the results from rigorous third-party evaluations of a year's worth of energy efficiency programs with both the energy savings projected for those programs and savings reported by utilities once the programs were done. The study had an unusually robust data set for an entire year's worth of energy efficiency programs: projected savings (estimated before the project was done), the utility reported savings (based on performance after the fact), and evaluated savings (third-party after-the-fact reviews of savings). California has an energy efficiency performance incentive mechanism where utilities are rewarded with increased profits the more energy savings they obtain, so the amount of verified energy savings has direct financial consequences for the utilities.[26]

[23] See, e.g., Louis-Gaetan Giraudet, Sebastien Houde, and Joseph Maher, "Moral Hazard and the Energy Efficiency Gap: Theory and Evidence," *Journal of the Association of Environmental and Resource Economists*, Vol. 5, No. 4 (October 2018), https://www.journals.uchicago.edu/doi/abs/10.1086/698446 (finding that for hard-to-observe efficiency measures like insulation, the gap between projected and actual energy savings is particularly pronounced when the efficiency measure is installed on a Friday, i.e., workers cut more corners on Fridays).

[24] See, e.g., Joshua A. Blonz, "The Welfare Costs of Misaligned Incentives: Energy Inefficiency and the Principal-Agent Problem," Washington: Board of Governors of the Federal Reserve System, Finance and Economics Discussion Series 2019-071, 34, https://doi.org/10.17016/FEDS.2019.071 (finding that many contractors intentionally misstated the age of replaced refrigerators in a large efficiency program, resulting in 50% less energy savings than claimed, and noting that such gaming of the system during implementation may be part of the explanation for the widespread failure of energy efficiency programs to deliver promised savings).

[25] Kaufman and Palmer, "Energy Efficiency Program Evaluations" (see *supra* note 11).

[26] Kaufman and Palmer, 245.

Kaufman and Palmer's research revealed that actual energy savings were 30 percent to 40 percent less than had been projected. In addition, utilities overstated the actual savings by 15 percent (electricity) and 53 percent (gas).[27] This study shows that not only were the projections for energy savings inflated—a point made in other studies—but the utilities were systematically overstating the savings even when they did their own after-the-fact reviews. This is exactly what a Next Gen analysis would predict in a setting where higher reported energy savings produces greater financial rewards.

These findings are particularly notable because they occurred in California, which has one of the strongest energy efficiency programs and one of the most rigorous evaluation, measurement, and verification systems in the country. And the utilities knew in advance they would be subject to third-party review. If utilities significantly overstate the energy savings when they expect regulators are strict and know in advance that an outside party will be carefully scrutinizing their work, what is likely happening when neither of those things is occurring?[28]

None of this is surprising. Anyone who has studied what has happened with environmental rules over the past decades would expect it. Hard-to-measure programs make it tough to ensure that the desired result is occurring. Everyone involved responds to what is in their best interest and makes close calls—and sometimes nowhere near close—in their own favor. Some companies will just cheat. When it is next to impossible to check, and everyone knows it, self-serving behavior will be widespread.

[27] Kaufman and Palmer, 250. The utilities' after-the-fact reports acknowledged that the actual savings were lower than had been projected in advance. But they still overstated the actual savings. For example, utilities reported that implemented programs only achieved 87% of the projected MWh savings. But the third-party evaluations found that those programs only achieved 67% of the savings that had been projected for those programs. Kaufman and Palmer, 252.

[28] Another interesting finding from the Kaufman and Palmer study was confirmation that third-party review does not necessarily ensure accuracy. The largest outside auditors were far more likely to find larger discrepancies in utility reports than smaller auditing companies did. Kaufman and Palmer, 259. The authors speculate that the larger firms have less financial dependence on individual clients, and thus a weaker incentive to please any single client, possibly resulting in more honest evaluations. That would be consistent with other studies finding that the financial incentives for third-party auditors selected by the regulated firm can distort audit findings. Esther Duflo, Michael Greenstone, Rohini Pande, and Nicholas Ryan, "Truth-Telling by Third-Party Auditors and the Response of Polluting Firms: Experimental Evidence from India," *The Quarterly Journal of Economics*, Vol. 128, No. 4 (September 26, 2013), https://doi.org/10.1093/qje/qjt024 (auditors selected and paid by the regulated firm are far more likely to report the plant in compliance); Jodi L. Short and Michael W. Toffel, "The Integrity of Private Third-Party Compliance Monitoring," *Administrative & Regulatory Law News*, Vol. 42, No. 1 (Fall 2016): 22–25 (describing factors that lead to third-party auditor bias in reporting), https://www.hbs.edu/faculty/Publication%20Files/ShortToffel_2016_ARLN_13fe8ba5-cb72-482b-b341-5c7632f7c164.pdf; Gillingham, Keyes, and Palmer, "Advances in Evaluating Energy Efficiency," 517; Justin Marion and Jeremy West, "Dirty Business: Principal-Agent Problems in Hazardous Waste Remediation" (December, 2019), http://conference.nber.org/conf_papers/f132544.pdf (finding that third-party hazardous waste site evaluators manipulated scoring to favor their clients, facilitating lower quality remediation of hazardous waste sites).

Why Does It Matter?

The net result is that we don't really know how effective energy efficiency is. We know it works—we have seen collective energy consumption go down despite growth in population, corresponding with implementation of energy efficiency programs.[29] But we can't really say with confidence how much energy is saved by specific energy efficiency programs. The band of uncertainty is wide. And with a lot of money to be made under a souped-up clean energy standard, the incentives that push against greater accuracy will get much worse.

That's why inserting energy efficiency into a clean energy standard creates such a big problem. This is how including it would work: an energy efficiency program projected to save a defined amount of energy would earn one clean energy credit. That credit could be used by the utility to meet its mandatory clean energy percentage. Its worth in the market is exactly the same as a clean energy credit from, say, solar energy. But actually, they aren't equal. We know how much clean energy is created by the solar project, but don't really know how much energy savings is reflected in the energy efficiency credit. There are some energy savings in the energy efficiency credit, probably, but it is almost certainly not as much as is being claimed. It might be a lot less. There's no way to know for sure. But that shaky energy efficiency credit is used to justify the release of more actual, real, we-know-it-is-happening CO_2. Because energy efficiency credits are likely a lot less expensive than renewable energy credits—that's one of the reasons that energy efficiency is so attractive as a climate solution—the market presses for more energy efficiency, leading to a plethora of dubious clean energy credits. The more of those doubtful credits there are, the less likely it is we will achieve the desired carbon reductions.

The conventional narrative that this book is seeking to overturn—that most companies comply, and that noncompliance can be handled though enforcement—is just obviously wrong here. Energy efficiency has all the hallmarks of a program where those assumptions will be potentially fatal. Nearly all the characteristics that led to widespread violations in the rules described in chapter 1 exist here. It is complicated to figure out, lots of activity goes on behind closed doors that is hard to monitor, checking on compliance is expensive and difficult, strong implementation requires both expertise and good judgment, and just about everyone involved benefits by overstating savings. In addition, the activities will be happening at millions of locations across the country. What are the chances that compliance will be good? Tiny. And the chances that enforcers

[29] See Kit Kennedy, "Lighting, Appliances, and Other Equipment," in Gerrard, *Legal Pathways*, 217, 218.

will be able to fix rampant noncompliance after the fact, never mind prevent it from occurring in the first place? Literally zero.

The fact that there will be widespread violations isn't something to bemoan and berate companies for; it is just acknowledging reality. Government accomplishes nothing by writing a rule that isn't compliance resilient and then focusing on who is to blame for violations. Regulators' obligation is to build a program that will achieve the goal. That is the animating principle of Next Gen: Will the program work in the real world, or not? It's on the regulators to make sure it does.

Enforcement can't possibly turn this around. In part, that's because the problem is inherently complex. Even among people of good will, trying their best, there is likely to be a mismatch between projected and actual efficiency savings. Lots of efficiency programs that fail to deliver will be tough to classify as violations. The complexity also means it will be incredibly time consuming to investigate even a single case, which requires checking all the data and the assumptions used in the evaluation and conducting field work to compare reports to reality. Many states don't have the expertise or interest in doing such investigations. None have the resources. On top of that, nearly all the interests of the many players involved push in the direction of overstating savings. These will run the gamut from outright fraud to looking on the bright side when making assumptions, but the incentives mean that it will be common, and inevitable, that lots of projected savings will never be realized. Enforcement needs to be part of the mix, but it cannot be our principal strategy for closing the gap between what is required and what actually happens. Pretending we can solve this problem through enforcement is just throwing in the towel.

These are the reasons that including energy efficiency in a clean energy standard will result in more carbon emissions than intended—possibly a lot more. That's the kind of insight that a Next Gen analysis, done in advance, contributes to policy. Here it tells us unambiguously that allowing energy efficiency credits into a clean energy standard will mean both more carbon emissions than desired and much greater uncertainty about how close we are to achieving the necessary declining trajectory. There may be ways to hedge, and modestly reduce that negative impact, but it can never be "fixed."[30]

[30] One strategy that has been considered is limiting the percentage of credits that can come from energy efficiency, although usually that is proposed to ensure that the push for renewable energy remains strong. See, e.g., C2ES, "Clean Energy Standards," 37. Another discussed option is reducing the relative value of EE credits (e.g., defining credits so that, for example, it takes three EE credits to equal one RE credit). Attempting to address the problem this way is really just an admission that we don't know how much EE credits are worth and is more about limiting the damage than solving the problem. We would still have clean energy credits in the market of uncertain value, and probably would see a lot more of them with an aggressive CES. But note that even without energy efficiency credits, a CES contains a natural incentive for utilities to promote energy efficiency; because CES defines the obligation as a percentage of the total, reducing demand through effective energy efficiency reduces the amount of clean energy needed to meet the CES obligations. See C2ES, "Clean Energy Standards," 32–33. One Next Gen advantage of relying on this natural incentive is that it puts

The same cautions, plus a few more, apply to including energy efficiency in a regulation-only approach to clean energy. That's because a regulatory strategy necessarily leaves all the implementation challenges to individual states. The existing 50-state strategy for energy efficiency evaluation has been described as a "mess."[31] Supercharging the existing problematic system, and then layering on state authority to create tradable energy efficiency credits—as the Clean Power Plan did, for example—makes the already overwhelmingly complex problem much worse.[32] We aren't doing a credible job of measuring energy efficiency savings now. Adding a market for credits created under 50 different systems encourages states to be even less demanding so their credits can compete in the energy market. Such a regulatory strategy has all the problems with energy efficiency measurement already described plus the drawbacks of operating extremely complicated programs through 50 different governments, all underresourced and many actively opposed to the climate objectives of the rule. That would be a Next Gen disaster waiting to happen: a near-impossible problem being exclusively administered by state governments whose every incentive is to look the other way. More carbon emissions are inevitable as utilities are allowed to increase emissions by purchasing overstated and unverifiable energy efficiency credits.

Some people who concede that the measurement challenges of energy efficiency are daunting nevertheless still push to include these programs in a clean power approach, whether that is legislative or regulatory, because allowing energy efficiency credits has another important objective: stimulating investment in energy efficiency. If utilities can buy energy efficiency credits to meet their clean energy compliance obligation, that's a source of cash for the vital work of energy efficiency. Some think this is actually the primary purpose of energy

the risk of energy efficiency not achieving its energy-reduction goals on the utilities, aligning the private interest of the utilities with the public interest in energy efficiency that actually reduces demand.

31 A survey of state approaches to energy efficiency evaluation confirmed that states have a "sometimes distressing amount of variability and inconsistency," and from a national perspective, the situation might be regarded as a "mess." Kushler, Nowak, and Witte, "National Survey of State Policies" (see *supra* note 14), 34, 39. Forty-four percent of states reported, for example, that they do not even have written rules for conducting evaluations. Kushler, 11.

32 The Clean Power Plan finalized in 2015 allowed use of deemed savings to develop energy efficiency credits, despite the reality that such projected savings are rarely realized. "Carbon Pollution Emission Guidelines for Existing Stationary Sources: Electric Utility Generating Units," *Federal Register*, Vol. 80 (October 23, 2015): 64662, 64909, https://www.govinfo.gov/content/pkg/FR-2015-10-23/pdf/2015-22842.pdf (describing EPA's draft guidance as containing best practices for evaluation, measurement and verification for energy efficiency under the Clean Power Plan). That draft guidance allowed use of deemed savings. "Draft Evaluation Measurement and Verification (EM&V) Guidance for Demand-Side Energy Efficiency (EE)," EPA, August 3, 2015, at 15–16, https://archive.epa.gov/epa/sites/production/files/2015-08/documents/cpp_emv_guidance_for_demand-side_ee_-_080315.pdf.

efficiency credits.[33] How can we achieve that critical investment in energy efficiency without undercutting the zero-carbon electricity plan?

Fortunately, states have shown us the way. Instead of mixing energy efficiency with electric generation, to the detriment of both, we can address energy efficiency directly, through an Energy Efficiency Resource Standard (EERS).[34] An EERS would direct utilities to accomplish the maximum achievable level of energy efficiency. Twenty-six states already have an EERS.[35] An aggressive federal energy efficiency standard could cause the leap forward in energy efficiency we need without putting it in competition with renewable energy. We can insist on both.

Separating energy efficiency from a clean power standard also makes it much easier to include other important goals in the work, like equity. Designing markets so they address environmental justice is not a simple problem. A standard solves that; it can require energy efficiency investment and specifically require that it occur in underserved communities.[36] In this case, as is so often true, it is far more straightforward to go directly at the desired outcome, rather than try to torque an ill-fitting method in an attempt to force it to achieve something that it is not well-designed to do.

A bill introduced in the US Senate in 2019 suggests ways this can work; it proposes a national standard, with states allowed to have a more stringent approach, and nationally consistent evaluation, measurement and verification requirements.[37] It directs the National Academy of Sciences to evaluate measurement methods and to incorporate more state-of-the-art experimental approaches, and directs that an evaluation database be made public.[38] Through

[33] The dual and often competing objectives—funding for desired projects and achieving the regulatory goals—are also what fuels the drive for offsets in larger carbon markets, with often disastrous result. See, e.g., Kenneth R. Richards, "Environmental Offset Programmes," in Kenneth R. Richards and Josephine van Zeben eds., *Policy Instruments in Environmental Law* (Edward Elgar Publishing, 2020), 325–51; "Climate Change Issues: Options for Addressing Challenges to Carbon Offset Quality," Government Accountability Office, GAO-11-345, February 2011. See also the discussion of carbon offsets in the conclusion.

[34] This is the strategy that the House Select Committee on the Climate Crisis recommended in June 2020. See House Committee, "Solving the Climate Crisis," 34.

[35] House Committee, "Solving the Climate Crisis," 34; See also Rachel Gold, Annie Gilleo, and Weston Berg, "Next-Generation Energy Efficiency Resource Standards," American Council for an Energy-Efficient Economy, Report U1905 (August 2019), iv, https://www.aceee.org/research-report/u1905#. For information on each state's EERS, see "Energy Efficiency Resource Standards," American Council for an Energy-Efficient Economy, https://database.aceee.org/state/energy-efficiency-resource-standards.

[36] See, e.g., Maryland's proposed legislation to add a low-income investment target to the state's energy efficiency resource standard. Deron Lovaas, "Energy Justice for Maryland's Low-Income Communities," Natural Resources Defense Council, February 5, 2021, https://www.nrdc.org/experts/deron-lovaas/energy-justice-marylands-low-income-communities.

[37] S. 2288, American Energy Efficiency Act of 2019, 116th Cong. (2019), https://www.congress.gov/bill/116th-congress/senate-bill/2288/text.

[38] Section 610(d) of Senate 2288. There are many potentially useful approaches that have not yet been applied at scale, like new technologies for real-time measurement and randomized controlled trials. See, e.g., "The Changing EM&V Paradigm," Northeast Energy Efficiency Partnership (NEEP),

strategies like this we can ramp up energy efficiency and figure out how to improve measurement, without undercutting the integrity of a clean power standard.[39] This approach also gives government the option to conclude, as seems eventually likely, that while measurement can greatly improve, it might never be possible to have real certainty about energy savings from energy efficiency. At some point it might not be cost-effective to obtain that last degree of clarity.[40] We are nowhere near that point yet. But keeping efficiency separate from other measures to cut carbon frees us from the necessity of pushing for energy efficiency measurement that isn't cost-effective and limits the damage to our climate change objectives that could result from mixing poorly measured actions with measurement-certain clean energy.

Energy efficiency shouldn't be included in a clean energy standard because it inserts a high degree of uncertainty into one of the central climate solutions we can otherwise be confident about achieving. For all the same reasons it should not be allowed as an offset under any other approach to cutting carbon in electric generation, whether that's a Clean Power Plan 2.0, a market approach like cap and trade or a carbon tax, or any other market-type strategy. Dramatic improvements in energy efficiency are indispensable for any climate plan, but the uncertainties of energy efficiency savings cannot be allowed to undercut the also absolutely essential reductions in CO_2 from electric generation. If we were starting this plan in 1980, maybe we would have been willing to gamble on creating more uncertainty in carbon reduction in exchange for possibly reducing costs. We don't have that luxury anymore. We have to cut carbon from electric generation to zero or close to it as fast as we can, and we can't use any strategy that makes achieving that goal significantly less certain.

It is worth remembering, when regulators make policy choices about ways to cut carbon to zero in electric generation, that this is the easy one. We can already see the pathway to zero. We know how to do this; we just need to summon the political will. Many of the other programs to achieve necessary carbon reductions will be much harder. There isn't as yet a clear solution for the industrial sector,

2015, https://neep.org/changing-emv-paradigm (potential for advanced data analytics and automated data availability to improve evaluation).

[39] Senate Bill 2288 also has a number of interesting Next Gen–type ideas for incentivizing energy efficiency measurement that is as accurate as possible, like excluding energy savings that are not adequately documented and setting fixed civil penalties for each unit of energy claimed but not delivered. S. 2288, American Energy Efficiency Act, § 610(e).

[40] See, e.g., discussion of the trade-offs between certainty and cost in energy efficiency evaluation in Gillingham, Keyes, and Palmer, "Advances in Evaluating Energy Efficiency," 528. See also Steven R. Schiller, Charles A. Goldman, and Elsia Galawish, "National Energy Efficiency Evaluation, Measurement and Verification (EM&V) Standard: Scoping Study of Issues and Implementation Requirements," State & Local Energy Efficiency Action Network (2011), v–vii, https://www7.eere.energy.gov/seeaction/system/files/documents/emvstandard_scopingstudy.pdf (discussing issue of "how good is good enough" in measuring energy efficiency savings).

agriculture, and making the cuts global, to name just some of the more formidable issues. Those are genuinely complicated and will require action in a highly uncertain environment. Injecting all the uncertainty and complexity of energy efficiency into a clean power strategy makes it both unnecessarily convoluted and less likely to succeed. We have to move fast on energy efficiency, but not by undercutting clean power. Zero-carbon electricity can be a sure thing. Let's keep it that way.

8
Don't Double Down on Past Mistakes with Low-Carbon Fuels

Any plan to address climate change has to put transportation emissions front and center. Transportation—passenger and freight vehicles including cars, trucks, planes, and ships—is currently the largest contributor to human-caused US greenhouse gas emissions.[1] The biggest share of that is from passenger cars and trucks.[2]

There is widespread agreement that the way to cut carbon from passenger vehicles is to shift them to electric power at the same time that we take action to make that electricity as low carbon as possible.[3] While we transition to electrification, make passenger vehicles as efficient as they can be. It isn't simple to do, but at least the path is clear.

Other types of transportation, like long-distance trucks, ships, and planes, are harder. At present, most of these still require some form of liquid fuel.[4] For these essential modes of transportation, most climate plans propose to ratchet up efficiency and at the same time reduce the carbon-intensity of fuels through low-carbon fuel standards.[5]

[1] "Fast Facts on Transportation Greenhouse Gas Emissions," EPA, 2019, https://www.epa.gov/greenvehicles/fast-facts-transportation-greenhouse-gas-emissions; "The Case for Climate Action: Building a Clean Economy for the American People," Senate Democrats Special Committee on the Climate Crisis, August 25, 2020, at 45, https://www.schatz.senate.gov/imo/media/doc/SCCC_Climate_Crisis_Report.pdf; Amy L. Stein and Joshua P. Fershée, "Light Duty Vehicles," in Michael B. Gerrard and John C. Dernbach eds., *Legal Pathways to Deep Decarbonization in the United States* (Environmental Law Institute, 2019), 353.

[2] EPA "Fast Facts"; Stein, "Light Duty Vehicles," 354; "Solving the Climate Crisis: The Congressional Action Plan for a Clean Energy Economy and a Healthy, Resilient, and Just America," Majority Staff of House Select Committee on the Climate Crisis, 116th Congress, June 2020, at 87, https://climatecrisis.house.gov/sites/climatecrisis.house.gov/files/Climate%20Crisis%20Action%20Plan.pdf. The next largest share is medium- and heavy-duty trucks, which contribute about 25%. EPA "Fast Facts."

[3] House Select Committee, "Solving the Climate Crisis," 86–100; Stein, "Light Duty Vehicles," 356, 373; "Resetting the Course of EPA: Reducing Air Emissions from Mobile Sources," Environmental Protection Network, August 2020, at 3, https://www.environmentalprotectionnetwork.org/reset/reducing-air-emissions-from-mobile-sources/.

[4] Senate Special Committee, "The Case for Climate Action," 57.

[5] House Select Committee, "Solving the Climate Crisis," 101; Andrea Hudson Campbell, Avi B. Zevin, and Keturah A. Brown, "Heavy-Duty Vehicles and Freight," in Gerrard, *Legal Pathways*, 384, 389.

Next Generation Compliance. Cynthia Giles, Oxford University Press. © Cynthia Giles 2022.
DOI: 10.1093/oso/9780197656747.003.0009

Climate isn't the only reason to push for changes in transportation emissions. The fossil fuel combustion that powers vehicles creates a lot of other dangerous air pollution too, so cutting climate emissions also addresses a key public health threat.[6] Vehicle pollution is a big part of the urgent calls for environmental justice because transportation emissions disproportionately affect already overburdened communities.[7]

What is the biggest compliance challenge in transportation emissions? Despite what you probably think, it isn't clean-car standards. Volkswagen's emissions fraud[8] certainly opened a lot of people's eyes to the problem of companies' flouting these standards; it is Exhibit A for the fact that big companies do cheat, the one-word response to the unsupportable but oft-repeated notion that big companies are all trying to comply. EPA learned a valuable lesson that its otherwise strong vehicle compliance program needed to block the pathways for fraud.[9] But with that change, the compliance provisions in the rules for new cars and trucks are actually quite resilient. The vehicle pollution limits need to be considerably strengthened; the Biden EPA has moved quickly to do that after progress was halted during the Trump administration.[10] Once we have more ambitious standards, though, the rules as presently designed can achieve good compliance. This industry has a strong profile for Next Gen effectiveness. There are a limited number of easily identified auto and truck makers who are sophisticated and knowledgeable. There are reliable tests to know how much pollution will be emitted from each vehicle, and these have been considerably improved by changes post-Volkswagen. EPA has world-class engineers and testing that easily go toe to toe with the manufacturers. Every new vehicle has to be pre-certified by EPA, based both on manufacturer and independent EPA testing.[11] If the vehicle

[6] Senate Special Committee, "The Case for Climate Action," 52.

[7] "Research on Near Roadway and Other Near Source Air Pollution," EPA, https://www.epa.gov/air-research/research-near-roadway-and-other-near-source-air-pollution.

[8] See chapter 2, note 56 for a brief description of the Volkswagen case.

[9] It's not just Volkswagen, of course. As a result of the investigation EPA did after the Volkswagen cheating was uncovered, other diesel-car manufacturers have also been tagged for having defeat devices. See EPA, "In Civil Settlements with the United States and California, Fiat Chrysler to Settle Allegations of Cheating on Federal and State Vehicle Emissions Tests," Press Release, January 10, 2019, https://archive.epa.gov/epa/newsreleases/civil-settlements-united-states-and-california-fiat-chrysler-settle-allegations.html; EPA, "U.S. Reaches $1.5 Billion Settlement with Daimler AG over Emissions Cheating in Mercedes-Benz Diesel Vehicles," Press Release, September 14, 2020, https://www.epa.gov/newsreleases/us-reaches-15-billion-settlement-daimler-ag-over-emissions-cheating-mercedes-benz.

[10] For an up-to-date history of emission standards for vehicles, including recent executive actions to strengthen those standards, see "Corporate Average Fuel Economy Standards/Greenhouse Gas Standards," Harvard Environmental and Energy Law Program Regulatory Tracker, https://eelp.law.harvard.edu/2019/09/corporate-average-fuel-economy-standards-greenhouse-gas-standards/.

[11] This is the big lesson of Volkswagen: don't trust the company's testing and don't let the companies know what testing EPA will do. EPA now does less predictable and more variable testing, which inspires companies to design their vehicles to achieve standards in actual operation in the real world, instead of "teaching to the test." See Cary Coglianese and Jennifer Nash, "The Law of the

doesn't clear the emissions hurdle, it isn't approved. This is the compliance gate that EPA used to eventually force Volkswagen to admit its deceit. Are there still violations? Yes. Can the regulations' compliance strategies be improved? Absolutely.[12] But overall, this is a solidly designed program. If it receives adequate funding, it can achieve quite good emissions compliance without constant enforcement attention.[13] New technologies hold promise to make it even better.[14]

Low Carbon Fuels

The big compliance challenge in transportation is low-carbon fuels. To call it a challenge understates; the compliance problems are daunting. The compliance landscape is global, extremely complex, and strewn with political land mines. There are compliance barriers in every direction, with huge implications for our ability to reduce climate-forcing emissions through low-carbon fuels. This is not for the faint of heart.

The theory of low-carbon fuels is very appealing. They are mostly made from plants, unlike the oil-based fuels we have traditionally used for transportation.

Test: Performance-Based Regulation and Diesel Emissions Control," *Yale Journal on Regulation*, Vol. 34 (2017): 33.

[12] Among the improvements that could strengthen compliance: stronger requirements to limit emissions deterioration over time, roadside monitoring of emissions to spot vehicle types that appear to have more emissions than allowed, and enhanced "in use" testing to find the vehicle models that show declining emissions performance over the life of the vehicle. We also need new regulatory strategies to address the scourge of the so-called after-market defeat devices, which eliminate or reduce emissions controls. These are not in the new vehicle when sold, but are added later, in plain violation of the law and with serious pollution consequences. More than 15% of diesel trucks in the United States are estimated to have had emission control systems removed through such tampering, adding 570,000 tons of nitrogen dioxide pollution to the air over the life of the vehicles. Coral Davenport, "Illegal Tampering by Diesel Pickup Owners Is Worsening Pollution, E.P.A. Says," *New York Times*, November 25, 2020, https://www.nytimes.com/2020/11/25/climate/diesel-trucks-air-pollution.html; "Aftermarket Defeat Devices and Tampering Are Illegal and Undermine Vehicle Emissions Controls," EPA Enforcement Alert, December 2020, https://www.epa.gov/sites/production/files/2020-12/documents/tamperinganddefeatdevices-enfalert.pdf. This is an important subject that calls out for Next Gen solutions but is beyond the climate focus of this chapter.

[13] Note that the small number of automakers and the extensive mandatory testing and reporting also make this a program where enforcement can make a big difference for compliance. The cases are large and complicated, but the significant chance of getting caught and the recent record of serious consequences for violations make enforcement a bigger deterrent in this program than it is in many others. Enforcement isn't practical as a systemic answer to other issues in this sector, but the stars align to make it one of EPA's powerful tools to deter violations of vehicle emission standards for new cars and trucks.

[14] See, e.g., Felipe Rodríguez and Francisco Posada, "Future Heavy-Duty Emission Standards: An Opportunity for International Harmonization," International Council on Clean Transportation White Paper, November 2019, https://theicct.org/sites/default/files/publications/Future%20_HDV_standards_opportunity_20191125.pdf.

When fossil fuels like gasoline and diesel are burned, they release a lot of carbon into the atmosphere that would otherwise have remained underground. That's why fossil fuels are the central problem for climate change. Plants, on the other hand, capture carbon as they grow.[15] They release carbon when they are used, but a lot of carbon can be recaptured when more plants are grown. That's why these are often called "renewable" fuels; the carbon is recycled, so we can use the fuels for transportation with far less carbon burden. That's the theory.

So what's the problem? Plants as fuel are only lower carbon if we don't chew up a lot of land to grow them.[16] Undisturbed lands, like forests and grasslands, store a lot of carbon, both above ground in the biomass of the plants, and below, in the soil. Cutting down a forest or a grassland and plowing that land for crops releases that carbon, and it keeps emitting carbon for decades.[17]

The quantity of carbon released when undisturbed land is converted to agriculture is surprisingly big. Carbon is emitted when the forests or grasslands are cut down and the vegetation either decays or is burned.[18] But the largest source of carbon from converting land to crops in the United States is the soil itself. Plowing under US grasslands releases a significant amount of carbon, 90 percent of which originates in the soil.[19] Carbon in biomass accumulates over years to decades, but soil carbon accumulates slowly, over decades to centuries; releasing the carbon in soils is thus effectively irreversible over human time scales.[20] The climate impacts of cutting down perennial vegetation and replacing it with annual commodity crops dwarf the other climate issues for biofuels.[21] That's why the National Academy of Sciences says that land-use changes can have profound effects on greenhouse gas emissions and that the carbon impact of biofuels depends on the changes to land use and land cover. The carbon released from land-use changes alone can wipe out any climate benefit from biofuels.[22]

[15] Timothy Searchinger et al., "Use of U.S. Croplands for Biofuels Increases Greenhouse Gases Through Emissions from Land-Use Change," *Science*, Vol. 319 (2008): 1238.

[16] Blake Hudson and Uma Outka, "Bioenergy Feedstocks," in Gerrard, *Legal Pathways*, 650. The energy used in production and transportation of biofuels also add to its carbon footprint, but induced land-use change is the largest source of greenhouse gas emissions associated with biofuels.

[17] National Research Council, *Renewable Fuel Standard: Potential Economic and Environmental Effects of U.S. Biofuel Policy* (The National Academies Press, 2011), 4, 191, https://doi.org/10.17226/13105.NRC.

[18] Seth A. Spawn, Tyler J. Lark, and Holly K. Gibbs, "Carbon Emissions from Cropland Expansion in the United States," *Environmental Research Letters*, Vol. 14 (April 2019): 1, https://iopscience.iop.org/article/10.1088/1748-9326/ab0399.

[19] Spawn, Lark, and Gibbs, "Carbon Emissions from Cropland Expansion," 5.

[20] Spawn, 7.

[21] National Research Council, "Renewable Fuel Standard," 4.

[22] National Research Council, "Renewable Fuel Standard," 4, 192, and 245 (in many cases land-use change is the variable with the greatest effect on greenhouse gas emissions from biofuels); "Biofuels and the Environment: Second Triennial Report to Congress," EPA, EPA/600/R-18/195, June 2018, at 20, 53, https://cfpub.epa.gov/si/si_public_record_Report.cfm?Lab=IO&dirEntryId=341491 (land-use change has been identified as one of the primary drivers affecting environmental impacts); Searchinger, "Use of U.S. Croplands for Biofuels," 1238; "Renewable Fuel Standard: Information

You might think, "OK, that's not an insurmountable barrier; just require that biofuels only be grown on land already in farming. Don't convert any undisturbed land for renewable fuel crops. Problem solved." Putting aside for a moment the practical problems with implementing that, it has a major conceptual flaw: many of the crops pushed off existing farmland in favor of biofuels will go somewhere else. Where? Obviously, on currently non-farmed land. The demand for biofuels increases the value of biofuel crops, and that itself can provide economic incentive for additional land-use change.[23] In the jargon of biofuels these are called indirect land-use changes; even if the biofuels themselves aren't grown on converted land, the demand for biofuels inevitably leads to undisturbed land being plowed under.[24] For this reason, requiring that biofuels be grown only on land already in farming doesn't solve the climate problem, it just pushes it around. The net effect of the demand for renewable fuels is more natural areas turned under for farming. More carbon released. Whether the effect is direct (forest and grassland land converted to grow crops for renewable fuels) or indirect (forest and grassland converted for other crops) doesn't matter for the climate. More land disturbance causes more carbon release causes more climate change.

The problem is complicated by the reality that efforts to tackle biofuels' exceedingly difficult land-use implications run headfirst into a political buzz saw. Land use in the United States has traditionally been a state or local governance issue.[25] Apart from federally owned lands, the federal government has not had a lot to say about land use. The few places where it has—for example, in protecting wetlands because of their central role in clean water—have been controversial.

on Likely Program Effects on Gasoline Prices and Greenhouse Gas Emissions," Government Accountability Office, GAO-19-47, May 2019, 22. Land-use change is the largest but not the only factor in assessing climate impacts of biofuels. Other factors include how the land is farmed and how much fertilizer is used. No-till farming has promise to reduce soil carbon losses, but its impact is uncertain (National Research Council, "Renewable Fuel Standard," 186), and some think it only matters when land is permanently no-till (Spawn, Lark and Gibbs, "Carbon Emissions from Cropland Expansion," 8–9, noting that intermittent tillage probably doesn't have much climate benefit and observing that permanent no-till management is relatively rare). Fertilizer is another significant climate issue, because fertilizer releases nitrogen dioxide, a potent greenhouse gas. Spawn, "Carbon Emissions," 9. Growing more corn leads to more nitrogen fertilizer applications, with their attendant climate impacts. Tyler J. Lark et al., "Environmental Outcomes of the U.S. Renewable Fuel Standard," *Proceedings of the National Academy of Sciences*, Vol. 119, No. 9 (2022): 3, https://doi.org/10.1073/pnas.2101084119. New land converted to grow crops is usually less productive, so also requires—surprise!—more fertilizer. EPA, "Second Triennial Report," 54. Climate is not the only environmental concern raised by biofuels. There are also many other environmental impacts of increasing demand for biofuels not addressed in this chapter, such as air and water pollution and wildlife habitat loss. For a description of these other impacts, see National Research Council, "Renewable Fuel Standard"; EPA, "Second Triennial Report," 65, x (finding that the environmental and resource conservation impacts of biofuels are, on balance, negative).

[23] National Research Council, "Renewable Fuel Standard," 5.
[24] National Research Council, "Renewable Fuel Standard" 5; EPA, "Second Triennial Report," 21.
[25] Hudson and Outka, "Bioenergy Feedstocks," 650.

Many local and state governments, which have the legal authority to impose restrictions to prevent low-carbon fuels from becoming carbon multipliers, don't have the interest or the political will to do that.[26] Using the authority of the federal government to prevent climate-damaging land-use changes is politically fraught.

It isn't just land use that is a political minefield. Farming—choosing which crops are grown where and in what way—has long been the third rail in environmental politics. Some of the most contentious policy debates in environmental protection resulted from the farming industry working hard to make it impossible to regulate farm activities.[27]

This is the tricky situation Congress confronted when it enacted the expanded Renewable Fuel Standard (RFS) as part of the Energy Independence and Security Act of 2007:[28] how to get the potential climate benefits of biofuels without causing new land-use change or detonating a political bomb.

The updated RFS law mandated that US transportation fuels, mainly gasoline and diesel, be blended with biofuels, made primarily from agricultural feedstocks. The statute and mandated annual EPA regulations specify how many gallons of biofuels must be included in transportation fuels sold in the United States each year. The law describes two basic types of biofuels: conventional and advanced. Conventional biofuels must be at least 20 percent lower carbon intensity than the petroleum-based fuels they replace.[29] Advanced biofuels, such as fuel from algae, must have a much greater carbon benefit: 50 percent or better

[26] See Hudson and Outka, "Bioenergy Feedstocks," 660, 662 (noting that state and local governments face pressure to keep land-use regulations flexible, even lax, so as not to drive out economic development).

[27] One recent example is the so-called Waters of the United States rule. See, e.g., EPA, "(Des Moines Register) EPA, Army Finalize Repeal of Controversial 'Waters of the U.S.' Rule," New Release, September 12, 2019, https://www.epa.gov/newsreleases/des-monies-register-epa-army-finalize-repeal-controversial-waters-us-rule. Here are the actual first two sentences of an article written by the Trump EPA administrator: "Today, EPA and the Department of the Army will finalize a rule to repeal the previous administration's overreach in the federal regulation of waters and wetlands. This action officially ends an egregious power grab and sets the stage for a new rule that will provide much-needed regulatory certainty for farmers, home builders, and property owners nationwide." Notice where the op-ed was placed. Farming is politics with a small p, except every four years when Iowa has the first voting of the presidential primary season, turning farming issues into national "capitol P" politics. The Renewable Fuel Standard has been a perennial player in Presidential elections, including 2020. See Rebecca Beitsch, "EPA Delivers Win for Ethanol Industry Angered by Waivers to Refiners," *The Hill*, September 14, 2020, https://thehill.com/policy/energy-environment/516364-epa-delivers-win-for-ethanol-industry-angered-by-waivers-to.

[28] Energy Independence and Security Act (EISA) of 2007, 42 U.S.C. § 17001 et seq. EISA replaced an earlier version enacted in 2005. The 2007 RFS statute is sometimes called RFS2.

[29] 42 U.S.C. § 7545(o)(2)(A)(i). The 20% threshold was significantly undermined elsewhere in the law; facilities that existed or were in construction by 2007 are exempted from meeting the 20% better-than-fossil-fuel requirement. 42 U.S.C. § 7545(o)(2)(A)(i). In 2017, 89% of all RFS blending volume was exempt from the 20% improvement standard. GAO, "Renewable Fuel Standard," at 7 n.7, and at 20 n.29; "Regulation of Fuels and Fuel Additives: Change to Renewable Fuel Standard Program," EPA Final Rule ("2010 RFS regulation"), *Federal Register*, Vol. 75 (March 26, 2010): 14670, 14677.

reduction in greenhouse gas emissions.[30] Congress directed EPA to analyze the total climate forcing emissions related to each fuel type; these lifecycle analyses determine which biofuels meet the mandatory percentage reduction threshold.

The hope and the goal of RFS was to spark a big increase in advanced biofuels. The vast majority of the expected climate benefits were projected to come from these advanced fuels.[31] Unfortunately, that isn't what happened. The program has produced several orders of magnitude less advanced biofuels than was set forth in the 2007 legislation.[32] Almost 90 percent of biofuels produced in the United States are conventional ethanol, which is made nearly entirely from corn; advanced biofuels comprise less than 10 percent, over half of which is biodiesel from soybeans.[33] This falls far short of the 60 percent share for advanced biofuels that Congress envisioned in the updated RFS.[34]

In an effort to ensure the climate benefits of biofuels, Congress included two provisions to limit conversion of undisturbed land to biofuel crops: (1) it directed EPA to consider land conversion in its lifecycle analysis of the climate impacts of biofuels, so land-use change would be included in EPA's decisions about which fuels meet the emissions reduction thresholds; and (2) it declared that biofuel crops grown on converted land were not eligible.[35]

As Congress instructed, EPA's lifecycle analyses contained a thorough evaluation of land-use change. Those analyses included robust consideration of how much land was likely to be converted as a result of the new standard and the expected carbon emission impacts of that conversion. EPA found that the land already being farmed in the United States as of 2007 was likely sufficient to support both the new biofuels and other crop products, so there would be no need to clear and cultivate additional land.[36] As a result, EPA predicted no additional

[30] 42 U.S.C. § 7545(o)(1)(B) Definition of Advanced biofuel. See also GAO, "Renewable Fuel Standard," 6–7. There are a variety of advanced biofuels, including fuels from algae or cellulose in crop residue. GAO, 7.

[31] Joseph E. Aldy, "Promoting Environmental Quality Through Fuels Regulations," in Ann Carlson and Dallas Burtraw eds., *Lessons from the Clean Air Act: Building Durability and Adaptability into U.S. Climate and Energy Policy* (Cambridge University Press, 2019), 159, 185.

[32] Aldy, "Promoting Environmental Quality," 198.

[33] EPA, Second Triennial Report, 7. "Biofuels Explained: Biomass-based Diesel Fuels," US Energy Information Administration, https://www.eia.gov/energyexplained/biofuels/biodiesel.php (percent of biodiesel from soybean oil). See also Jonathan Lewis, "Biofuels, Part 2," *Harvard CleanLaw podcast*, September 30, 2019, transcript at 2, http://eelp.law.harvard.edu/wp-content/uploads/CleanLaw-26-Joe-Jon-Lewis-biofuels-2.pdf.

[34] "Legislated Renewable Fuel Standard (RFS) Volume Requirements," U.S. Dept. of Energy, Alternative Fuels Data Center, last updated May 2020, https://afdc.energy.gov/data/10421.

[35] 42 U.S.C. § 7545 (o) (1)(H) (definition of lifecycle greenhouse gas emissions, including direct and indirect emissions from land-use change); 42 U.S.C. § 7545 (o) (1)(I) (definition of renewable biomass, which includes only crops and other feedstocks from land in farming before the law was enacted in December 2007).

[36] EPA, "2010 RFS regulation," 14682.

climate-forcing emissions from domestic land-use change.[37] However, EPA also found that the RFS would cause shifts in agricultural markets worldwide, leading to significant greenhouse gas emissions associated with land conversion in other countries; EPA's analysis attributed 40 percent of the total lifecycle greenhouse gas emissions for corn ethanol to international land-use change.[38] Some commenters suggested that international land-use change was too unpredictable to be included. EPA had the opposite reaction. Although predicting land conversion is inherently uncertain, the impact of land-use change on greenhouse gas emissions is large. Therefore, EPA concluded, if it's impossible to determine that carbon emissions from likely land-use change are small enough to meet the threshold, then the fuel would have to be excluded from the program.[39]

Based on significant carbon emissions predicted from land-use changes overseas, but virtually no land-use change emissions here at home, the dominant biofuel—corn ethanol—barely squeaked by the 20 percent threshold, with a 21 percent benefit rating for new ethanol plants.[40]

What about the second prong of the law's attempt to limit emissions from land conversion: Congress's prohibition on using biofuel crops grown on newly farmed land? EPA rule writers wrestled with how to ensure that biofuels weren't produced on land not in farming as of 2007 (the year the law was enacted), as the statute required. EPA's proposed regulation suggested putting the burden of making sure lands met the 2007 cutoff on the biofuel producer or importer.[41] Producers and importers would be required to certify that the crops they used didn't come from newly farmed land, and to maintain records to support that claim, including maps or electronic data identifying the boundaries of the land where each type of feedstock was produced, and other documents, traceable to the specific land used, proving that each such tract of land was eligible under the rules.[42]

There are limited options for how this proposed compliance regime could play out: (1) growers admit to the biofuels producer or importer that the crops the farmers are selling violate the 2007 cutoff; (2) the records required to prove

[37] EPA, "2010 RFS regulation," 14788 (corn ethanol); EPA, "2010 RFS regulation," 14789–790 (soy biodiesel).

[38] EPA, "2010 RFS regulation," 14788.

[39] EPA, "2010 RFS regulation," 14679.

[40] EPA, "2010 RFS regulation," 14786–788. The reason EPA only analyzed the lifecycle greenhouse gas (GHG) emissions from new ethanol plants is that biofuels from pre-2007 ethanol plants are exempt from the 20% reduction standard. See discussion accompanying notes 67 to 71 in this chapter.

[41] "Regulation of Fuels and Fuel Additives: Changes to Renewable Fuel Standard Program," EPA Proposed Rule (EPA, "Proposed RFS regulation"), *Federal Register*, Vol. 74 (May 26, 2009): 24904, 24911.

[42] EPA, "Proposed RFS regulation," 24933 (recognizing that "it may be difficult" to determine qualification with the 2007 cutoff using some of the documents identified as supporting compliance determinations).

compliance are missing or wrong so it is difficult to impossible to figure out if the crops complied with the rule; or (3) the producer or importer says everything is hunky dory and EPA attempts to verify that claim by looking at decades-old records about what was in farming in prior to 2007 and brings an enforcement case if it turns out some crops were ineligible. Everyone who thinks number 1 is likely, raise your hand. Here's what would really happen: EPA would find that fuel producers' or importers' records are incomplete or inaccurate, so EPA couldn't determine if the crops they used complied with the 2007 cutoff. Where records do exist, EPA would be faced with the impossible task of verifying that the specific crops sold were actually grown on the precise land claimed and that land was eligible.[43] Many of the (usually paper) records would likely be missing or ambiguous. And these investigations would provoke a firestorm of political pushback. See politics 101 above. Not to mention that EPA doesn't have the staff to conduct such inquiries at even one in a thousand locations; there are about 90 million acres of corn in the US, about 40 percent of which are dedicated to biofuels.[44] If by some miracle EPA did identify a few violators, the biofuel made with ineligible crops would be long since sold. In the unlikely chance that a violation by a producer or importer were proved, they could easily, and probably truthfully—since knowing is not in their interest—claim they had no idea the purchased crops weren't compliant. As a result, the chance of persuading a court to impose penalties that have deterrent punch would be extremely low. The entire enterprise is totally unworkable. File this under enforcement: your worst nightmare.

Anyone tracking the Next Gen indicia for rules likely to have widespread violations will recognize them in EPA's proposal. The regulated companies have strong incentives not to hassle their suppliers with demands for hard-to-find records or to scrutinize those records too closely. The farmers who are supposed to provide the records are regulation resistant and have little interest in meeting what likely feel like intrusive requests for business information. And both know that it is unlikely to impossible that EPA will be able to sort out which feedstocks meet the 2007 cutoff requirement.

Here's the creative idea that EPA finalized in 2010 in place of the initial proposal. EPA would monitor land in farming in the United States through satellite data and other means to see if there had been a net increase in farmland since

[43] Note that this is not as simple as looking at a satellite image and comparing pre- and post-2007 land use. For example, land in conservation status is eligible too, even if not being actively farmed. Determining eligibility is therefore a records-intensive exercise, not knowable from just looking at a photograph. There are many other compliance challenges too, including that the same eligible land could be used repeatedly as the claimed origin of crops that actually were grown other places. That couldn't be spotted without a complete database for all the fuel producers and importers that could compare every piece of land claimed for compliance to every processor, a very demanding task.

[44] EPA, "Second Triennial Report," 10 (acres of US land in corn production in 2016), and 53 (40% of corn grain produced nationally goes to biofuels).

2007. If not, EPA would assume that the land-use restrictions were being met and none of the recordkeeping obligations would be triggered.[45] EPA called this "aggregate compliance." In theory it assured that biofuels would not cause undisturbed lands to be farmed—protecting the climate benefits of biofuels—at the same time it reduced implementation burden for EPA and for industry. As of 2020, EPA continues to find that there has been no net increase in farmland in the United States, so government is not checking if newly farmed land in the United States is the source of biofuel feedstock, and industry doesn't have to keep any records.[46] Unfortunately, EPA's creative idea didn't work.

The prediction that land conversion would be limited in the United States and the backstop of the aggregate compliance approach have both been proven wrong. In fact, a huge amount of additional land has come under the plow in the United States since 2007. EPA estimates that by 2012, 4 to 7.8 million acres of new farmland were added—an area the size of New Jersey.[47] The RFS is responsible for 5.2 million acres of additional US land farmed, all land that but for RFS would have been busy storing carbon but is instead emitting carbon in large quantities.[48] The amount of carbon being released from converted land in the United States eliminates the modest carbon savings that EPA had predicted for

[45] 40 C.F.R. § 80.1454(g) (aggregate compliance approach). See 40 C.F.R § 80.1454(c) (importers), and (d) (domestic producers), 40 C.F.R. § 80.1451(g) (records required if EPA determines that total land in farming has increased and record keeping obligations are therefore triggered); 40 C.F.R. § 80.1401 Definition of existing agricultural land (what records are required to demonstrate that land was in farming as of 2007).

[46] See "Renewable Fuel Standard Program: Standards for 2020 and Biomass Based Diesel Volume for 2021 and Other Changes," EPA, *Federal Register*, Vol. 85 (February 6, 2020): 7053–54.

[47] EPA, "Second Triennial Report," 37–38, 44. There is clear scientific consensus on the extensive cropland expansion that has occurred in the United States over the last decade. Tyler J. Lark, Seth A. Spawn, Matthew Bougie, and Holly Gibbs, "Cropland Expansion in the United States Produces Marginal Yields at High Costs to Wildlife," *Nature Communications*, Vol. 11 (September 2020): 4–5, https://doi.org/10.1038/s41467-020-18045-z; EPA, "Second Triennial Report," 39. Corn is the predominant crop planted on newly cultivated land. Lark, "Cropland Expansion," 3. How can there be so much post-2007 land conversion to farming in the United States but also no net increase in farmed land? One reason is development. If I am farming 100 acres, sell 10 of those acres to a developer for a subdivision, and start farming 10 new acres, I have not increased the net land in farming; it is still 100 acres. But the subdivision isn't storing carbon, and my 10 newly farmed acres are releasing a lot of carbon, producing a net increase in carbon emissions even though there is no net increase in acres farmed. That is happening all over American today. Between 2001 and 2016, 11 million acres of farmland and ranchland were lost to development. Julia Freedgood, Mitch Hunter, Jennifer Dempsey, and Ann Sorensen, "Farms Under Threat: The State of the States," American Farmland Trust (2020), 4, https://s30428.pcdn.co/wp-content/uploads/sites/2/2020/09/AFT_FUT_StateoftheStates_rev.pdf.

[48] Lark, "Environmental Outcomes of RFS," 2. The study estimates the carbon flux from the changes in land use caused by RFS at 398 MMT CO2e. Lark, 3. The cause-and-effect relationship of RFS to US land conversion is supported by a study showing that the rate of land conversion for corn production is significantly higher close to facilities making biofuels. See EPA, "Second Triennial Report," 54, xi (noting that there are "strong indications" that biofuel feedstock production is responsible for some of the observed changes in land used for agriculture since enactment of RFS).

corn ethanol back in 2010.[49] The bottom line is that neither strategy—lifecycle analysis that included land-use change or prohibitions on using biofuels grown on land newly converted to farming—prevented significant land conversion and the resulting large carbon emissions.

Citing this evidence, some have called for EPA to withdraw the aggregate compliance approach and enforce the prohibition on biofuels grown on post-2007 farmland.[50] Is it possible to ensure compliance with that obligation? And if we could, would that help? No and no.

It is completely impractical to implement the check-every-parcel idea using paper records that are increasingly unavailable or ancient, for all the reasons already explained. Next Gen teaches that regulatory obligations running directly counter to a company's self-interest—like expecting biofuel producers and importers to scare away their suppliers through rigorous compliance checks—won't happen unless there are robust real-time countermeasures that make that hard to avoid. Relying on the dubious proposition of enforcement after the fact will never work. Next Gen, and a huge body of compliance evidence, tells us that in a situation like this, where all the incentives line up against compliance and it is virtually impossible to check, serious violations will be widespread. Holding the line for the 2007 cutoff isn't feasible.

Even if it were possible to perfectly enforce the prohibition on biofuels grown on farmland converted after 2007, that doesn't solve the problem. At best it forces biofuels onto existing cropland and pushes other crops to go elsewhere. Undisturbed land is still plowed under, resulting in significant carbon releases.[51] The climate doesn't care if we label land-use conversion as direct or indirect. The relentless logic of science can't be gamed in that way. The demand for biofuels is causing—direct, indirect, call it whatever you want—more carbon emissions from land-use change.

If you think it is hard to sort this out in the United States, consider how much more complicated it is for imported renewable fuels, made from crops grown in other countries. The largest sources of imports to the United States have been

[49] Lark, "Environmental Outcomes of RFS," 3 (concluding that the carbon emissions associated with land-use change caused by RFS eliminate the climate advantage that EPA predicted for ethanol and make corn ethanol significantly worse for climate than conventional fossil fuel gasoline).

[50] See, e.g., National Wildlife Federation at al., "Amended Petition to the United States Environmental Protection Agency to Amend Its 'Aggregate Compliance' Approach to the Definition of 'Renewable Biomass' Under the Renewable Fuel Standard in Order to Prevent the Conversion of Native Grassland," January 18, 2019, https://earthjustice.org/sites/default/files/files/Amended-Aggregate-Compliance-Petition_1-18-19.pdf.

[51] Searchinger, "Use of U.S. Croplands for Biofuels," 1240: "Because emissions from land use change are likely to occur indirectly, proposed environmental criteria that focus only on direct land-use change would have little effect. Barring biofuels produced directly on forest or grassland would encourage biofuel processors to rely on existing croplands, but farmers would replace crops by plowing up new lands." See also National Research Council, "Renewable Fuel Standard," 5.

Brazil, Argentina, and Indonesia.[52] The same discouraging data about expansion of farming in the United States are evident around the world. Cropland expansion and deforestation have been documented in these major exporters of biofuels to the United States.[53] EPA's 2018 assessment acknowledges that increased biofuel production has contributed to these international land-use changes, including in the three countries that are the major exporters of biofuels to the United States.[54] Unlike in the United States, where the vast majority of new farmland is converted from grasslands, in other countries more of the newly disturbed area is at the expense of even more intensively carbon-storing land, like forests. Increased biofuel production for the US market has contributed to these climate-damaging international land-use changes.[55]

The sheer complexity of the biofuels land-impact problem makes it much harder to solve. Lots of things affect farming choices, including crop prices, international trade agreements, development pressures, and, ironically, changing weather due to climate change.[56] Because there are so many factors that influence when, where, and how crops are grown, there isn't a straight line from the RFS to changes in farming. It's not possible to "measure" the land-use impacts of biofuels. The only way to try to figure that out is through a combination of land-use data and sophisticated economic models that try to separate the causal patterns.[57] Anyone who would rather not take this on, or wants to challenge government's conclusions, will find ample cover behind the curtain of uncertainty and real or manufactured difference of opinion. As the evidence mounts that this well-intentioned program might actually be making things worse, the intricacies of modeling give lots of room for claims of conflicting evidence and ambiguity.

Land-use change turns out to be the Achilles heel of the push for biofuels as a climate solution. For good reason, the National Academy of Sciences describes land-use change as the variable with the highest uncertainty and greatest effect.[58]

Why the long recitation of the sorry state of land-use change in the biofuels program in a chapter about compliance?

The first reason is to explain why enforcement cannot ride to the rescue. The problem is not that we are unable to restrict biofuels to lands in farming as of 2007, although we are unable to do that. It's that even if we could, all it would

[52] EPA, "Second Triennial Report," 48.

[53] EPA, "Second Triennial Report," 48.

[54] EPA, "Second Triennial Report," 48, 52, 111. EPA also notes that some recent studies showing international corn ethanol land-use change trending downward are based on models that tend to understate land-use change. EPA, "Second Triennial Report," 50.

[55] EPA, "Second Triennial Report," 115.

[56] EPA, "Second Triennial Report," 45.

[57] EPA, "Second Triennial Report," 51–52; National Research Council, "Renewable Fuel Standard," 190; GAO, "Renewable Fuel Standard," 19.

[58] National Research Council, "Renewable Fuel Standard," 245.

do—after huge regulatory and political resources were devoted to a probably hopeless task—is change which crops are planted on newly disturbed land. Just as much land would be plowed under, just as much carbon released. Preventing that kind of indirect land-use change is unachievable for a compliance program. There is no way to design anything enforceable that would accomplish that. In the United States, it is impossible, and in rest of the world, it is impossible squared. Doubling down on the 2007 cutoff would only be pretending to do something.

The second is to underscore that we cannot expect to solve the climate challenge by piling new demands for biofuel production on top of the existing rickety structure. We are not getting the job done now. Ratcheting up the demand and raising the economic stakes, as many climate proposals suggest we do, is likely to do far more harm than good under RFS as it is designed today.[59] No one wants this to be true. We all hoped we had found a climate solution that was also a political winner. There is, alas, no free lunch.

The third is to show how stunningly difficult this is. Smart, well-intentioned people tried their best to find a way through the thicket. Yet they failed. Some problems are so daunting that they demand a little humility. In that spirit, what might we do to make low-carbon fuels a reality, so that transportation that can't be electrified can still significantly cut carbon emissions?

Reset the level of carbon benefit required to qualify as a biofuel

Given the levels of ambiguity that are inherent in determining the impact of low-carbon fuel standards on land-use emissions and the limits of our ability to quantify those impacts, encouraging biofuels that only seek to achieve a 20 percent improvement over fossil fuels is not sufficient. The band of uncertainty is so wide that what regulators think is a 20 percent benefit could in reality be causing harm.[60] Many scientists think we are in negative territory already, pursuing a policy that is making the climate worse rather than better.[61] If we set our sights

[59] See, e.g., House Select Committee, "Solving the Climate Crisis" (see n.2) 102, 131, 137 (recommends growth for low-carbon fuels "with guardrails to prevent conversion of any non-agricultural lands into cropland").

[60] See National Research Council, "Renewable Fuel Standard," 192 (noting that the range of estimates for greenhouse gas emissions from indirect land-use change is wide so the precise value is highly uncertain), and 202 ("Food based biofuels such as corn-grain ethanol have not been conclusively shown to reduce GHG emissions and might actually increase them.") See also Lewis, "Biofuels Part 2," 10–11 (noting that we should discount the biofuel options for which benefits are highly uncertain).

[61] See, e.g., GAO, "Renewable Fuel Standard," 19 (noting that half the experts interviewed for their report thought RFS had a net negative impact on greenhouse gas emissions, and half said it was net positive), and at 22 (scientists who think that corn ethanol does not meet the 20% emission reduction standard almost all pointed to land-use change as the reason). Lark, "Environmental Outcomes of RFS," 3 (due to land-use change, corn ethanol worse for climate than fossil fuel gasoline). See also Hudson and Outka, "Bioenergy Feedstocks," 656–57; National Research Council, "Renewable Fuel

higher—requiring all biofuels to achieve at least a 50 percent reduction in carbon emissions—we give ourselves a fighting chance of knowing that we are helping rather than hurting. The only way not to risk going deeply into the red on our carbon accounts is to bet on renewable fuels that are much further away from the break-even line.

Redo the cost-benefit and lifecycle analyses

We have learned a lot since 2010. About land-use change, direct and indirect, and about many other facets of biofuels. Let's apply that knowledge to the law. The 2010 EPA analyses, despite best efforts, don't align with what we know now. EPA can revisit the lifecycle analyses so that, despite the many flaws and unbridgeable gaps, EPA can bring the current findings up to date with science since 2010. Congress should consider requiring an independent body—like the National Academies—to verify the analyses, to ensure that the outcome has the external credibility to withstand the inevitable political firestorm. And they need to be regularly updated, so that new science can be incorporated. Keeping current with the science supports investment in the lowest carbon strategies, because close-to-the-line approaches won't be economically attractive. And it encourages new rigorous science about the impacts of biofuels because scientists know their work will be used.

Focus on the truly carbon-improving biofuels

The bright spot in the gloom is that some researchers think there may be truly renewable biofuels. Sustainable grasses and waste biomass, for example, when grown on marginal lands, in the right way, might be close to the holy grail of significantly reduced carbon fuels.[62] But—remember the no free lunch problem?—they cost a lot of money.[63] Way more than fossil fuels or the vast majority of

Standard," 199 (noting that corn ethanol might not have lower greenhouse gas lifecycle values than petroleum and that there are plausible scenarios in which the greenhouse gas emissions from corn ethanol are much higher than those of petroleum-based fuels); and 201 (finding that EPA's own analysis suggests that RFS2 might not achieve the intended greenhouse gas reductions). See also Gal Hochman and David Zilberman, "Corn Ethanol and U.S. Biofuel Policy 10 Years Later: A Quantitative Assessment," *American Journal of Agricultural Economics*, Vol. 100 (February 13, 2018): at 570, 582 (finding a minuscule 0.23% benefit for ethanol compared to gasoline based on a meta-analysis of studies quantifying GHG impacts of ethanol).

[62] See EPA, "Second Triennial Report," 108; National Research Council, "Renewable Fuel Standard," 202; Searchinger, "Use of U.S. Croplands for Biofuels," 1240.

[63] See GAO, "Renewable Fuel Standard," 10.

plant-based fuels used today. We haven't cracked that nut yet, perhaps because so far there has not been sufficient sustained incentive to do so. The economics of these potentially climate-beneficial fuels are tough, and so are the politics, but we haven't committed sufficient economic or policy resources to know if the problems are solvable.[64]

Consider shifting from a volume standard to a carbon-intensity standard

California has adopted a different strategy for reducing carbon in fuels, called a low-carbon fuel standard. Instead of the approach used in the RFS—defining renewable fuels based on how their emissions compare to fossil fuels (20 percent, 50 percent, 60 percent emissions reduction from comparable fossil fuels) and mandating volumes—a low-carbon fuel standard sets standard for carbon intensity in fuels, with the standard getting increasingly stringent over time. There are many theoretical benefits to this approach, including that it continues to incentivize improvements beyond the static thresholds in RFS.[65] However, the foundational problem of defining the carbon impact of the fuel remains; both RFS and the California low-carbon fuel standard require a lifecycle analysis to figure out how much carbon a fuel is responsible for emitting. If a low-carbon fuel standard causes a lot of land-use change, it isn't helping. That's why restricting eligibility to fuels with less potential for climate damage and rigorous science-driven lifecycle analysis will continue to be essential no matter which regulatory model is used.[66]

Eliminate exemptions

The 20-percent-less-than-fossil-fuel standard is already far too low; we can't be waiving the obligation to meet even that modest expectation for improvement. The RFS law exempted the existing production facilities from the 20 percent

[64] See Lewis, "Biofuels Part 2," 3, 8 (noting that incentives for advanced biofuels have to be durable and certain to incentivize investment). The other benefit of lower carbon biofuels is that they generally also have fewer of the other damaging environmental impacts. See EPA, "Second Triennial Report," 108.

[65] Hudson and Outka, "Bioenergy Feedstocks," 665–66; James M. Van Nostrand, "Production and Delivery of Biofuels," in Gerrard, *Legal Pathways* (see n.1), 692, 698–99; Lewis, "Biofuels Part 2," 9–10.

[66] See Van Nostrand, "Production and Delivery of Biofuels," 700–01 (describing lifecycle analysis under California's Low Carbon Fuel Standard). The dangers of establishing an accurate carbon intensity for biofuels, given huge but challenging-to-measure induced land-use change, are a warning flag to states that are now moving toward what they hope will be clean fuel standards as part of their push toward lower carbon transportation. See, e.g., the New Mexico Clean Fuel Standard Act, S.B. 11, 55th Leg., Reg. Sess. (NM 2021).

standard.[67] In 2017, 89 percent of the RFS blending volume was exempt from the obligation to achieve a 20 percent improvement in greenhouse gas emissions.[68] As bad as this is in the United States, it is even worse internationally. For example, EPA refused to approve palm oil as a renewable fuel feedstock because it doesn't clear the 20 percent climate improvement hurdle, but because of the exemption, palm oil–based biofuels continue to be imported to meet RFS compliance.[69] The exemption for existing facilities not only reduced climate benefits from ethanol but also likely contributed to depressing investment in lower carbon-intensity fuels.[70] The waiver provisions have another downside too: they make already difficult compliance and enforcement strategies that much harder. Even if regulators spot what appears to be ineligible fuel, they still have to engage in a complicated and paperwork-heavy research project to determine if the fuel is nevertheless eligible because it is exempt. Enforcement of these rules is hard enough without deliberately inserting more confusion. All exemptions should be phased out.[71]

Use Next Gen ideas to ensure that we only use truly renewable fuels

After we clear away the obstacles that hobble our ability to ensure compliance with the lower carbon purpose of biofuels, through the steps outlined in this chapter , we will confront the challenge of ensuring that the approved biofuels meet the standards. The only way to assure that the eligible fuels are actually climate beneficial requires that we be sure about where the feedstock comes from and how it is grown. How can we know if what's claimed is what actually happens? Some of that is inherent in choices about qualifying feedstock; algae

[67] 42 U.S.C. § 7545(o)(2)(A)(i). See also 40 C.F.R. § 80.1403(b), (c), and (d) (biofuels from facilities that commenced construction by 2007 are exempt). See also EPA, "2010 RFS regulations," 14682; Aldy, "Promoting Environmental Quality," 196.

[68] GAO, "Renewable Fuel Standard," at 7 n.7, and at 20 n.29 (amount of biofuel exempt from the 20% standard).

[69] See EPA, "Second Triennial Report," 104, 107 (noting that Indonesia is the United States' second largest biodiesel import country of origin, where palm oil is the dominant feedstock, and stating that although EPA found that palm oil biodiesel didn't meet the 20% threshold, because of the exemption palm oil biofuels are nevertheless eligible). See also "EPA Renewable Fuel Standard Program—Standards for 2020 and Biomass-Based Diesel Volume for 2021 and Other Changes: Response to Comments," EPA-420-R-19-018, December 2019, at 71 (noting that increased demand for feedstocks in the biodiesel market likely has increased use of palm oil in other markets, evidenced by the dramatic increase in imports of palm oil to the United States since 2007); EPA, "Notice of Data Availability Concerning Renewable Fuels Produced from Palm Oil Under the RFS Program," *Federal Register*, Vol. 77 (January 27, 2012): 4300, 4313 (finding that land-use change emissions account for over half the GHG emissions associated with palm oil biofuels).

[70] Aldy, "Promoting Environmental Quality," 196.

[71] See also Hudson and Outka, "Bioenergy Feedstocks," 665.

and waste oils aren't grown on crop land, so we know without having to check that undisturbed land wasn't farmed to make them. This feature makes such biofuels much more attractive than others. But some fuels will need an origin-confirmation strategy. The fact that the same verification methods used in the United States also have to work around the globe means that the compliance plan will have to be both automatic and unusually resilient. It will take hard work to figure out if a minimally acceptable compliance verification strategy is possible for a redesigned low-carbon fuel program. It certainly would require innovation and maximum use of technology and satellite imagery, and a relentless focus on simplicity and no exemptions. But here's the catch: low-carbon fuel producers and importers would have to accept a level of accountability that is much higher than they have now. It can't be a make-the-government-chase-it-down, labor-intensive way of doing business. If companies want to be in this industry, they will have to accept that real-time verification will be required because tracking it down after the fact will not prevent the harm and is in any event impossible. Even with these improvements, there is likely to be a lot of fraud, as we have already seen in this sector, as discussed further later in this chapter. Fraud reduces the emissions benefits of the law and undermines the market. Our best bet to keep it within reasonable bounds is creative and rigorous strategies that effectively shift the burden to industry to show that they are compliant before they can register or sell the fuel, making them partners in finding a workable solution.

Save biofuels for the sectors with no other option

Given the huge inherent uncertainty about the climate benefits of biofuels, it makes no sense to use them in sectors that have a known and verifiable low/zero-carbon alternative. Like passenger vehicles. Electrification is a much more attractive approach for this largest portion of transportation emissions. Biofuels, if the approach is changed as noted here, might help reduce carbon emissions for transportation sectors that can't be electrified, like aviation.[72] Because biofuels have the potential of causing more harm than good, for climate as well as many other issues, we should reserve them for the climate challenges where we really have no other choice.

[72] House Select Committee, "Solving the Climate Crisis," 102; Hudson and Outka, "Bioenergy Feedstocks," 650, 654; Van Nostrand, "Production and Delivery of Biofuels," 694, 710; Jonathan Lewis, "Biofuels, Part 1," *Harvard CleanLaw podcast*, August 16, 2019, transcript at 9, http://eelp.law.harvard.edu/wp-content/uploads/Joe-and-Jon-Biofuels-1-transcript.pdf; Lewis, "Biofuels Part 2," 4–5. See also "The Long-Term Strategy of the United States: Pathways to Net-Zero Greenhouse Gas Emissions by 2050," US Department of State and US Executive Office of the President, November 2021, at 18, https://www.whitehouse.gov/wp-content/uploads/2021/10/US-Long-Term-Strategy.pdf.

Renewable Fuel Standard Fraud

Sometimes companies deliberately cheat. It isn't that they just don't try hard enough or fail to make compliance a priority. They intentionally and deliberately violate. They cross the line into clear criminal territory. The RFS unfortunately provides a powerful example.

Why include criminals in a Next Gen story? Aren't they just bad guys who deserve prosecution? They are, but focusing on defendants' moral culpability is looking for a solution in the wrong place. We know companies cheat. It happens all the time. Lift your head up and look around. If regulators design a program where significant money can be made and there is virtually no check on fraud, criminals will run rampant. It isn't a question of maybe or it's possible, it's definite: fraud will happen. Let's not pretend we don't know that when we design rules.

Market programs have the biggest fraud risk. That's because markets are trading something of value, that is, there is money to be made, and often the thing being traded is separated from the thing regulators really care about. That's what happens in the RFS. So not only is RFS fraud something to focus on for any future low-carbon fuels program, it is also illustrative of the challenge of using markets for environmental problems.

So far in this discussion of RFS the focus has been on the renewable fuels themselves, and whether those fuels—assuming they are made exactly as claimed—do or don't achieve the climate benefits ascribed to them. But what if the whole thing is a scam and there is actually no fuel made at all? That's the fraud issue.

To understand the story of fraud in biofuels, you need the short version of how we get from wet biofuels to the numbers on a computer that are actually traded, which is where the fraud occurs. RFS is set up as a requirement to blend biofuels into transportation fuels. Compliance isn't achieved by making biofuels; RFS only cares about substituting biofuels for traditional oil-based fuels, because that's where the greenhouse gas emissions savings are supposed to occur. There are a lot of players in the system for producing transportation biofuels: farmers who grow the crops, companies that produce biofuels, blenders who combine biofuels with petroleum fuels, and companies that sell the finished product at a gas station near you, along with many brokers and dealers up and down the chain. The choice in RFS was to place the legal obligation on the relatively small number of oil companies that sell finished transportation fuels—a smart Next Gen choice—and create a market for buying and selling biofuel credits.[73]

[73] For a description of the biofuels program and an explanation of the market that was created, see "Overview for Renewable Fuel Standard, Compliance Program Basics," EPA, https://www.epa.gov/renewable-fuel-standard-program/overview-renewable-fuel-standard; "The Renewable Fuel Standard (RFS): An Overview," Congressional Research Service, R43325, April 14, 2020, at 4, https://crsreports.congress.gov/product/pdf/R/R43325.

Each gallon of biofuels that is blended into transportation fuel creates credits, called Renewable Identification Numbers (RINs). At the end of the year the large oil companies have to buy and retire the number of RINs that the RFS requires. The theory is each RIN reflects an actual gallon of biofuel blended into transportation fuels, so if you have the right number of RINs in total, the fuel supply contains the expected quantity of biofuels.

At the initial stages, a gallon of wet biofuel is "attached" to its RIN. They travel together. But once that gallon of biofuel is blended into petroleum-based fuel, through the magic of the market, the gallon and the RIN are separated. The RIN has value, because the oil companies must have a defined number of them every year, but the wet gallon of biofuel is somewhere else. Maybe. In this system, RINs have all the value, biofuels have very little. The criminally minded among you will immediately see the problem: life would be much easier, not to mention more lucrative, if a company could just produce RINs for sale and avoid the expense and mess of actually making fuel. And that's what happened.

What kind of fraud went on? Here's a sampling:

Skip making the biofuel, and just print money. Some fraudsters didn't make any biofuel at all. They just pretended to make biodiesel, claimed the RINs, then sold them. A wide variety of sham transactions helped to hide the truth. The title of an article about one of these companies says it all: "The Fake Factory that Pumped Out Real Money."[74]

Pretend your fuel is biodiesel when it isn't. Instead of going to the bother of producing eligible fuel, make a cheaper one instead, but claim it's biofuel and sell the RINs. One example: the nearly 5 million invalid biodiesel RINs sold by New Energy Fuels and Chieftain Biofuels.[75]

Reuse and recycle. Why make three batches when one will do the trick? Some companies created or purchased biodiesel, sold the RINs, and then through complicated paperwork maneuvers, claimed that the same fuel was newly made, generated new RINs, and sold that second set of RINs. Rinse and repeat. Here's one example: Gen-X sold 60 million RINs for which there was no actual fuel, or the same fuel was "reprocessed."[76]

[74] Mario Parker, Jennifer A. Dloughy, and Bryan Gruley, "The Fake Factory that Pumped Out Real Money," *Bloomberg Businessweek*, July 13, 2016, https://www.bloomberg.com/features/2016-fake-biofuel-factory/.

[75] "Notice of Violation to New Energy Fuels and Chieftain Biofuels," EPA, July 28, 2015, https://www.epa.gov/enforcement/notice-violation-new-energy-fuels-inc-and-chieftain-biofuels-llc.

[76] "Summary of Criminal Prosecutions" (search for Gen X or Richard Estes 2017), EPA, https://cfpub.epa.gov/compliance/criminal_prosecution/index.cfm. For another example of this type of RINs fraud, see EPA, "NGL Crude Logistics, LLC Agrees to Pay Civil Penalty of $25 Million and to Retire $10 Million in Renewable Fuel Production Credits Under Settlement with United States," Press Release, September 27, 2018, https://archive.epa.gov/epa/newsreleases/ngl-crude-logistics-llc-agrees-pay-civil-penalty-25-million-and-retire-10-million.html.

Export the fuel, keep the RINs. When biodiesel is exported, the producer is required to retire the RINs, because that is fuel no longer available to the US market. But you can make more money by unlawfully selling the RINs in the United States instead, a practice so common that it even has its own name: "strip and ship." That's what Chemoil did, to the tune of over 70 million RINs.[77]

These are prime examples of why this market was viewed as "rife with fraud."[78] A former head of the EPA Criminal Investigation Division explained that the publicly known fraud was only a fraction of the total and that the initial fairly crude fraud in the early years of the program has given way to much more complex fraud schemes, with signs that organized crime is becoming involved.[79] Congress held a hearing specifically on the topic of RINs fraud.[80]

All this fraud was possible because of a design decision made at the outset of the program. At the urging of the oil companies and many other players, the RFS was set up as a market program. Instead of fixing a direct blending requirement, the program allowed trading of RIN credits. The RINs market, like most markets in environmental programs, was intended to increase compliance flexibility and reduce costs.[81] How would this market ensure the integrity of the credits? EPA embraced an economist's approach to compliance as well. The rule was crystal clear that it was up to the obligated parties to make sure the RINs they used for compliance were valid. If the obligated party—generally the oil company—didn't check before buying, too bad for them. They bore the risk if it turned out the credits were no good; they would be required to replace the bad RINs and pay

[77] DOJ, "Chemoil Agrees to Pay Civil Penalty of $27 Million and to Retire a Total of More Than $71 Million in Credits from Renewable Fuels Market Under Settlement with United States," Press Release, September 29, 2016, https://www.justice.gov/opa/pr/chemoil-agrees-pay-civil-penalty-27-million-and-retire-total-more-71-million-credits. For a summary of some of the other RINs fraud cases, see Doug Parker, "White Paper Addressing Fraud in the Renewable Fuels Market and Regulatory Approaches to Reducing this Risk in the Future," *E&W Strategies*, September 4, 2016, at 7–10, https://www.earthandwatergroup.com/wp-content/uploads/2016/09/Expert-Report-Fraud-in-the-RFS-9-4-16-.pdf.

[78] "U.S. Says Glencore Unit to Pay Record $27 Million for Biofuels Compliance," *Reuters*, September 29, 2016, https://www.reuters.com/article/chemoil-usa-idAFL2N1C521I. See also Mike Newman, "The Problem of Invalidated RINs in the Renewable Fuel Standard," *Stillwater Associates*, November 14, 2018 (stating that although EPA has prosecuted millions of dollars' worth of fraudulent RINs transactions, fraud continues to be a significant risk), https://stillwaterassociates.com/the-problem-of-invalidated-rins-in-the-renewable-fuel-standard/; American Fuel and Petrochemical Manufacturers, "The RINsanity Continues," Press statement, October 7, 2016, https://www.afpm.org/newsroom/blog/rinsanity-continues.

[79] Parker, "White Paper," 8.

[80] "RIN Fraud: EPA's Efforts to Ensure Market Integrity in the Renewable Fuels Program," Hearing before the H. Comm. on Energy and Commerce, 112th Cong. (2012), https://www.govinfo.gov/content/pkg/CHRG-112hhrg81890/html/CHRG-112hhrg81890.htm.

[81] "The EPA Should Improve Monitoring of Controls in the Renewable Fuel Standard Program," EPA Office of the Inspector General, Report No. 13-P-0373, September 5, 2013, at 1–2, https://www.epa.gov/sites/production/files/2015-09/documents/20130905-13-p-0373.pdf.

a penalty to boot.[82] Appropriately, this strategy was known as "buyer beware." EPA would not be checking on the validity of the credits; that was up to the obligated parties. Proceed at your own risk. The theory was that the obligated parties would at least do some minimal checking because they faced potentially significant financial losses should the credits prove invalid, and that would keep the market honest.

Except they didn't check. As a result, the fraudulent companies looking to cash in knew that the chances of being discovered were small and so they went to town. When EPA did uncover invalid RINs, it demanded that the obligated parties replace the bad RINs and pay a penalty, just as the rules clearly stated. Then came the uproar. The obligated parties complained that they didn't know the RINs were bad. The market seized up because now everyone was nervous about buying RINs. And the small biofuel producers had trouble finding buyers, who were electing to limit their purchases to companies with deep pockets to cut financial risk.

Everyone was hollering and demanding that EPA do something. In response, EPA promulgated the Quality Assurance Plan (QAP) regulations.[83] The QAP regulations set up a voluntary private program designed to provide more assurance of RIN integrity. Biofuel producers could elect to have an independent third-party auditor monitor their facilities to ensure these facilities were producing qualifying renewable fuel and generating valid RINs. Companies who wanted some financial insurance could buy those certified RINs and reduce their liability should the RINs turn out to be bad.

Did that help? Well, it turns out the assurance of being fraud-free that QAPs were supposed to provide wasn't so reassuring: the QAP auditor Genscape certified Gen-X's fraudulent RINs, for which Gen-X's president was criminally convicted. EPA found that the RINs auditor Genscape "ignored and failed to follow through on glaring signs of RIN fraud," including trucks that visited alleged delivery locations without unloading fuel, or locations alleged to be fuel suppliers that had no equipment. Genscape ignored these obvious indicia of fraud and verified the RINs anyway.[84]

[82] 40 C.F.R. § 80.1431(b)(2): "Invalid RINs cannot be used to achieve compliance with the Renewable Volume Obligations of an obligated party or exporter of renewable fuel, regardless of the party's good faith belief that the RINs were valid at the time they were acquired." See also OIG, "EPA Should Improve," 1–2.

[83] "RFS Renewable Identification Number (RIN) Quality Assurance Program," EPA, *Federal Register*, Vol. 79 (July 18, 2014): 42078. See also "Quality Assurance Plans under the Renewable Fuel Standard Program," EPA, https://www.epa.gov/renewable-fuel-standard-program/quality-assurance-plans-under-renewable-fuel-standard-program (describing the QAP program).

[84] See "Final Determination in the Matter of Genscape, Inc., Option A Quality Assurance Plan Auditor Under the Renewable Fuel Standard Program," EPA, May 31, 2019, at 2–3 (attached as Exhibit A to Genscape's Petition for Review, filed with the Sixth Circuit Court of Appeals on July 26, 2019, https://www.epa.gov/sites/production/files/2019-07/documents/genscape_19-3705_pfr_07262019_0.pdf); Todd Neeley, "Genscape Must Replace RINs," *Progressive Famer*, August 7, 2019,

Why was there so much fraud in RFS? The Department of Justice itself inadvertently identified the heart of the problem in announcing one RINs fraud criminal conviction: "Like many government programs, the EPA's renewable fuel initiative was designed with the assumption that people would act in good faith and actually produce renewable fuel before collecting the subsidy for it."[85] As one law firm explained: "[T]he entire system floats on a sea of good faith."[86] This is the deeply flawed compliance assumption that is at the heart of so many rules with terrible compliance performance. As was previously explained in all too painful detail, the assumption that most companies will comply is unfounded. A rule that defines a compliance obligation and then just hopes for the best will not succeed.

EPA has also learned the hard way in the RINs fraud experience that it is risky to put the burden of keeping the system honest on the regulated parties. Yes, that's a great market-embracing idea in theory, but when push comes to shove, companies aren't willing to shoulder the accountability that comes with market flexibility. Companies liked the idea of buyer beware for its cost-cutting potential, until the beware part became real.[87] Then they wanted EPA to regulate.[88] A significant lesson from RFS is that an ad hoc private system for deterring fraud in a complicated market is unlikely to withstand the inevitable political pressure from companies wailing about the consequences. And it isn't likely to work.[89]

So what if companies cheat? Why do we care? At the values level, we care because a democratic society depends on the rule of law. That's what sets us apart from autocracies. Widespread flouting of the law undercuts this foundational premise of our society. But you don't have to get so lofty to see why fraud matters at a much more practical level: if lots of companies are not doing what the law

https://www.dtnpf.com/agriculture/web/ag/news/business-inputs/article/2019/08/07/biodiesel-fraud-identified-early-epa; Parker, "White Paper," 6 (noting that the QAP program has not been successful at preventing fraud).

[85] DOJ, "Owner of "Clean Green Fuel" Convicted of Scheme to Violate EPA Regulations and Sell $9 million in Fraudulent Renewable Fuel Credits," Press Release, June 25, 2012, https://www.epa.gov/sites/production/files/2015-09/documents/20120625-clean_green_fuel.pdf.

[86] Andrew S. Levine, "Criminal Liability for Abuse of Renewable Fuel Credits," Stradley Ronon, June 26, 2019, https://www.stradley.com/insights/publications/2019/06/white-collar-insider-june-2019.

[87] See "'Buyer Beware' Is EPA's Response to Fraud in the Renewable Identification Number (RIN) Market," Seyfarth Shaw LLP, October 12, 2012, https://www.seyfarth.com/news-insights/buyer-beware-is-epa-s-response-to-fraud-in-the-renewable-identification-number-rin-market.html.

[88] See "Amended Second Interim Enforcement Response Policy—Violations Arising from the Use of Invalid 2012 and 2013 Renewable Identification Numbers," EPA, February 2014, at 1, https://www.epa.gov/sites/production/files/2014-02/documents/amendedsecondierp020514_0.pdf (noting that after the RINs fraud came to light, the regulated community wanted EPA to impose regulations to assure that RINs entering commerce are valid).

[89] See "Renewable Fuel Standard: Program Unlikely to Meet Its Targets for Reducing Greenhouse Gas Emissions," Government Accountability Office, GAO-17-94, November 2016, at 16, https://www.gao.gov/assets/690/681252.pdf.

requires, we aren't getting the health and environmental benefits. In the case of biofuels, that's carbon reductions that will slow the advance of climate disruption. We can debate whether some kinds of biofuels actually accomplish that, as the preceding section of this chapter does. But this is definite: when no biofuels are produced at all, there's no climate benefit. Once we cross the threshold of a nontrivial amount of fraud, as we have in RFS, we aren't going to achieve the purpose of the law.

That is why regulators in all programs—but especially market programs—need to ask themselves before finalizing a regulation: Can someone willing to flout legal standards readily get around this? If the answer is yes, watch out. Why especially market programs? For the reasons that are evident in RINs fraud. Most environmental markets attempt to increase efficiency and liquidity by trading a piece of paper that is supposed to reflect a ton of pollutants or a gallon of biofuel or whatever is being regulated. If oil companies only got compliance credit when they purchased an actual gallon of biofuel, they would not be easily deceived. Try selling them an empty tanker and see how far you get. It's the separation of the thing being regulated from what's being traded that introduces a host of problems. Compliance fails when the traded thing does not reliably reflect the thing regulators actually care about. The market in the Acid Rain Program worked so well primarily because an ingeniously designed regulatory program made it a sure thing that a traded credit reflected an actual ton of emission reductions.[90] That's where the RINs market fell apart. A RIN might reflect a real gallon of biofuel, or it might not. Once the RIN went forth into the world on its own, stripped of the wet fuel, there was virtually no way to tell. Policing each and every gallon of biofuels to be certain is impossible. There is no chance that enforcement can discover and rectify every incident of fraud, or even a significant fraction. Private policing might just introduce new avenues for collusion, as the RINs example proves.

Zero fraud is usually not a reasonable goal. There isn't a way to prevent every possible violation, nor can we afford that in most cases. Some bad actors will find a way. That's one reason we have to have a strong criminal enforcement program. But fraud has to be rare. When it does happen, criminal enforcement is a way to say, "Hey we're not kidding." We punish the malefactors both because they deserve it and because it signals to the rest of the players that compliance is a good choice. But the measure of a solid compliance program is how good compliance will be even if there is little enforcement. What is the default setting? Designing a program so the default setting is strong compliance has a double benefit: it pushes toward actually achieving the goals that were the reason for the program

[90] See the discussion of the Acid Rain Program design in chapters 1 and 6.

in the first place, and it makes it easier to find the now much smaller number of serious violators and take enforcement action.

I am not suggesting that markets don't work for environmental problems. They certainly can, and have. But they create new and difficult issues that need to be addressed. Fraud is one of them. Rules get implemented here in the real world. In the real world there is fraud. Let's spend less time being shocked and more time figuring out how to make it close to impossible. If there really isn't a way to do that, maybe a market isn't the best strategy. Some market challenges may be solvable, but only before the market is launched. Once fraud permeates the market, it will be extremely difficult to root it out or even know how extensive it is. RINs fraud is both a problem that must be fixed in any redesigned low-carbon fuels program and a cautionary tale as we consider creating additional environmental markets.

9

Innovative Strategies Are the Only Way to Cut Methane from Oil and Gas

Climate change is usually thought of as a carbon problem. It results from the rapidly increasing concentration of carbon in the atmosphere, which is largely the result of the combustion of fossil fuels releasing carbon dioxide.[1] Fossil fuels contribute to climate change in another way too; they result in releases of methane, which is a far more potent greenhouse gas than carbon dioxide. Methane in the atmosphere traps 84 times as much heat as carbon dioxide over the first 20 years after it is released.[2] So even though there is less methane than carbon dioxide emitted each year, methane packs a big climate punch. That's why cutting methane is urgent for addressing climate change in the near term.[3] The largest source of anthropogenic methane in the United States is fossil fuel production and transportation.[4] For this reason, a climate strategy must include controlling methane releases from oil and gas.

Methane, the main component of natural gas, is naturally under pressure in the tight spaces underground. When a well is punched deep below ground by drilling for gas or oil, it brings that gas to the surface. There are many ways that methane can be released into the air during production, processing, and transportation. This section focuses on methane emissions during production, because that is where the bulk of the releases happen.[5]

There are three main ways methane gets into the air during production of oil and gas. The first is venting. That means the company just lets the pressurized

[1] James H. Williams et al., "Technical and Economic Feasibility of Deep Decarbonization in the United States," in Michael B. Gerrard and John C. Dernbach eds., *Legal Pathways to Deep Decarbonization in the United States* (Environmental Law Institute, 2019), 21, 22.

[2] Steven Ferrey and Romany M. Webb, "Methane and Climate Change," in Gerrard and Dernbach, *Legal Pathways to Deep Decarbonization*, 879; EPA, "Standards of Performance for New, Reconstructed, and Modified Sources and Emissions Guidelines for Existing Sources: Oil and Natural Gas Sector Climate Review," Proposed Rule, *Federal Register*, Vol. 86 (November 15, 2021): 63129 ("EPA 2021 Proposed Methane Rule").

[3] UN Environment Programme, "Global Assessment: Urgent Steps Must Be Taken to Reduce Methane Emissions This Decade," Press Release, May 6, 2021, https://www.unep.org/news-and-stories/press-release/global-assessment-urgent-steps-must-be-taken-reduce-methane ("Cutting methane is the strongest lever we have to slow climate change over the next 25 years").

[4] "EPA 2021 Proposed Methane Rule," 63129. A close second is agriculture, and third is landfills. Proposed Methane Rule, 63130.

[5] Daniel H. Cusworth et al., "Intermittency of Large Methane Emitters in the Permian Basin," *Environmental Science and Technology Letters*, Vol. 8, No. 7 (June 2021): 571, https://doi.org/10.1021/acs.estlett.1c00173.

Next Generation Compliance. Cynthia Giles, Oxford University Press. © Cynthia Giles 2022.
DOI: 10.1093/oso/9780197656747.003.0010

gas escape into the air. It isn't captured or stored; they just let it go. That puts the methane directly into the atmosphere, along with its 84-times-carbon heat-capture ability. The second is flaring. Sometimes oil and gas companies direct the escaping methane to a structure with a flame, where the methane is burned, just like on a gas stove. When it works perfectly, this converts the methane to carbon dioxide. Of course, carbon dioxide is also a greenhouse gas, so flaring exchanges the more powerful greenhouse gas of methane for another one that lasts much longer in the atmosphere. Needless to say, flaring doesn't always work perfectly, so flares often end up releasing methane along with the carbon dioxide. And sometimes the flare doesn't work at all and ends up being identical to venting.[6] The third pathway for release of methane is leaking. Even after the methane is captured, all the equipment to hold, store, and transport it can leak. Undersized tanks, leaky relief valves, hatches left open, cracks in pipes—the list of places from which leaks can happen is long. And they do leak. A lot.

There is mounting evidence that all this venting, flaring, and leaking is much worse than initial estimates guessed.[7] Conclusively quantifying the amount of methane released is challenging, but here's one thing that is definite: a lot more methane is being emitted than government reports claim.[8] Recent studies show

[6] "With Initial Data Showing Permian Flaring on the Rise Again, New Survey Finds 1 in 10 Flares Malfunctioning or Unlit, Venting Unburned Methane into the Air," Environmental Defense Fund, July 22, 2020, https://www.edf.org/media/initial-data-showing-permian-flaring-rise-again-new-survey-finds-1-10-flares-malfunctioning (new aerial survey finds that in the Permian Basin more than one in every 10 flares surveyed were either unlit—venting uncombusted methane straight to the atmosphere—or only partially burning the gas they were releasing). See also Cusworth, "Intermittency of Large Methane Emitters," 571; Jonah M. Kessel and Hiroko Tabuchi, "It's a Vast, Invisible Climate Menace. We Made It Visible," *New York Times*, December 12, 2019, https://www.nytimes.com/interactive/2019/12/12/climate/texas-methane-super-emitters.html.

[7] See, e.g., Yuzhong Zhang et al., "Quantifying Methane Emissions from the Largest Oil-Producing Basin in the United States from Space," *Science Advances*, Vol. 6, No. 17 (April 22, 2020): 4, https://advances.sciencemag.org/content/6/17/eaaz5120 ;Ramón A. Alvarez et al., "Assessment of Methane Emissions from the U.S. Oil and Gas Supply Chain," *Science*, Vol. 361 (July 2018): 186, https://doi.org/10.1126/science.aar7204 ;Jeffrey S. Rutherford et al., "Closing the Methane Gap in US Oil and Natural Gas Production Emissions Inventories," *Nature Communications*, Vol. 12 (August 2021), https://doi.org/10.1038/s41467-021-25017-4. See also "Methane Advisory Panel Technical Report," the report of the technical committee convened by New Mexico Environment Department and New Mexico Energy, Minerals and Natural Resources Department, December 19, 2019, at 38–39, https://www.env.nm.gov/new-mexico-methane-strategy/wp-content/uploads/sites/15/2019/08/MAP-Technical-Report-December-19-2019-FINAL.pdf ("NM Methane Advisory Panel Report").

[8] Yuanlei Chen et al., "Quantifying Regional Methane Emissions in the New Mexico Permian Basin with a Comprehensive Aerial Survey," *Environmenal Science and Technology*, Vol. 56, No. 7 (March 2022): 4321, https://doi.org/10.1021/acs.est.1c06458 (finding that total methane emissions from upstream and midstream oil and gas activities in the New Mexico Permian Basin are 6.5 times EPA's estimates); Zhang, "Quantifying Methane Emissions," 1, 3 (measured methane emissions in the largest producing basin are more than twice EPA estimates); Alvarez, "Assessment of Methane Emissions," 186 (annual methane emissions from the oil and gas supply chain are 60% higher than official estimates); "NM Methane Advisory Panel Report," 38–40; Nicholas Kusnetz, "Is Natural Gas Really Helping the U.S. Cut Emissions?," *Inside Climate News*, January 30, 2020, https://insideclimatenews.org/news/30012020/natural-gas-methane-carbon-emissions ;Carlos Anchondo, "Study: Methane Emissions in Central U.S. Twice EPA Estimates," *E&E News*, January 30, 2020, https://www.eenews.net/stories/1062205045. The same problem happens in countries around the world. See Chris Mooney et al., "Countries' Climate Pledges Built on Flawed Data, Post Investigation Finds," *Washington Post*, November 7, 2021, https://www.washingtonpost.com/climate-environment/interactive/2021/greenhouse-gas-emissions-pledges-data/.

that actual methane emissions from oil and gas production are at least 50 percent to 100 percent higher than EPA data suggest.[9]

The amount of methane released to the air during gas production and transportation is at the heart of the debate about whether natural gas can be an effective "transition" fuel in the fight against climate change. When natural gas is burned at a power plant, it produces less carbon dioxide than would be generated making the same amount of power with coal.[10] That's why the switch from coal to gas has reduced greenhouse gas emissions from electric power generation.[11] But if a significant chunk of powerfully climate-altering methane escapes into the air during production and transportation, the climate benefits of gas-fired power are lost. Experts say that the crossover point is about 3.2 percent; if more than that percentage of total gas production escapes due to venting and leaks, then the climate impacts of natural gas will be worse than coal.[12] Some experts believe we have already crossed that threshold; others say we are close.[13] Either way, it is incredibly bad news.

And it gets worse. Methane emissions and flaring are both increasing dramatically.[14] The United States is among the worst offenders globally.[15]

[9] Rutherford, "Closing the Methane Gap," 4715 (reviewing recent data from multiple studies). See also sources cited in note 8 *supra*.

[10] "How Much Carbon Dioxide is Produced When Different Fuels Are Burned?," US Energy Information Administration, https://www.eia.gov/tools/faqs/faq.php?id=73&t=11.

[11] Gas also produces less pollution of other types when combusted, another benefit to public health. Emanuele Massetti et al., "Environmental Quality and the U.S. Power Sector: Air Quality, Water Quality, Land Use and Environmental Justice," Oak Ridge National Laboratory, ORNL/SPR-2016/772, January 4, 2017, at 7, https://info.ornl.gov/sites/publications/files/Pub60561.pdf.

[12] Ferrey and Webb, "Methane and Climate Change," 883. Others say the crossover point is lower. See Benjamin Storrow, "Is Gas Really Better Than Coal for the Climate?," *E&E News*, May 4, 2020, https://www.eenews.net/stories/1063041299 (EDF says that 2.7% leakage negates the climate benefits of natural gas).

[13] Ferrey and Webb, "Methane and Climate Change," 883. Zhang, "Quantifying Methane Emissions," 4 (finding that 3.7% of the gas produced in the Permian Basin—the highest producing basin in the country—was being released to the atmosphere through venting and flaring); Alvarez, "Assessment of Methane Emissions," 186–88 (estimates percent leakage at 2.3%). The authors of the Alvarez 2018 study note that the 2.3% estimate was based on data from companies willing to cooperate and may be an underestimate because worse performers would have likely opted out of the study. Steven Mufson, "Methane Leaks Offset Much of the Climate Change Benefits of Natural Gas, Study Says," *Washington Post*, June 24, 2018.

[14] Robert B. Jackson et al., "Increasing Anthropogenic Methane Emissions Arise Equally from Agricultural and Fossil Fuel Sources," *Environmental Research Letters*, Vol. 15, No. 7 (July 2020): 6, https://iopscience.iop.org/article/10.1088/1748-9326/ab9ed2; "Natural Gas Annual 2020," US Energy Information Administration, September 30, 2021, at 1, https://www.eia.gov/naturalgas/annual/ (significant increase in reported oil and gas venting and flaring in 2018, 2019 and 2020 compared to prior years); New Mexico Environment Department, "Significant Emission Increases from Oil and Gas Operations Confirm Need for Stronger Rules and Enforcement, Greater Industry Compliance," News Release, December 20, 2020 (noting that leak rates in New Mexico's Permian Basin increased 250% in 12 months), https://www.env.nm.gov/wp-content/uploads/2020/12/2020-12-21-Flyovers-reveal-high-leak-rates-in-Permian.pdf ;Kristina Marusic, "Babies Born Near Natural Gas Flaring Are 50 Percent More Likely to Be Premature: Study," *Environmental Health News*, July 16, 2020, https://www.ehn.org/fracking-preterm-births--2646411428.html (United States responsible for highest number of flares of any country); Hiroko Tabuchi, "Despite Their Promises, Giant Energy Companies Burn Away Vast Amounts of Natural Gas," *New York Times*, October 16, 2019, https://www.nytimes.com/2019/10/16/climate/natural-gas-flaring-exxon-bp.html (flaring and venting both significantly increasing, despite oil and gas companies' public commitments to address climate change).

[15] Jackson, "Increasing Anthropogenic Methane Emissions," 4; Marianne Kah, "Columbia Global Energy Dialogue: Natural Gas Flaring Workshop Summary," Columbia Center on Global Energy

We know what to do to dramatically reduce methane releases and cut back flaring. The solution isn't uncertain or untested, or the cost astronomical. We could make big cuts very quickly using known, already deployed technologies.[16] Experience in the Permian Basin that straddles Texas and New Mexico makes this crystal clear. Operators limit flaring in New Mexico, where regulations are more stringent, but flaring is skyrocketing in Texas, where regulations are lax.[17] Oil and gas companies know how to cut greenhouse gas emissions using currently available technologies; they just aren't doing it.

This situation has led to the consensus that quickest and most direct way to dramatically reduce methane emissions and flaring in the oil and gas industry is through strong federal regulations. Tough new regulations were proposed in 2021, with commitments to strengthen them further.[18] Some of the country's methane problems—like emissions from agriculture, the second largest source of human-caused methane—don't have such clear-cut answers and will take a little more time.[19] Oil and gas isn't like that; as climate challenges go, this is one of the easier ones. We have to move fast on these obvious and near-term opportunities, so emissions decline quickly while we figure out what to do for the tougher problems. This section focuses on the compliance challenges for EPA rules to significantly reduce climate-forcing emissions from oil and gas production.

Policy, April 30, 2020, https://www.energypolicy.columbia.edu/research/global-energy-dialogue/columbia-global-energy-dialogue-natural-gas-flaring-workshop-summary (United States is the fourth largest source of flared gas in the world). The Trump administration made this bad situation worse by rolling back rules to cut methane and other pollutants from oil and gas production. Coral Davenport, "Trump Eliminates Major Methane Rule, Even as Leaks Are Worsening," *New York Times*, August 13, 2020.

[16] Ferrey and Webb, "Methane and Climate Change," 883–84; "NM Methane Advisory Panel Report," 59–60 (methane-reduction strategies known and cost effective); Storrow, "Is Gas Really Better Than Coal"; "Comprehensive Control Standards," Environmental Defense Fund, https://www.edf.org/nm-oil-gas/ComprehensiveControl.pdf (showing that standards to dramatically cut emissions from new and existing wells are already in place in a number of oil and gas producing jurisdictions).

[17] See Jennifer Hiller, "Crossing State Lines? Oil Firms Flare Texas Gas as Investors Vent on Climate," *Reuters*, March 12, 2020, https://www.reuters.com/article/us-climate-change-flaring-analysis/crossing-state-lines-oil-firms-flare-texas-gas-as-investors-vent-on-climate-idUSKBN20Z23C (some drillers flare at six times the rate in Texas as they do in New Mexico; regulation is the difference). Forty percent of flaring in the Permian Basin in 2025 is avoidable at no cost to operators. "Permian Basin Flaring Outlook," Rystad Energy and EDF, January, 2021, slide 38, http://blogs.edf.org/energyexchange/files/2021/01/20210120-Permian-flaring-report.pdf.

[18] EPA 2021 Proposed Methane Rule; Lisa Friedman, "Biden Administration Moves to Limit Methane, a Potent Greenhouse Gas," *New York Times*, November 2, 2021.

[19] Ferrey and Webb, "Methane and Climate Change," 890–95.

The Technical Challenges Are Comparatively Easy; the Compliance Challenges Are Not

Unfortunately, regulating methane and flaring from oil and gas production presents the classic situation in which compliance is likely to be bad. There are well over a million oil and gas wells in the United States, with new ones being drilled all the time.[20] The wells are scattered around the country and are often in remote places. (Of course, they are also sometimes right next to where people live, presenting a serious pollution threat to neighborhoods.[21]) Once a well is completed and is producing, there are usually no personnel routinely on site, so no one to keep a daily eye on failing or leaking equipment.

And, just to make things more complicated, emissions are unpredictable. Sometimes a site has small methane releases at a fairly constant rate; these still matter because even small emissions at a fraction of a million or more sites quickly add up. But a gigantic share of the methane comes from eye-popping big methane releases at a relatively small number of sites with malfunctions or other abnormal operating conditions.[22] Robust surveys have shown that methane emissions in oil and gas have what's known in statistics world as a "fat tail": a tiny portion of the sites are responsible for a huge share of the total emissions.[23] It would make life easier if we could just find the super-emitters and consider it done, but no: the really big emitters change over time. On any given day, different sites can be the super-emitters, and it has proven close to impossible to predict which ones it will be.[24]

[20] EPA 2021 Proposed Methane Rule, 63153.

[21] See Julia Rosen, "Study Links Gas Flares to Preterm Births, with Hispanic Women at High Risk," *New York Times*, July 22, 2020, https://www.nytimes.com/2020/07/22/climate/gas-flares-premature-babies.html (study finds that pregnant women who lived near frequent flaring had 50% greater chance of preterm birth); Janet Currie, Michael Greenstone, and Katherine Meckel, "Hydraulic Fracturing and Infant Health: New Evidence from Pennsylvania," *Science Advances*, Vol. 3, No. 12 (December 2017), https://advances.sciencemag.org/content/3/12/e1603021 (negative impacts on infant health when mothers live close to fracking sites).

[22] Alvarez, "Assessment of Methane Emissions," 187, 188; Cusworth, "Intermittency of Large Methane Emitters," 570, 571; Daniel Zavala-Araiza et al., "Super-emitters in Natural Gas Infrastructure are Caused by Abnormal Process Conditions," *Nature Communications* Vol. 8 (January 2017), https://doi.org/10.1038/ncomms14012 ;Adam R. Brandt, Garvin A. Heath, and Daniel Cooley, "Methane Leaks from Natural Gas Systems Follow Extreme Distributions," *Environmental Science and Technology*, Vol. 50, No. 22 (October 2016), https://doi.org/10.1021/acs.est.6b04303.

[23] Cusworth, "Intermittency of Large Methane Emitters," 570, 571; Brandt, "Methane Leaks from Natural Gas Systems," 12514–515 (largest 5% of leaks responsible for over 50% of leaked methane); Alvarez, "Assessment of Methane Emissions"; "NM Methane Advisory Panel Report," 37–39 (at any one time roughly 90% of emissions come from 10% of sites). See also Kessel and Tabuchi, "It's a Vast, Invisible Climate Menace."

[24] "NM Methane Advisory Panel Report," 39; Zavala-Araiza, "Super-emitters in Natural Gas Infrastructure," 4; Cusworth, "Intermittency of Large Methane Emitters," 568, 571; EPA 2021 Proposed Methane Rule, 63129–130.

An additional challenge for controlling methane from oil and gas production sites is that the pollution doesn't come from one discrete place, like pollution from a stack. Huge emissions can happen at many different places: an undersized tank, an open hatch, and malfunctioning equipment such as valves, pumps, or flares.

In a situation like this—a gigantic number of potential sources at which emissions are collectively huge but individually sporadic and unpredictable—what you would want is a robust monitoring system that could continuously keep an eye on things, quickly find the serious problems, and prompt companies to immediately fix them. Alas, we don't have that yet. Methane is invisible to the naked eye, so even massive releases can't be spotted without specialized monitoring equipment. The most dependable monitoring in wide use today is a person on-site with the appropriate equipment.[25] That's a definitive and reliable way to not only find leaks but also determine exactly where they are coming from so that action can be taken. Of course, a person with sophisticated monitoring equipment can't be present at each of the over 1 million locations every day. And that's just the well pads: there are also plenty of other places between well pad and consumer, including about 3 million miles of pipelines, that also have to be monitored for leaks.[26] A great deal of effort is being directed at this daunting monitoring challenge, and there are hopeful signs that improvements may be coming soon—through low-cost screening monitors at the well pad, wide-area scanning monitors, and mobile monitors on drones, vehicles, aircraft, and even satellites. But we are not there yet.[27]

The economic incentives also push against compliance. Avoiding venting, flaring, and leaking costs money. The technology to do this exists, and we know it works. The good news is that the natural gas that is captured can be sold, so many of these strategies pay for themselves or are low cost. But equipment must be purchased and installed, people have to track how well it is working, and it needs to be monitored, maintained, and operated correctly. Sometimes the lowest cost

[25] "NM Methane Advisory Panel Report," 40; David Lyon, Aileen Nowlan, and Elizabeth Paranhos, "Pathways for Alternative Compliance," Environmental Defense Fund and Environmental Council of the States Shale Gas Caucus (April 2019): 13–14, https://www.edf.org/sites/default/files/documents/EDFAlternativeComplianceReport_0.pdf; "EPA 2021 Proposed Methane Rule," 63190 (EPA proposes to find that the Best System of Emission Reduction (BSER) for most sites is quarterly onsite monitoring using Optical Gas Imaging (OGI) equipment).

[26] "Natural Gas Explained: Natural Gas Pipelines," Energy Information Administration, https://www.eia.gov/energyexplained/natural-gas/natural-gas-pipelines.php (miles of natural gas pipelines in the United States).

[27] Mike Lee, "The Key for EPA Rules? Inside the Methane Tech Revolution," *E&E News*, October 25, 2021, https://www.eenews.net/articles/the-key-for-epa-rules-inside-the-methane-tech-revolution/; Brady Dennis, "How Satellites Could Help Hold Countries to Emissions Promises Made at COP26 Summit," *Washington Post*, November 9, 2021, https://www.washingtonpost.com/climate-environment/2021/11/09/cop26-satellites-emissions/; "EPA 2021 Proposed Methane Rule," 63147 (innovative methane detection technologies discussed at a workshop EPA held in August 2021).

way to capture the gas at the wellhead—putting it into a pipeline—isn't available, so the company would have to defer production until a gathering line is installed, probably not their first choice. The challenge is that often the cheapest approach from the oil or gas company's perspective is dumping the "waste" gas or burning it off. They save money by making it the public's problem.

This regulatory challenge has been almost entirely in the hands of the states.[28] That has played out exactly as it has for many other environmental problems. Some states stepped up to the plate and made serious attempts to address it. Colorado, for example, adopted the first methane rules in the nation; California also has applicable regulations, and New Mexico has recently proposed rigorous new standards.[29] Some other states have been lax. Texas is infamous for having among the weakest rules in the nation despite being by far the largest producer.[30]

This is our current situation: over a million widely dispersed sources; emissions that are hard to observe or measure; industries that know government's chances of figuring out they are in violation are low; and many states that are unwilling to hold operators accountable. You know what happens. If we set out to create a situation in which violations would be rampant, it would look a lot like this. Even under the current less stringent standards, violations are common.[31] Methane emissions and flaring are on the rise.[32] Even industry recognizes that this is an untenable situation. So long as this irresponsible behavior continues, the largest companies understand that their social license to operate is in jeopardy.[33]

[28] For a short history of EPA's regulatory actions for methane emissions from oil and gas wells, and Congress's recent action to restore EPA's authority, see "EPA 2021 Proposed Methane Rule," 63134–137.

[29] Matt Garrington, "Colorado Adopts Stronger Rules to Protect Health and Climate from Oil and Gas Pollution," Environmental Defense Fund, December 19, 2019, https://www.edf.org/media/colorado-adopts-stronger-rules-protect-health-and-climate-oil-and-gas-pollution; "EPA 2021 Proposed Methane Rule," 63137 (brief description of some state actions on emissions from oil and gas).

[30] Hiller, "Crossing State Lines"; Lee, "The Key for EPA Rules" (the two biggest oil-producing states, Texas and North Dakota, have notoriously lax rules about oil field emissions). See "US Overview, State Total Energy Rankings," US Energy Information Administration (2018), https://www.eia.gov/state/?sid=US (Texas is largest producer).

[31] See EPA, "New Owner Clean Air Act Audit Program for Upstream Oil and Natural Gas Exploration and Production Facilities, Questions and Answers," March 29, 2018, at 1, https://www.epa.gov/sites/production/files/2018-06/documents/qaoilandnaturalgasnewownerauditprogram.pdf (EPA has seen "significant excess emissions and Clean Air Act noncompliance" at upstream oil and natural gas exploration and production facilities). See also two compliance and enforcement alerts issued by EPA about widespread violations in the oil and gas sector: "EPA Observes Air Emissions from Controlled Storage Vessels at Onshore Oil and Natural Gas Production Facilities," EPA Compliance Alert, September 2015, https://www.epa.gov/sites/production/files/2015-09/documents/oilgascompliancealert.pdf; "EPA Observes Air Emissions from Natural Gas Gathering Operations in Violation of the Clean Air Act," EPA Enforcement Alert, September 2019, https://www.epa.gov/sites/production/files/2019-09/documents/naturalgasgatheringoperationinviolationcaa-enforcementalert0919.pdf.

[32] See sources cited in note 14 *supra*.

[33] Some oil companies have acknowledged that the extensive flaring in the Permian Basin has given the oil industry a "black eye." "Parsley Energy CEO Calls Out Industry for Shale Gas Flaring,"

All of these challenges, and some states' well-established aversion to serious action to address them, have led to the obvious conclusion that federal rules are necessary. Those rules must mandate significant cuts in methane and other health-threatening pollution. We can save the more nuanced incentive programs for issues that don't have an obvious answer; for methane from oil and gas production, we know what has to happen and we know who has to do it. There are probably many creative new approaches that could make the capture of gases at the wellhead and along the transportation chain cheaper and easier. Those aren't being explored now because there is little economic incentive to do so with regulation being so lax. Tough new regulations will provide that incentive. That's what the Biden EPA has commenced doing.[34]

Given the compliance-challenged situation of oil and gas, how should federal rules be designed to be as effective as possible at curbing emissions? There are three overarching issues to consider in making these choices.

First, if ever there were a context for remembering the central ideas of Next Gen, this is it. The widespread and faulty assumptions that most companies comply, and that enforcement can take care of the rest, are obviously incorrect here. The belief that most companies comply is wrong even in the programs that have tough regulations, a limited number of sources, good monitoring, and vigilant regulatory agencies.[35] If it is incorrect under these favorable conditions, you know it's wrong when noncompliance is close to impossible to discover and the sources number over a million. Widespread violations are going to be the norm unless we take deliberate action to prevent that. It is equally obvious that a handful of government regulators can't force compliance on literally millions of sources and miles of pipelines, many of which are far from roads or human observation. That's especially true when violations are intermittent, unpredictable, and currently impossible to monitor continuously. Pretending otherwise is laughably unrealistic.

Second, noncompliance is going to get much worse. Today's regulatory environment for oil and gas is relaxed. The incentives to cut corners are comparatively modest because the expectations are already so low. Which isn't to say that companies aren't currently avoiding regulatory obligations.[36] When we

Reuters, February 5, 2020, https://www.reuters.com/article/us-parsley-egy-flaring/parsley-energy-ceo-calls-out-industry-for-shale-gas-flaring-idUSKBN1ZZ2ZK.

[34] "EPA 2021 Proposed Methane Rule." In addition to proposing tough new standards for both new and existing oil and gas wells, EPA announced in its proposal that it intends to suggest additional action in a supplemental proposal and finalize these rules by the end of 2022. "EPA 2021 Proposed Methane Rule," 63115.

[35] See the discussion of widespread noncompliance in many different programs in chapter 2.

[36] See, e.g., EPA, "EPA Settlement with Gulfport Energy to Reduce Emissions from Oil and Natural Gas Operations by 313 Tons Per Year," Press Release, January 22, 2020, https://www.epa.gov/newsreleases/epa-settlement-gulfport-energy-reduce-emissions-oil-and-natural-gas-operati

ramp up expectations through more stringent regulation, and real money is on the table, the pressure to find another way out will increase dramatically. That's how it almost always works. It's a reminder that the regulatory structure has to be designed to be resilient to far more compliance pressure than it faces today.

Third, more reliable and continuous monitoring is the lynchpin. Even with today's intermittent-at-best monitoring, we can cut emissions across the board much more than we are doing now.[37] But we cannot achieve all the necessary pollution reductions without a much better way to quickly spot violations, especially the super-emitters. There are many promising monitoring strategies currently being explored, from on-site 24/7 methane monitors, to mobile monitors that can scan larger areas more rapidly, to systems that merge satellite monitoring with big data analytics to find the largest sources.[38] These are already demonstrating their value today, but they are likely years away from at-scale deployment. And no single one of these is an all-purpose solution; it is far more likely that it will take a combination of all of these, plus on-the-ground close-up monitoring, to provide a consistent enough answer. Government should be exploring all these options and others, as quickly as possible, because a longer-term solution depends on it. Once industry appreciates government's resolve, they might be motivated to help.

Next Gen Strategies for Federal Oil and Gas Regulations

In the meantime, we need to finalize federal regulations that will be effective at dramatically cutting emissions, using today's technologies. The following list has some Next Gen ideas that might make those rules more likely to succeed.

ons-313-0 (Ohio); EPA, "The U.S. Government and Pennsylvania Settle with MarkWest for Air Emission Violations at Natural Gas Facilities," Press Release, April 24, 2018, https://www.epa.gov/enforcement/reference-news-release-us-government-and-pennsylvania-settle-markwest-air-emission; EPA, "Slawson Exploration Company, Inc., to Make System Upgrades and Undertake Projects to Reduce Air Pollution in North Dakota," Press Release, December 12, 2016, https://www.epa.gov/enforcement/reference-news-release-slawson-exploration-company-inc-make-system-upgrades-and; EPA, "Noble Energy Inc. Agrees to Make System Upgrades and Fund Projects to Reduce Air Pollution in Colorado," Press Release, April 22, 2015, https://archive.epa.gov/epa/newsreleases/noble-energy-inc-agrees-make-system-upgrades-and-fund-projects-reduce-air-pollution-0.html. See also note 31 *supra*.

[37] See "NM Methane Advisory Panel Report," 34–67 (describing evidence from multiple states that use of existing monitoring technologies significantly and cost-effectively reduces methane emissions). See also "EPA 2021 Proposed Methane Rule," 63188 (EPA estimates that emissions could be cut by 80% through quarterly monitoring surveys and required repair obligations, although EPA notes that this does not include larger super-emitter emission events).

[38] See sources cited in note 27 *supra*.

Aim for clarity and simplicity

When a regulation is clear and opportunities to obfuscate or avoid complying are few, compliance will be better. The fewer exceptions and special conditions it contains, the less likely a regulation is to give companies a chance to confuse the matter and thereby evade or delay compliance. Obligations that depend on individual discretionary judgment on a site-specific basis create loopholes that undercut compliance. Numeric, straightforward, measurable obligations are likely to produce better environmental results than more nuanced and theoretically stringent requirements that are not actually implemented. Simplicity is an underappreciated powerhouse in the regulatory toolbox. Regulators should ask themselves this question: "Are there a lot of ways to avoid complying?" If there are, there will be a lot of violations. Compliance simplicity can be completely consistent with technical complexity; it just requires regulatory design that translates complicated issues into an easy-to-understand and hard-to-avoid compliance answer.[39]

Minimize exemptions

In any regulatory context, smaller businesses will object to the costs of regulation, claiming that they are less affordable for more cash-strapped companies. That manifests for a methane rule in pressure to exempt so-called "low-production" wells. Next Gen teaches us that whenever regulators draw a line and say on this side you are regulated and on that side you aren't, it creates powerful incentives for more companies to find a way to be—or claim to be—on the un-/less-regulated side of the line.[40] Exempting lower-producing wells is also tough to justify from a pollution control perspective; low-production wells can leak just as much as higher producing ones.[41] It creates a compliance black hole

[39] Simple does not mean simplistic. The Acid Rain Program is an excellent example of a complex program that had a simple overall compliance design. See discussion of the Acid Rain Program in chapters 1 and 6. The 2021 proposed methane rule adopts a simplicity frame in proposing for some types of equipment, e.g., pneumatic controllers, that the standard is zero emissions. "EPA 2021 Proposed Methane Rule," 63202.

[40] Small quantity generators of hazardous waste, for example, face fewer regulatory requirements than do larger generators. EPA's experience is that many companies therefore claim to be small quantity generators when in reality they are not. See discussion of small quantity hazardous waste generators in chapter 2. This kind of regulatory line drawing will likely inspire widespread violations that are almost impossible to find, especially when it is based on factors for which regulators lack easily verifiable information.

[41] "EPA 2021 Proposed Methane Rule," 63187; Jean Chemnick, "Trump's Climate Dismantling Complete with Methane Rollback," *E&E News*, August 14, 2020, https://www.eenews.net/stories/1063711683 (quoting Peter Zalzal from the Environmental Defense Fund that most wells in

by motivating companies to improperly claim the exemption, while at the same time eliminating the monitoring and reporting that would allow regulators to know what is going on. Multiply that by over a million wells and you see why this kind of exemption creates both pollution and compliance trouble. EPA's 2021 proposed methane rule takes a significant step toward a better compliance outcome by varying monitoring obligations based on equipment at the site—which is a physical fact that is much easier to verify than production records—and by requiring more monitoring at sites that have equipment most likely to leak.[42]

Require frequent monitoring

The best widely available monitoring today is on-site use of photoionization equipment, or comparable technology, which allows the operator to "see" emissions of VOCs and methane that are invisible to the naked eye. Aiming those monitoring devices at individual pieces of equipment allows the operator to identify not only that there is a release at the site but where it is coming from, along with a rough idea of the rate of release. This kind of monitoring is presently the only way to achieve the level of granularity necessary to remedy the problem. You can't fix a leak if you don't know its source. If you only do such inspections every six or twelve months, you might miss a significant emissions event that occurred between visits. The more frequent these visits are, the more likely it is that serious problems will be found, and the less time that they can continue until repaired. It isn't economically feasible to have people present on-site every day—that's the kind of regular screening that satellite data or other innovative technologies might eventually make feasible—but once a quarter is currently best practice for wellhead monitoring in state rules, and it's what EPA has proposed for all but the lowest emitting sites.[43] As monitoring improves it may be possible to require instead much more frequent screening level monitoring, with an obligation to fix the emissions problems that screening identifies.

the United States are low producing, and they have emission rates that are as high, or higher, than high-producing wells); Carlos Anchondo, "Study: Low-Producing Oil Wells Cause 50% of Methane Emissions," *E&E News*, April 21, 2022, https://www.eenews.net/articles/study-low-producing-oil-wells-cause-50-of-methane-emissions/

[42] "EPA 2021 Proposed Methane Rule," 63187–188. The most disastrous types of regulatory line drawing depend on information that is in the exclusive control of the regulated or can only be confirmed through individualized investigation.

[43] "NM Methane Advisory Panel Report," 66. "EPA 2021 Proposed Methane Rule," 63190.

Embrace innovation

A giant share of methane emissions from oil and gas production are from a very small number of super-emitters. The quarterly inspection strategy of prior and recently proposed methane rules won't identify most of those, particularly the intermittent leaks, and certainly not on a schedule that aligns with the dangerous impacts such large emission events pose. That's why monitoring innovation is essential for solving the methane problem. Technological progress in recent years holds promise for frequent, even continuous, monitoring at the site level. So in addition to requiring today's state-of the-art monitoring, new rules should encourage innovative monitoring alternatives. Rules need to create a pathway for innovators to bring rigorously tested new approaches to market.[44] EPA's 2021 proposed methane rule explicitly supports the deployment of alternatives to quarterly monitoring that are the key to both lower emissions and lower costs.[45]

Crowdsource compliance monitoring

Spotting super-emitter events is extremely challenging because they are intermittent, often unpredictable, and can't be seen without specialized equipment. It's looking for a needle in a haystack when the needle is invisible. Nevertheless, finding and rapidly stopping these large emission events is essential for combatting the outsize impact of methane on climate. Crowdsourcing the search for super-emitters is one possible solution. Advances in methane monitoring from aircraft and satellites are occurring at a breakneck pace. These techniques have already been deployed by researchers who have informed our quickly expanding knowledge about methane leaks. Many more monitoring tools are on the way. Regulation can turn that accelerating monitoring know-how into actionable obligations. Once an expert from government or the private sector has identified a large emissions source, a rule can require the source to promptly fix the problem. It makes sense to have this separate strategy for super-emitters because they are not a good fit for the quarterly monitoring strategy designed for more routine problems. Super-emitters are decidedly not routine; they are usually the result of malfunctions and abnormal conditions that should not be occurring at a well-operated site.[46] EPA has announced that it is considering a plan of this type for its methane regulations.[47]

[44] That's what Colorado's alternative approved instrument monitoring methods (AIMM) strategy does. See Lyon, Nowlan, and Paranhos, "Pathways for Alternative Compliance," 9.

[45] "EPA 2021 Proposed Methane Rule," 63175–177 (section titled "Alternative Screening Using Advanced Measurement Technologies").

[46] "EPA 2021 Proposed Methane Rule," 63177.

[47] "EPA 2021 Proposed Methane Rule," 63115, 63177.

Shift the burden of proof

Part of what makes enforcement particularly fraught in this sector is the reality that the number of potential violators dwarfs regulators' capacity to investigate. The crowd sourcing idea that could be a game changer for super-emitters might also work at a smaller scale, sidestepping the impossible-to-execute necessity of having a government inspector on-site with the necessary equipment at the exact moment that unlawful emission events occur. Because emissions from a well-controlled site that is complying with the EPA regulations that will be finalized in 2022 should not be significant, detection of significant emissions can create a presumption of a violation that it is up to the company to refute.[48] It is logical that the firm operating the site should have to figure out the reason for the emissions and what they are doing wrong: they're the ones that have access to the site and to the necessary information. The simple regulatory tool of shifting the burden creates a virtuous feedback loop; outside experts who know how to look for pollution will be motivated to do it when it is more likely to lead to corrective action, and regulated firms will figure out that they are not so invisible as they may have thought, so will put more effort into complying.

Automate everything that can be

One way to avoid serious problems is to be less dependent on human intervention to spot and fix them. So make some operations automatic. Automated pilot lights, which reduce the times when there is no flame at the flare for combustion of the gas, and automated thief hatches, so they aren't accidentally left open, are two examples.[49]

Require electronic reporting

We are in the twenty-first century. All reporting should be electronic. It is faster, more reliable, and less prone to errors. A federal rule should create a consistent reporting format, with data shared between EPA and states. We have seen the uneven and unreliable data that are generated when essential information comes

[48] New Mexico is proposing to adopt a rule with this presumption shifting approach. See New Mexico Environment Department, "Oil and Gas Sector—Ozone Precursor Pollutants," Proposed Rule, May 6, 2021, § 20.2.50.127, https://www-archive.env.nm.gov/air-quality/wp-content/uploads/sites/2/2021/03/Proposed-Part-20.2.50-May-6-2021-Version.pdf.

[49] See NM proposed oil and gas ozone precursor rule, § 20.2.50.115(C) (flares), and § 20.2.50.123(B) (thief hatches).

in only through the states; the way to go is direct electronic reporting by the regulated companies to a common data system.[50] Electronic reporting makes the information more accessible for the companies, increasing the chances that they use it to improve their operations. And it allows for automated compliance checks to avoid incomplete or obviously impossible answers. More importantly, it makes the information available to regulators, and potentially also the public, in as close to real time as possible, helping them to spot problems and making companies wonder if their violations are more likely to be noticed. The more likely they think that is, the better for compliance.

Impose data-substitution requirements

Conducting on-site inspections is a hassle for companies and also costs money. Cost-cutting zeal, or just personnel problems, could result in companies not doing the required inspections. Usually, the regulatory consequences for failing to inspect or report are less severe than the consequences of admitting a violation of emissions obligations. That kind of regulatory setup can create perverse incentives, motivating companies to skip inspections when the results are expected to be unfavorable.[51] One way to bypass this tangle is to create a powerful motivation to do inspections. A rule can require the company to assume that missed inspections would have produced negative results and impose the consequences that go along with that. EPA's air office has employed these kinds of data substitution requirements to good effect in other air emissions programs.[52] Automatic data-substitution requirements can inspire greater adherence to inspection and reporting obligations.

[50] EPA's 2021 proposed rule requires electronic reporting. "EPA 2021 Proposed Methane Rule," 63185. See also EPA, "National Pollutant Discharge Elimination System (NPDES) Electronic Reporting Rule," *Federal Register*, Vol. 80 (October 22, 2015): 64063, https://www.federalregister.gov/documents/2015/10/22/2015-24954/national-pollutant-discharge-elimination-system-npdes-electronic-reporting-rule (example of requiring electronic reporting to a system shared by EPA and states, in this instance for Clean Water Act dischargers). For discussion of data quality problems with state reporting in two programs, see chapter 2, section titled "For Some Important Programs, EPA's Understanding of Noncompliance Is Wrong" (drinking water and stationary source air pollution).

[51] That appears to be happening in air and water pollution programs today. Yingfei Mu, Edward A. Rubin, and Eric Zou, "What's Missing in Environmental (Self-)Monitoring: Evidence from Strategic Shutdowns of Pollution Monitors," *National Bureau of Economic Research*, Working Paper 28735 (April 2021), https://doi.org/10.3386/w28735 (statistical evidence that local governments skip pollution monitoring when air quality is expected to be poor); Daniel Nicholas Stuart, "Strategic Non-Reporting Under the Clean Water Act," chapter in "Essays in Energy and Environmental Economics," PhD diss., Harvard University 2021, https://nrs.harvard.edu/URN-3:HUL.INSTREPOS:37368502 (nonreporting increases when water pollution discharge levels are expected to exceed permit limits).

[52] See EPA, "Plain English Guide to the Part 75 Rule," section 9.0 Missing Data Substitution Procedures at 78–84 (June, 2009), https://www.epa.gov/sites/production/files/2015-05/documents/plain_english_guide_to_the_part_75_rule.pdf.

Tap the power of public accountability

Electronic reporting makes it easier to increase transparency. Posting emissions and compliance data online creates multiple pressure points for better performance. Companies, especially publicly traded companies, are loathe to look bad in the eyes of the public, so transparency about violations motivates them to do better. Competitors might be inspired to either learn from high-performing companies or spot the ones whose public reports are not credible and let regulators know. Companies claiming to be green, a quality that investors increasingly value, will be encouraged to have results that match their assertions. Enterprising citizens might be motivated to dig into the data and compare it to other sources of information to test its validity or to see how production sites in one state compare to sites elsewhere. Neighbors certainly are keenly interested in keeping a close eye on sources that affect them. Creating the opportunity for all these interested parties to shine a light can help.

Set the stage for robust data analytics

With data in common formats received electronically, government can perform data analytics to see what comparisons across reports reveal and spot anomalies that require more investigation. Incoming electronic reports can also be compared to external sources of information, including other government reports, like royalty payments or SEC filings, for example, to identify problematic information that suggests the need for follow up. It may eventually be possible to use the data to develop predictive analytics to flag in advance the locations most likely to cause serious problems. The rules themselves shouldn't mandate specific government analytics, but likely analytic approaches need to be investigated before regulations are finalized, so the rules require the right information in the right format to make such analytics possible.[53]

Require an engineer to certify the design

Underdesign of emissions control equipment has been a ubiquitous problem in the oil and gas industry, leading to significant emissions.[54] If the tank isn't

[53] As one example, the rules could require data that make it easy for government to match satellite images with permitting data bases, to help spot unpermitted facilities that are evading the rules.

[54] See EPA Compliance Alert, "EPA Observes Air Emissions from Controlled Storage Vessels at Onshore Oil and Natural Gas Production Facilities."

big enough or the pipe is too small for the expected pressures, that gas has to go somewhere so it is released to the air. Requiring engineer-certified plans reduces the chances that the company is unaware they have a flawed design.[55] Independent engineering reviews are best—because research confirms intuition that independent experts provide more accurate reports than auditors or employees who have an incentive to agree with the regulated company.[56]

Design reports to highlight violations

Sending in reams of spreadsheets to regulators from which one could, with investigation and calculations, deduce that there was a violation doesn't have much deterrent kick. This is Regulation 101. A form that requires companies to state, for the most important data, "is this a violation, yes or no? Check here [cannot be left blank]" carries more clout than a report of numbers only. And it's more likely to get the attention of company management. There's some incentive to answer honestly because a false answer is another violation. Reporting obligations should make it hard to obfuscate and easy to spot serious problems.

Create automatic consequences

One of the reasons that enforcement lacks credibility as a principal compliance driver for oil and gas rules is the glaring mismatch between the huge number of regulated facilities, companies, and activities and the tiny number of enforcers. Using traditional approaches, companies know that government has to catch them, prove a case likely to involve highly technical evidence that is held almost entirely by the company itself, and prevail in a lengthy enforcement process in which companies can throw up procedural roadblocks. The enterprise

[55] EPA's 2021 proposed rule includes an engineer certification for some design decisions. See, e.g., "EPA 2021 Proposed Methane Rule," 63162 (engineer certification required for determination that it is technically infeasible to route emissions from a pneumatic pump to a control device or process and for design to ensure no detectable methane emissions from closed vent systems).

[56] See Esther Duflo, Michael Greenstone, and Nicholas Ryan, "Truth-Telling by Third-Party Auditors and the Response of Polluting Firms: Experimental Evidence from India," *The Quarterly Journal of Economics*, Vol. 128, No. 4 (2013): 1499–1545 (auditors selected and paid by the regulated firm are far more likely to report the plant in compliance); Jodi L. Short and Michael W. Toffel, "The Integrity of Private Third-Party Compliance Monitoring," *Administrative & Regulatory Law News*, Vol. 42, No. 1 (Fall 2016), 22–25 (describing factors that lead to third-party auditor bias in reporting), https://www.hbs.edu/faculty/Publication%20Files/ShortToffel_2016_ARLN_13fe8ba5-cb72-482b-b341-5c7632f7c164.pdf. See also Noah Kaufman and Karen L. Palmer, "Energy Efficiency Program Evaluations: Opportunities for Learning and Inputs to Incentive Mechanisms," *Energy Efficiency*, Vol. 5 (June 2011): 259 (auditor bias in energy efficiency audits).

is doomed before it begins. A rule that short-circuits this process by imposing automatic consequences stands a better chance. Fixing high, and automatic, penalties for serious self-reported violations is one idea, but there are others that could be considered, such as limits on transferring ownership after self-reported violations. As long as the consequences are undesirable from the companies' perspective, proportional to the violation, and most importantly, imposed without the need for government intervention, they could be expected to provide more motivation than the uncertain to unlikely prospect of eventual individualized enforcement.

These ideas are not mutually exclusive: they can be combined into a single rule. The goal of all these ideas is to increase the alignment between the companies' incentives and the public interest in reduced emissions. There will always be a certain level of cat and mouse between regulators and the regulated. The idea of Next Gen is to cut that way back by building compliance drivers into the rule, so that the system is as close to self-implementing as it can be.

The Role of Enforcement in Oil and Gas Compliance

This book argues that rule design is the most important driver of compliance outcomes. A tightly designed rule can deliver strong compliance results if it makes compliance the path of least resistance. That is especially true where there are millions of geographically scattered sources and the worst violations are intermittent.

The unavailability of reliable and continuous measurement for oil and gas emissions mean that the best and most effective regulatory compliance solutions aren't currently possible. We can make significant improvements—and the preceding suggestions include ways to do that—but we cannot get all the way there through a regulation alone. It's unavoidable that this is one problem where enforcement will have to play a higher profile role. Government resources are extremely limited, so we can't have many problems where enforcement must be front and center. Oil and gas production is an exception because it is essential to the climate imperative, and the options to avoid a central role for enforcement aren't likely to be wholly effective due to the monitoring gap.

What that means for oil and gas regulation is that it has to create a bigger deterrent by making enforcement easier. When violations are hard to find, and enforcement is long, drawn out, and complicated—as happens now—enforcement loses its power to motivate. Cases cannot be brought against every violator, or even one in a thousand. Assuming regulators can even identify the violators. The threat of enforcement will only motivate better behavior if companies think it is likely they will get caught and that the consequences will be swift. That underscores how essential Next Gen provisions are in the rule. The regulation must require currently state-of-the-art monitoring to spot the worst problems, regular electronic and common format reporting that quickly flags the worst issues, public availability of data to make the most of crowdsourcing the search for violators, and automatic consequences that avoid opportunities for delay. Strategies that can be included in regulations to increase deterrence punch include shifting the burden of proof, limiting the number of compliance options available to companies so enforcement doesn't get bogged down in complicated compliance or applicability determinations, and imposing mandatory minimum penalties so companies know the consequences in advance and time spent negotiating is reduced. Rules could also do a lot to increase accountability by limiting the constant churn in ownership and operation of wells that can create ambiguity about legal responsibility and send enforcement down the rabbit hole of forensic accounting. Enforceability is important for every rule, but never more so than in situations like this, where the implementation challenges dwarf regulators' resources but a credible deterrence presence is nevertheless necessary.

The Problem of Abandoned Wells

Once a well stops producing oil or gas, the emissions problem is not over. The owner/operator may lose interest, but the well can keep belching methane and other pollutants into the air (and water) until it is carefully plugged.[57] This kind of structural mismatch—the action needed to protect the public is a low priority for the responsible company, and the government has virtually no leverage to insist—underlies some of our most vexing environmental problems. It is what created the Superfund program. And it challenges government's ability to control pollution from no longer operating oil and gas wells.

There are millions of abandoned wells in the United States.[58] Often there is no one for government to pursue to fix abandoned wells; all the players with money have walked away or gone bankrupt. The amount of methane released from abandoned wells is significant. A researcher in one state estimated that

[57] Government Accountability Office, "Bureau of Land Management Should Address Risks from Insufficient Bonds to Reclaim Wells," GAO-19-615, September 2019, at 1–2, 6; Mary Kang et al., "High Methane-Emitting Abandoned Oil and Gas Wells," *Proceedings of the National Academy of Sciences*, Vol. 113, No. 48 (November 29, 2016): 13636, https://www.pnas.org/content/pnas/113/48/13636.full.pdf.

[58] EPA estimates that there are about 3.4 million abandoned oil and gas wells in the United States, over 2 million of which are unplugged. EPA, "Report on Inventory of U.S. Greenhouse Gas Emissions and Sinks 1990–2019," EPA 430-R-21-005 (April 2021), at 3-111, https://www.epa.gov/ghgemissions/inventory-us-greenhouse-gas-emissions-and-sinks-1990-2019. EPA estimates that the number of abandoned oil wells in 2019 increased by 28% since 1990, while the number of abandoned gas wells increased by 84%. "EPA 2021 Proposed Methane Rule," 63240.

abandoned wells were the source of between 5 percent and 8 percent of the state's total annual human-caused methane emissions.[59] EPA estimates it at 263,000 metric tons of methane nationwide in 2019.[60] Since methane has 84 times the global warming power of carbon dioxide over the next 20 years, that's a lot of climate change.[61] There is a robust discussion about how to fund the plugging of all these abandoned wells through taxing systems and government programs; that isn't addressed here.[62] The compliance issue is how to avoid creating new ones.

State and federal regulations require oil and gas companies to properly close their wells and to set aside the funds necessary to accomplish that. The challenge is that the obligation to plug the well may not arise until decades after the well is originally permitted.[63] By that time, the well may be years past its productive period, and parties viable enough to perform the shutdown may have disappeared.[64] So yes, there is a regulatory obligation to conduct proper plugging and other remediation, but long experience with the oil and gas industry shows that many companies don't do that. That's how we end up with what are called orphan or abandoned wells.

Assuring widespread compliance is challenging in any circumstance, as the data proving extensive violations for all kinds of environmental programs shows,[65] but particularly so when the problem government is seeking to prevent does not go away when the company closes down. Government's leverage in many environmental protection programs comes from the company's desire to continue to operate. If the environmental problem continues after the company shuts down, government's leverage is gone. The company saves money by walking away, and the public is left holding the bag. Under those circumstances Next Gen predicts that violations will be common.

[59] Kang, "High Methane-Emitting Abandoned Oil and Gas Wells," 13640.

[60] EPA, "GHG Emissions and Sinks 1990–2019," 3-111. This estimate is acknowledged to be highly uncertain. EPA, "GHG Emissions," 3-313 to 3-314. It could be substantially more. See Nichola Groom, "Millions of Abandoned Oil Wells Are Leaking Methane, a Climate Menace," *Reuters*, June 16, 2020, https://www.reuters.com/article/us-usa-drilling-abandoned-specialreport/special-report-millions-of-abandoned-oil-wells-are-leaking-methane-a-climate-menace-idUSKBN23N1NL.

[61] See "Understanding Global Warming Potentials," EPA, https://www.epa.gov/ghgemissions/understanding-global-warming-potentials (describing current knowledge on the global warming potential of methane, including a range of 84–87 for the 20-year methane multiplier).

[62] The recently enacted infrastructure law provides $4.6 billion in funding to close abandoned oil and gas wells. See Mike Lee, "'Remember Solyndra.' Will Feds' Oil Cleanup Plan Work?," *E&E News*, November 23, 2021, https://www.eenews.net/articles/remember-solyndra-will-feds-oil-cleanup-plan-work/. It's a good down payment, but nowhere near enough to close all of the existing unplugged abandoned wells. Lee, "Remember Solyndra."

[63] GAO, "BLM Should Address Risks from Insufficient Bonds," 6.

[64] GAO, "BLM Should Address Risks from Insufficient Bonds," 10. See also Hiroko Tabuchi, "Fracking Firms Fail, Rewarding Executives and Raising Climate Fears," *New York Times*, July 12, 2020, https://www.nytimes.com/2020/07/12/climate/oil-fracking-bankruptcy-methane-executive-pay.html.

[65] See the evidence and examples from many regulatory programs in chapter 2.

Regulators know that the point of maximum ability to prevent abandoned wells is at the time the well is originally permitted. The company wants something of value (the permit) so is motivated to address the long-term problem if that's a condition of obtaining a permit. The obvious solution, and one that most states and the federal government have adopted, is to require bonds or similar assurance before a well can be permitted. Once the well is properly plugged, the bond or other assurance is released back to the company. The theory is that the company will be motivated to remediate the well and thereby recoup its financial instrument, but if it abandons the well, regulators can still protect the public by using the financial assurance to pay for the well to be properly closed.

It is a great idea, but it isn't working. The reason is that the value of most required bonds is nowhere near sufficient to cover the cost of plugging the well. For oil and gas leases on federal lands, 84 percent of the bonds, covering over 99 percent of the wells, would not cover closure costs even at the low end of the possible range; less than 1 percent of bonds would be sufficient if closure costs turned out to be higher.[66] States are facing large shortfalls to address the wells already abandoned, with many more wells at serious risk of becoming orphaned.[67] Even this dire scenario might be understating the risk.[68]

This system sets up the wrong incentives. The company isn't motivated to close the well properly to get its bond back, because the cost of the bond is

[66] GAO, "BLM Should Address Risks from Insufficient Bonds," 15. GAO found that over 99% of wells on federal lands were covered by bonds worth less than $20,000 per well. GAO, 15. GAO notes that the regulatory minimum bond—which is still what BLM uses today—has not been adjusted for inflation since the 1960s. GAO, 16. See also "Report on the Federal Oil and Gas Leasing Program," Department of Interior, November 2021, at 9, https://www.doi.gov/pressreleases/interior-department-report-finds-significant-shortcomings-oil-and-gas-leasing-programs.

[67] See, e.g., Cathy Bussewitz and Martha Irvine, "Forgotten Oil and Gas Wells Linger, Leaking Toxic Chemicals," *ABC News*, July 31, 2021, https://abcnews.go.com/US/wireStory/forgotten-oil-gas-wells-linger-leaking-toxic-chemicals-79188255#; "Reclaiming Orphaned Oil and Gas Wells: Creating Jobs and Protecting the Environment by Cleaning Up and Plugging Wells," Virtual Forum Before the Subcommittee on Energy and Mineral Resources, US House Committee on Natural Resources, 116th Cong (2020) (statement of Adrienne Sandoval, Oil Conservation Division Director, New Mexico Energy, Minerals & Natural Resources Department), 3; Ellie Potter, "State Officials Call for Federal Funds to Plug Orphaned Wells Amid Pandemic," *S&P Global Market Intelligence*, June 1, 2020, https://www.spglobal.com/marketintelligence/en/news-insights/latest-news-headlines/state-officials-call-for-federal-funds-to-plug-orphaned-wells-amid-pandemic-58874155.

[68] The average plugging cost per well is over $70,000 in some well-intensive states like North Dakota and Pennsylvania. "Idle and Orphan Oil and Gas Wells: State and Provincial Regulatory Strategies," Interstate Oil & Gas Compact Commission (2019), 25, https://iogcc.ok.gov/sites/g/files/gmc836/f/2020_03_04_updated_idle_and_orphan_oil_and_gas_wells_report_0.pdf. Carbon Tracker thinks that the cost for plugging more recent vintage wells is substantially more; they estimate that the cost of plugging a typical 10,000-foot shale well is about $300,000. Robert Schuwerk and Greg Rogers, "It's Closing Time: The Huge Bill to Abandon Oilfields Comes Early," *Carbon Tracker*, June 18, 2020, https://carbontracker.org/reports/its-closing-time/. And even supposedly plugged wells can continue to leak methane. Kang, "High Methane-Emitting Abandoned Oil and Gas Wells," 13639.

substantially below the actual cost of plugging the well. The company's best financial move is giving up the bond and walking away, because that is far cheaper than paying to close the well properly.[69] Some companies with a large number of wells know that government might find a way to come after them if they fail to plug, and they may continue to want new permits to drill from those regulators, so conclude that it makes financial sense to do the right thing. But for many companies, it doesn't.

This problem is likely to get a lot worse. Newer wells are much deeper and use horizontal drilling, making them more complicated and expensive to close.[70] Many experts are predicting that we will see a surge in abandoned wells as a result of the current difficult financial circumstances in the oil and gas industry, which will only get worse as the country shifts toward renewable energy.[71]

This is not rocket science. Oil and gas is a boom-and-bust business.[72] It has ever been thus. We see the evidence of that in front of us at the present minute. Blithely counting on compliance, despite the extensive evidence to the contrary, has gotten us to the plight we are in today. Enough already. It is time to require financial assurance that covers the actual costs of properly shutting down the well.[73] In God we trust, all others pay cash.

[69] GAO, "BLM Should Address Risks from Insufficient Bonds," 14–15. In many states it is too easy for operators to insulate themselves from liability by selling aging wells before they are officially abandoned. See IOGCC, "Idle and Orphan Oil and Gas Wells," 21; Leanna First-Arai, "Will Taxpayers Bear the Cost of Cleaning Up America's Abandoned Oil Wells?," *Guardian*, September 21, 2021, https://www.theguardian.com/environment/2021/sep/21/infrastructure-bill-taxpayers-oil-cleanup-costs. There is often no fail-safe; usually it is up to overworked and underresourced government staff to find the companies likely to bail and try to take action to increase bonds before it is too late. That's a structure that experience shows doesn't work. See GAO, "BLM Should Address Risks from Insufficient Bonds," 3.

[70] GAO, "BLM Should Address Risks from Insufficient Bonds," 17.

[71] Tabuchi, "Fracking Firms Fail"; Matt Egan, "A Wave of Oil Bankruptcies Is on the Way," *CNN Business*, April 2, 2020, https://www.cnn.com/2020/04/02/business/oil-crash-bankruptcies-whiting/index.html; GAO, "BLM Should Address Risks from Insufficient Bonds," 17–18; Carbon Tracker, "It's Closing Time" (explaining the incentives that drive companies to delay the day of reckoning and to sell older wells to weaker companies); Heather Richards, "'Game Changer'? Deal on Orphaned Wells Sparks Debate," *E&E News*, August 9, 2021, https://www.eenews.net/articles/game-changer-deal-on-orphaned-wells-sparks-debate/.

[72] GAO, "BLM Should Address Risks from Insufficient Bonds," 1.

[73] There are encouraging signs that government may start to address the insufficient bonding problem. See "EPA 2021 Proposed Methane Rule," 63240–242 (EPA considering regulatory standards that require demonstration of financial capability to close the well); DOI, "Report on the Federal Oil and Gas Leasing Program" (BLM should increase minimum bond amounts for well closure); Infrastructure Investment and Jobs Act of 2021, Public Law 117-58, November 15, 2021, § 40601 (includes financial incentives for states to improve orphan well programs, including through financial assurance reform).

Conclusion to the Climate Chapters

This and the prior two chapters focus on three top priority climate regulatory topics for EPA—electric generation, transportation, and oil and gas production—and examine the most prominent compliance issues each presents. All of these regulatory areas have many more challenging implementation design problems than are discussed here. The purpose of this analysis isn't to provide an exhaustive list of either the compliance difficulties or the solutions for these sectors. It's to show how complicated and important those issues are. And to explain why implementation and compliance cannot be an afterthought, appended to the rule after the design is completed.

Implementation is the foundation. If a beautiful-on-paper rule doesn't cut it in the real world, we have failed. Next Gen design must be front of mind when the rule is being crafted and seen as a central obligation throughout. When Next Gen analysis shows regulators that the approach they had in mind when they started out has little to no chance of success once it meets the rough and tumble of gritty reality, rule writers have to be willing to reconsider. And they have to fight just as hard for the necessary-for-effective-implementation provisions as they do for the standards the rule sets. There will be strong external resistance from companies that recognize that Next Gen approaches mean they might actually have to comply. There are also many internal government barriers to building compliance into regulations. Sticking by the implementation-necessary provisions and rejecting popular ideas that will not get us there will require tenacity and commitment.

We are out of time to address climate change. The rules EPA develops now have to work. The famous sage Yoda's advice should be our touchstone for climate rule effectiveness: "There is no try. There is only do or not do."

10
Updating Federalism

Most federal environmental laws are set up in a similar way. EPA translates the mandate from Congress into standards that apply nationally. States can elect, and virtually all do, to implement the federal standards by passing comparable state laws and regulations and getting authorized by EPA. State requirements cannot be less rigorous than federal law—the federal standard is the "floor" that states cannot go below—but states may adopt more stringent standards if they wish. The federal government is supposed to ensure that everyone everywhere has the benefit of the federal standards, while state governments are charged with making that a reality in the context of their specific circumstances. EPA retains an obligation to oversee state implementation of federal law, including enforcement, where EPA and states usually have concurrent authority to address violations.

No federal standard that applies nationwide will fit perfectly with every individual state's situation or preferences.[1] States differ in the expertise and resources dedicated to environmental protection. Every state has distinct types and sizes of industries, environmental threats, geography, and population. State philosophy about compliance and enforcement also varies. Some states see it as mission central, some are actively hostile, and there is just about every flavor in between.[2] Attitudes diverge even within the same state; a state with a strong enforcement program for air can have a lackluster or worse enforcement system for drinking water or hazardous waste. Many states don't have meaningful capacity to prosecute environmental crimes. Viewpoints can also shift dramatically with the political winds. We have seen that happen at both the state and the federal level in recent years.

Federal oversight of state implementation leads to friction between state and federal regulators.[3] Feds try to ensure the same protection everywhere,

[1] In addition to the relationship with states, EPA also has responsibilities to address environmental issues in Indian Country, dealing with Tribes as sovereign nations. That is an important and complex topic of its own, but this chapter is limited to cooperative federalism and the role of EPA and states.

[2] See, e.g., Tim McLaughlin, "Three Exxon Refineries Top the List of U.S. Polluters," *Reuters*, May 28, 2021 (sources noting the "surprising amount of unevenness among states in enforcing pollution limits").

[3] EPA's Inspector General continues to list oversight of state environmental programs as one of EPA's top management challenges. EPA OIG, "EPA's FYs 2020–2021 Top Management Challenges," EPA OIG Report #20-N-0231 (July 21, 2020) (noting that oversight of delegated environmental

Next Generation Compliance. Cynthia Giles, Oxford University Press. © Cynthia Giles 2022.
DOI: 10.1093/oso/9780197656747.003.0011

while states assert their prerogative to make their own decisions. Those two perspectives are often not an easy fit. The natural tension between the state and federal viewpoints is often at its height in interactions about enforcement.[4]

Next Gen argues for channeling that tension in a more productive direction. The perspective that drives Next Gen, and the modern technologies that fuel Next Gen ideas, can also change our political dynamic. I hasten to add that no approach will eliminate conflict. That will always be with us. Congress intended that both the federal and the state perspective play a role, knowing it would lead to disagreement and disputes. Over time this creative conflict has made the program better. If regulators ever get comfortable, they are doing something wrong. The trick is to use the tension to drive progress rather than fighting to a standstill.

Although there is a lot that could be said about why the cooperative federalism strategy adopted in the 1970s isn't working for us today, I want to focus here just on three main ideas. (1) We cannot have a system where states control information on sources, pollution, and compliance. That has never worked, and it isn't going to start working now. Modern technologies provide a solution. (2) We need to take a lot more advantage of states as laboratories. States have long agitated for more ability to try new approaches and they are right. Breaking up the information logjam can open the door to innovation. (3) Next Gen ideas in federal rules need to be written with states' central role in mind. Strategies to ensure effective execution should be included as an essential part of the federal floor. At the same time, Next Gen ideas offer hope of shared solutions across federalism's ideological divide.

The 1970s Federalism Model for Controlling Information Is Holding Us Back Now

There may be national standards, but there isn't national information about how we are doing meeting them. In most programs information about source compliance is reported to or generated by states, and states decide how much they share with EPA. Yes, the regulations require states to tell EPA about violations, but there is a mountain of evidence that many states don't do that.

programs is central to EPA's core functions and that OIG continues to uncover problems with EPA's oversight of state environmental programs and has multiple investigations on that topic underway).

[4] Environmental Law Institute (ELI), "The Macbeth Report: Cooperative Federalism in the Modern Era," *ELI* (2018), 27; GAO, "EPA-State Enforcement Partnership Has Improved, but EPA's Oversight Needs Further Enhancement," GAO-07-883 (July 2007).

States don't tell EPA about significant air pollution violators

The EPA Inspector General found that while one industrialized state claimed that only a fraction of 1 percent of its major air pollution sources were in serious violation, the actual percentage was at least 25 percent.[5] The EPA Office of Inspector General (OIG) then expanded its review to five more states and found the same huge gap between what was in the states' files and what states were telling EPA. States with well over 1,000 major air sources were reporting fewer than four sources with significant violations. Some industrialized states even claimed zero serious violators. That wasn't even superficially credible. Today the picture isn't much better: many states claim to have serious violator rates below 1 percent, and the national average is about 3 percent.[6] It is inconceivable that EPA's most complex program has by far the lowest major source serious violation rates.[7]

States don't tell EPA about drinking water violators

There have been many investigations into the accuracy of national drinking water violation data. They all find that states don't tell EPA about significant numbers of serious violations; somewhere between 20 percent and 40 percent of known health-based violations are not disclosed.[8] For lead in drinking water, the most recent in-depth look revealed that a stunning 92 percent of health-based violations are not reported to EPA.[9] States are similarly failing to notify EPA about monitoring and reporting violations by drinking water systems: 84 percent of those violations are not reported, leaving EPA in the dark about how safe the drinking water is.[10] Because the national data is based almost entirely on the information provided by states, EPA's public numbers are nowhere close to reality.[11]

[5] See the in-depth discussion of this topic, including citations for the OIG studies, in chapter 2, section titled "For Some Important Programs, EPA's Understanding of Noncompliance Is Wrong," subsection titled "Stationary Sources of Air Pollution."

[6] EPA, ECHO (select topic: Analyze Trends: State Air Dashboard, select Classification: major, Box 4 (High Priority Violations, select % Facilities (Majors) with HPVs) (data 2015 through 2021).

[7] See chapter 2, section titled "For Some Important Programs, EPA's Understanding of Noncompliance Is Wrong," subsection titled "Stationary Sources of Air Pollution."

[8] See chapter 2, text accompanying notes 76 to 82.

[9] EPA, "2006 Drinking Water, Data Reliability Analysis and Action Plan for State Reported Public Water System Data in the EPA Safe Drinking Water Information System/Federal Version (SDWIS/FED)," EPA 816-R-07-010 (2008), i, 19.

[10] GAO, "Unreliable State Data Limit EPA's Ability to Target Enforcement Priorities and Communicate Water Systems' Performance," GAO-11-381 (2011), 16–17.

[11] For a more complete description of the gap in state drinking water reporting, see chapter 1, section titled "Programs with Pervasive Violations: Four Examples," subsections titled "Drinking Water: Pathogens" and "Drinking Water: Lead" and chapter 2, section titled "For Some Important Programs, EPA's Understanding of Noncompliance Is Wrong," subsection titled "Drinking Water."

States don't tell EPA about water discharge violators

When states reported only summary data about water violators to EPA, they claimed a record that was more than three times better than that of states with verified data.[12] Government Accountability Office (GAO) documents that state nonreporting on water pollution discharge violators continues to this day.[13]

There is always a risk of painting with too broad a brush when making blanket statements about how things work across many different environmental laws. Not in this case. States have made it plain that they are not going to give EPA complete or reliable information about violations. Why?

States across the political spectrum can agree on one thing: they would really like EPA to butt out. Once a source is listed in the federal data system as being in noncompliance, states can expect hassle from EPA. Some states don't want pressure to do something about noncompliance that they don't think is important or just don't have capacity to address. Other states think EPA's involvement will just slow things down.[14] Many states don't want to take the more assertive enforcement action that EPA will likely prefer, and definitely don't want EPA to start its own enforcement case. States of all political stripes see EPA's involvement as interference. Even a state with a strong commitment to enforcement often doesn't see an advantage to EPA looking over its shoulder. Plus, many states are well aware that if the actual rate of noncompliance were known, the public would demand more action. And it isn't just states that would come under fire; EPA managers know that high numbers of serious violators create blowback for them too. Added to these challenges is the fact that putting violation information into the federal database takes time and effort. State agencies are already strapped for resources; completing federal reports isn't at the top of their list.

All of these factors contribute to the bad to horrible compliance data quality evident in many programs today. And they explain why enforcement is so often a flash point in the federal and state relationship. National consistency—making sure every community has the same protections from federal law and

See also Cynthia Giles, "Comments on the Agency's Proposed Revisions to Its Lead and Copper Rule in the National Primary Drinking Water Regulations," submitted February 4, 2020, https://www.regulations.gov/comment/EPA-HQ-OW-2017-0300-1003.

[12] Water dischargers with verified data reported serious violation rates of 60%. States that provided only summary data claimed their serious noncompliance rate for the same kinds of sources was 18%. "U.S. EPA Annual Noncompliance Report (ANCR) Calendar Year 2015," EPA Office of Enforcement and Compliance Assurance (2016), 6, https://echo.epa.gov/system/files/2015_ANCR.pdf.

[13] GAO, "Clean Water Act: EPA Needs to Better Assess and Disclose Quality of Compliance and Enforcement Data," GAO-21-290, July 2021, at 22, 24.

[14] Although it is rare for states to say on the record that they don't tell EPA about violations because they don't want EPA to be involved, a few were unusually candid in their responses to the IG in a report examining state failure to disclose air violations. EPA OIG, "Consolidated Report on OECA's Oversight of Regional and State Air Enforcement Programs," September 25, 1998, at 10–12.

that businesses have a level playing field across the nation—is usually not a state priority. The time and effort such consistency demands from states is not only a diversion from higher state priorities, it can cause real disruption as companies object and state administrative agencies are forced to deal with attorneys general and other state political power centers. State regulators are sometimes secretly pleased by federal pressure that forces a solution to tough problems, but that won't stop them from vigorously complaining in the public realm.

The push for credible national data about pollution and violations isn't just the feds being demanding. Not knowing what's happening, or having the illusion that EPA knows when actually it doesn't, matters for protecting people's health and the environment on which that health depends. Here's why.

Without reliable national data, no one knows if the standards meant to protect health are working. Five percent of sources significantly violating is completely different from 70 percent violation rates. At 5 percent, it seems likely the rule is on the right track. At 70 percent, not so much. The Safe Drinking Water Act emphasizes this point by requiring EPA to prepare an annual nationwide report on violations by public water systems. But if states aren't telling EPA about 25 percent to 80 percent of the drinking water violations states know about, that national report—especially when it doesn't emphasize how huge the data gaps are—isn't really serving that function. The giant data holes mean it is impossible to say how well our safe drinking water rules are being implemented.

Bad data leads to bad policy. A mostly accurate picture of noncompliance nationwide isn't just of academic interest. Information about violations—especially serious violations—drives national policy choices. Do rules need to be amended? Does national attention need to be directed more to problem A and less on problem B? Are new compliance drivers necessary? These are all questions that national compliance data helps to answer. If EPA doesn't know what the compliance status is, EPA doesn't have eyes on the problems and isn't working on solutions. Big issues can fester unnoticed.

Serious violations may not get enforcement attention. There are many national policies requiring enforcement action for significant violations. When a history or pattern of violations poses a serious health risk, or widespread reporting gaps prevent government from knowing what's going on, government is supposed to act. States take the vast majority of the enforcement cases, but if the states don't move quickly or forcefully enough, sometimes EPA steps in. That's especially true for the largest, most complicated cases, and companies that are violating across the country or affecting people in other states. If states aren't revealing violations, the system to prioritize noncompliance for enforcement attention doesn't work.

We can't have a level playing field with uneven violation reporting. One of the inequities in the system as it operates today is that states that have tougher standards, look hard for violators, and meet the requirement to inform EPA of violations—exactly what you would hope they would be doing—can look comparatively worse. On the public scorecards, it looks like those states have more than their share of violators, even though a big contributor is other states' failure to disclose. Companies that operate in those more stringent and more transparent states understandably feel like they are getting the short end of the stick. Responsible states and companies are made to look bad; not the best way to encourage them to do the right thing.

I spend a lot of time in this book explaining why mandating that facilities meet certain standards is not by itself enough to ensure that most comply. States are no different. Requiring them to notify EPA about violators isn't making that happen. A Next Gen analysis of the compliance drivers would predict exactly that outcome. Telling EPA about violations is almost all downside for states: wasted time, more hassle, less autonomy, more public pressure, and demand for action that the state doesn't want and doesn't have the resources to do. On the other hand, there are almost no consequences for failing to report. Yes, reporting is legally required, but just about the only thing EPA can do in response to failures to report is to threaten to withhold some of the state's funding. For many states that threat doesn't carry a lot of weight. And it can feel illogical to penalize an under-resourced state by reducing its budget further. This unsatisfactory situation has persisted for decades.

Today's technologies provide a way out. Federal regulations can require that sources report directly to an electronic system, and both EPA and states can have access to the data in real time. The route-everything-through-the-states system was devised in the 1970s, when reporting was on paper. Now that electronic reporting is not only possible but easier than the on-paper method, the data can be shared with the state and EPA at the same time. That cuts out the data entry costs for the states and provides a much more accurate and comprehensive picture of pollution and compliance nationwide. It isn't a complete answer because it only captures source self-reporting and not violations found through state inspections, but it is a big start.

That's what EPA did for the water discharge program through the NPDES e-reporting rule.[15] Instead of the old system where state engineers had to transfer data from paper reports to the federal electronic database, regulated sources now have to report electronically to a data system that is shared by EPA and the states. EPA estimates that the NPDES e-reporting rule will save states $24 million a year

[15] EPA, "National Pollutant Discharge Elimination System (NPDES) Electronic Reporting Rule," *Federal Register*, Vol. 80 (November 22, 2015): 64063, 64065 ("NPDES E-reporting Rule").

when fully implemented.[16] It will provide a much more complete picture of water compliance nationwide and make the data available to the public.

An e-reporting system will only succeed if it is mandatory. Voluntary electronic reporting doesn't work—not for sources, not for states. That's one thing EPA consistently heard from states that had tried voluntary electronic reporting and is what EPA has itself experienced.[17] Regulated entities that haven't moved into the electronic age need the mandate to finally do it, and those that have been hiding behind unread paper reports won't readily move out into the light. It takes some initial investment in computer systems at both the state and federal level, and those won't happen if no one is requiring it. Plus, about half the states have a state law saying that they cannot have rules more stringent than EPA's, so if electronic reporting isn't federally mandated, in many states it can't happen.[18]

Moving to mandatory e-reporting with data shared by EPA and states sets the stage for the other advances that Next Gen can bring for compliance. Yes, it helps to fix the problem of states failing to report violations they know about. But its virtues for driving better results go well beyond that, as is explained in detail in chapter 5. States understand that too. The original vision of E-Enterprise, a joint federal/state initiative that was started by the Obama EPA, was to seize the potential of the electronic age to get things done that have been insoluble up until now. The promise was to make the entire environmental protection enterprise work better, for regulators at the state and federal level, the regulated and the public.[19] It deliberately takes a practical, work together, solution-oriented stance, sidestepping the ideological debates of decades past. Most states continue to support it.[20] That's a nonpartisan platform on which Next Gen can build.

States as Laboratories: Unlocking Innovation Potential

One of the big advantages of the federalism system is the opportunity for states to try new things. Whatever the outcome—success or failure—we can all learn

[16] EPA, "NPDES E-reporting Rule," 64065. It isn't fully implemented though. A July 2021 GAO report finds that 15 of the 17 states reviewed didn't meet the standards for accuracy and completeness for their NPDES data required by the rule. GAO, "EPA Needs to Better Assess," 24.

[17] That's why EPA proposed that a national e-reporting requirement be adopted for drinking water systems. EPA "November 2016 Drinking Water Action Plan" (November 2016), 8.

[18] See GAO, "Unreliable State Data," 34; Environmental Law Institute, "State Constraints: State-Imposed Limitations on the Authority of Agencies to Regulate Waters Beyond the Scope of the Federal Clean Water Act," *Environmental Law Institute* (May 2013) (cataloguing the 28 states that have adopted laws prohibiting or limiting state environmental rules that are "more stringent than" federal regulations).

[19] EPA, "E-Enterprise for the Environment," https://www.epa.gov/e-enterprise.

[20] "Cooperative Federalism 2.0," Environmental Council of the States (ECOS), June 2017, https://www.ecos.org/news-and-updates/cooperative-federalism-2-0/.

from their experience. States have long complained that EPA is resistant to innovation in compliance work. What's going on?

Here's how the compliance system usually works. EPA's national rules frequently don't include a robust measurement system to track what the rule most cares about. Some do require self-reporting, but often the compliance information goes only to states, or it is demonstrably undependable, or both. The lack of adequate monitoring and reporting means the program depends on inspections, and even those can be hit or miss at discovering violations. In the uncertain terrain created by the unavailability of dependable outcome measures, EPA's enforcement program has mostly imposed process-type compliance obligations on delegated states: required percentages of facilities to be inspected, mandatory reporting of violations to EPA, expected enforcement response for the most serious violations. Usually those are just for the so-called "major" facilities; the vastly greater number of smaller sources generally don't have comparable obligations or federal tracking.

EPA's oversight of state enforcement programs focuses heavily on the process metrics: whether states are doing the required number of inspections, properly identifying violations, entering the data into national data systems, and taking timely and appropriate enforcement action in response to (reported) serious violations.[21] Despite the acknowledged shortcomings of relying on these means-to-an-end obligations, they do provide some assurance, however imperfect, that many serious problems will be spotted and addressed.

States chafe at EPA's process requirements. There are regular calls for more outcome-oriented compliance measures, like compliance rates or measures of unlawful pollution, and repeated criticism about EPA's inability to say how well the enforcement programs are working. There have been some well-intentioned efforts to advance the ball in that direction over the past decades, but today's inadequate data have usually not been equal to the challenge.

States that want to try an innovative compliance idea quickly encounter the reality that close to all of their enforcement staff are completely occupied—if not already overwhelmed—fulfilling the state's existing obligations. Doing something new requires cutting back elsewhere. So states that want to try something else end up wanting to reduce the number of required inspections or the expectations for enforcement. That's often how EPA first hears about the state's new idea: a request to do less of what EPA is demanding.

In the alternative universe where we had pretty good outcome data, where it was possible to keep an eye on compliance overall or measure pollution as it

[21] For a description of the measures EPA uses for oversight reviews, developed under a joint federal and state effort to bring more consistency and predictability to the process, see "State Review Framework, Compliance and Enforcement Program Oversight," EPA, https://www.epa.gov/compliance/state-review-framework.

was happening, trying that new idea could be intriguing. The possibility of better results without new resources is something everyone can get behind. That's not usually the proposal though. Instead of inspections and enforcement, the state proposes to do something else, but the effectiveness of that something else isn't measurable. Or the suggested measurement is obviously dubious, like proposing to evaluate effectiveness by asking companies if they found the new approach useful. So here's how it usually stacks up: the idea presented to EPA is to cut back on things known to work, in favor of things that the state hopes will work but lacks the ability to reliably measure.

EPA's understandable skepticism about this proposed trade-off has been amplified by states' nearly single-minded focus on compliance assistance as the most suggested alternative. Enforcement contributes to better compliance overall through general deterrence, the idea that taking enforcement against some motivates many others to comply to avoid the downsides of sanctions. Extensive research proves that general deterrence works.[22] Proponents of compliance assistance argue that government should help companies comply, asserting that it is a lower resource, and less confrontational, method to get to the same result. Compliance assistance ideas usually suffer from the same overall lack of data that plagues most compliance programs, plus a few more: (1) the empirical data so far, although far from complete, isn't convincing that compliance assistance causes better compliance; (2) compliance assistance can't work for companies that don't participate; unlike general deterrence for enforcement, compliance assistance has no theory about how it improves compliance for the many facilities that will never interact directly with government; and (3) some states promoting compliance assistance alternatives have made it very clear that their primary goal isn't better compliance, it's doing less enforcement.

This is how we arrived at the innovation stalemate so frequently experienced today. Many state proposals have only one thing that's for sure: reduced inspections and enforcement. Whether there will be improved performance, or even any defensible data about the outcome, is iffy at best. A disproportionate number involve different flavors of compliance assistance, despite the paucity of verifiable data suggesting it is a better alternative. And EPA's skepticism meter is high for state ideas that are grounded in opposition to enforcement.[23]

Notwithstanding these problems, states have launched a number of creative and thoughtful compliance innovations. Here are just a few examples.

[22] Wayne B. Gray and Jay P. Shimshack, "The Effectiveness of Environmental Monitoring and Enforcement: A Review of the Empirical Evidence," *Review of Environmental Economics and Policy*, Vol. 5, No. 1 (2011): 3–24.

[23] This discussion is based on how it has usually worked over the decades, not including the Trump administration years, when EPA was generally willing to defer to states entirely, unless states wanted to be more stringent than federal standards.

Colorado set up a mandatory self-certification checklist for small hazardous waste generators, which improved compliance by over 50 percent, as confirmed by random follow-up inspections.[24] Maryland deployed a simple but effective strategy to prevent the too-common situation where stormwater permittees fail to submit a required stormwater management plan: it required the management plan to be part of the application—no plan, no permit.[25] Tennessee put up a remotely operated radiation detection monitor at a government disposal location for low-level radioactive waste to monitor the over 100,000 entering trucks; loads that trigger a higher than normal reading are pulled over for further investigation, preventing unlawful disposal of waste for which the site isn't permitted.[26] California—an innovation leader—is field-testing an inventive strategy to cut unlawful emissions from heavy-duty trucks.[27]

Here's where things stand today. States have proven that front-line innovation can be a powerful part of the solution to our compliance dilemmas, and we need to encourage that to happen more. At the same time, measurement of compliance end points is often poor to nonexistent, so backing away from enforcement in favor of other ideas will often not have a verifiable result. Compliance might be better; it might be worse; usually we won't know. Giving up the sure thing of enforcement in favor of the unknowable thing of an alternative—especially when state commitment to compliance, and to assuring reliable compliance information, is at best uneven—seems like heading down the path of no accountability. This innovation impasse isn't in anyone's interest, but the way out has proven elusive.

Next Gen ideas, developed to drive better compliance, might also address the innovation dilemma. The foundation of the federalism conflict between new ideas and same-old is lack of data. We don't know what's happening with pollution, we don't reliably know how it is going with compliance, so we don't know how many violations there are or how much risk they are creating. You can't have innovation to achieve better outcomes if you don't have measurable outcomes at all. Giving up the definite of enforcement in favor of the untested alternative won't feel like gambling with people's health if outcome data is reliably available.

[24] See Colorado Department of Public Health & Environment, "2014 Annual Report to the Colorado General Assembly: Status of the Hazardous Waste Control Program in Colorado," February 1, 2015, at 9, https://spl.cde.state.co.us/artemis/heserials/he171318internet/he1713182014internet.pdf.

[25] EPA, "Compendia of Next Generation Compliance Examples in Water, Air, Waste, and Cleanup Programs," NPDES Compendium, 3, https://www.epa.gov/compliance/compendia-next-generation-compliance-examples-water-air-waste-and-cleanup-programs.

[26] EPA, "Compendia of Next Generation Compliance Examples in Water, Air, Waste, and Cleanup Programs," RCRA Compendium, 14, https://www.epa.gov/compliance/compendia-next-generation-compliance-examples-water-air-waste-and-cleanup-programs.

[27] University of Chicago Energy and Environment Lab, "Reducing Heavy-Duty Truck Emissions in California," https://urbanlabs.uchicago.edu/labs/energy-environment.

That's why the strategy that makes sense for stronger compliance—required actual measurement reported electronically in real time via shared data systems—could contribute to more state innovation too. Once it is possible to have a pretty good fix on what's happening on the ground, EPA will find it much more attractive to lighten up on the process controls and embrace ideas that show promise. If a proposal has a convincing strategy, and a credible way to figure out if it is working, let's give it a run. Real-time information means that it is possible to pull the plug if it becomes obvious that it isn't working, and realistic to monitor for the individually catastrophic situations that can't wait.

At the same time, we should open our collective eyes to experimentation with a much broader range of options for driving better performance. Like the creative idea tried in India, where random assignment of third-party auditors significantly improved the accuracy of compliance data and reduced pollution.[28] Or the wide array of evidence that requiring continuous monitors by itself improves performance.[29]

As more robust and reliable information becomes available through these Next Gen approaches, compliance will be better, which will help advance environmental justice. Minority and low-income communities have historically suffered the most from poor compliance, so moving that needle will make a big difference. Ideally, many state innovations will focus on addressing these disparities. But, as a range of new outcome metrics becomes viable, we need to remain ever mindful about adopting superficially neutral measures that disadvantage already overburdened communities. Metrics like compliance rates, for example, can disguise disproportionate impacts that shift attention away from the places that need it most.[30]

States as laboratories of innovation was part of Congress's vision of federalism. We need that more than ever now, given our unparalleled challenges and episodic federal leadership. But the lack of solid compliance foundation in federal regulations has made it too difficult for states to achieve that vision. We can do a lot better—and increase public health protection at the same time—by embracing Next Gen measurement as an expected element of every program.

[28] Esther Duflo, Michael Greenstone, and Nicholas Ryan, "Truth-telling by Third-Party Auditors and the Response of Polluting Firms: Experimental Evidence from India," *The Quarterly Journal of Economics*, Vol. 128, No. 4 (2013): 1499–1545.

[29] See discussion in chapter 5, section titled "Monitoring," subsection titled "Continuous Monitoring Has the Most Compliance Power."

[30] Elinor Benami et al., "The Distributive Effects of Risk Prediction in Environmental Compliance: Algorithmic Design, Environmental Justice, and Public Policy," *FAccT'21*, Virtual Event, Canada (March 3–10, 2021).

Implementing Next Gen Ideas within the Federalism System

For most environmental programs, EPA and states are joined at the hip. There are some federal-only programs—like car standards and toxic chemical approvals—but most don't work that way. After EPA sets the national standard, the state decides whether it wants to assume delegation of the program, and if so, to stand up state rules, permits, compliance, and other implementation efforts and seek EPA's approval. The state is free to adopt more protective standards, although—at least in theory—they cannot be less stringent than EPA's rules. But, to the states' continual irritation, delegation of federal programs comes with caveats: EPA has a responsibility to oversee the states' implementation, which usually includes authority to object to permits the state issues and insist on revisions, and concurrent authority to enforce the law. Cooperative federalism is the term usually used to describe this structure.[31]

The fact that federal environmental rules are mostly implemented through the states needs to be part of Next Gen thinking. A terrific Next Gen strategy in a federal regulation won't matter much if state implementers don't include it.

Federal rule writers used to be able to ignore a lot of the mess and complexity of 50-plus implementation programs through the convenient fiction that compliance would just happen and if it didn't enforcement (leaving it vaguely unclear whether this would be state or federal enforcement) would take care of it. That dodge isn't available anymore. Next Gen reveals that rule design controls the outcome. It can't be left to someone else to sort out later. Given that implementation—what actually happens in real life—is largely up to rule writers, how should they design rules with state implementers in mind?

In some ways this is a lot simpler than it seems. It just requires a Next Gen shift of perspective. Rule writers should stop thinking of rules as divided into (1) standards (mandatory) and (2) everything else (negotiable). Instead, the implementation drivers—the provisions that are there to make sure the rule actually happens—are part of the standard. They go together. A standard means nothing without reliable execution. If you don't have rule structure that makes compliance close to the default, you don't have a standard. You have a hope.

Looked at in this way, it is obvious what to do. The provisions that are essential for achieving the goal are part of the standard and thus mandatory. They are part of the federal floor. We know from long and painful experience, described throughout this book, that without these provisions, compliance will be bad to

[31] ECOS, "Cooperative Federalism 2.0" (states have assumed more than 96% of the delegable authorities under federal law). For a concise description of the benefits of cooperative federalism, see Clifford Rechtschaffen and David L. Markell, "Reinventing Environmental Enforcement and the State/Federal Relationship," *Environmental Law Institute*, (2003): 15–35.

terrible. That means the rule won't achieve the standard. The regulation will have mandated something, but not really.

What would that mean in practice? Measurement is a great example. Next Gen teaches that without dependable measurement, it is impossible to know if we are getting there. Direct measurement, rather than estimating, is the best, and continuous monitoring has by far the most compliance clout. The measurement method is therefore part of the standard. We also know that requiring monitoring does not reliably make it happen, never mind happen correctly, so monitoring needs to include drivers to ensure it does occur, like data substitution, and other provisions that ensure data quality. These would also be part of the standard.

The same goes for reporting. The most powerful compliance-forcing designs require reporting of facts, formats that make it impossible to avoid admitting violations and inducements to make reporting a priority, like automatic consequences for reporting failures. For many problems, reporting in real time, required third-party verification, and third-party information reporting can dramatically improve the reliability of reported information and motivate companies to do it right. All of these depend on electronic reporting, which can also help prevent errors and even make some violations impossible.

These ideas, and many others discussed in chapter 5, are the foundation of rules that will actually work. They aren't extras, do them or not. They are what ensures that the federal floor actually is the floor and not the beginning point for cutting back through weak implementation.

The necessity of approaching it this way is underscored by the many state legislatures that have limited regulators' ability to go even one inch above the federal floor. More than half the states have laws that prohibit or strongly constrain states from adopting any rule that is "more stringent than" the federal standard.[32] In these states, any variation from the federal regulation can only work one way. Under these strict limits, state implementation isn't a creative process to see how the federal program can be made to work better or more effectively in their state. Nothing beyond the federal mandate is possible without a gigantic effort to surmount the huge barriers placed in front of state regulators. Therefore, essential Next Gen provisions have to be part of the federal mandate. It isn't just that state regulators may not want to adopt the Next Gen ideas. Many literally cannot unless those ideas are mandated in the federal rule.

Next Gen is built on the understanding that regulations telling companies what they have to do doesn't mean they will actually do it. The same insight

[32] ELI, "State Constraints"; National Association of Clean Air Agencies, "Restrictions on the Stringency of State and Local Air Quality Programs," NACAA, December 8, 2014. See also James M. McElfish Jr., "Minimal Stringency: Abdication of State Innovation," *Environmental Law Reporter*, Vol. 25, No. 1 (1995): 10003.

applies to federalism. Just because states have an implementation obligation doesn't mean it happens. Repeated investigations have underscored this reality, as the following examples demonstrate.

- State air pollution permits have significant issues with practical enforceability, including vague permit language, insufficient monitoring provisions, and incomplete annual compliance certifications, with a broad range of variance in permit quality across states.[33]
- State reporting of water pollution compliance information continues to be inadequate; 15 of the 17 states EPA has assessed on this topic between 2018 and 2021 did not meet expectations for accuracy and completeness.[34]
- States are not catching errors in companies' air pollution monitoring stack tests or noticing that necessary data and documentation are missing, increasing the chances that violations persist for a year or more before they are detected.[35]
- Most state-issued Greenhouse Gas Prevention of Significant Deterioration permits don't contain a specific CO_2 control technology or technique and instead include only qualitative and vague standards.[36]
- States aren't doing the required inspections to determine if there are violations.[37]
- States don't meet regulatory obligations to inform EPA about violations in many programs.[38]

EPA lacks the means to insist on these essential implementation measures so that protection is the same for communities across the country, because everyone knows that EPA's only tool with power is to take the program back from the state. There is no conceivable way EPA can do that.[39] EPA doesn't have the resources to

[33] EPA OIG, "Substantial Changes Needed in Implementation and Oversight of Title V Permits If Program Goals Are to Be Fully Realized," EPA OIG Report No. 2005-P-00010 (March 9, 2005). See, e.g., EPA, "Order re BP Amoco Chemical Company, Petition No VI-2017-6" (July 20, 2021), 17, 20, https://www.epa.gov/system/files/documents/2021-07/bp-amoco-order_7-20-21.pdf. The same problems exist with state minor source air permits. EPA OIG, "EPA Should Conduct More Oversight of Synthetic-Minor-Source Permitting to Assure Permits Adhere to EPA Guidance," EPA OIG Report No. 21-P-0175 (July 8, 2021).

[34] GAO, "EPA Needs to Better Assess," 22, 24–25.

[35] EPA OIG, "More Effective EPA Oversight Is Needed for Particulate Matter Emissions Compliance Testing," EPA OIG Report No. 19-P-0251 (July 30, 2019), 11–12.

[36] Matt Haber and Seema Kakade, "Revitalizing Greenhouse Gas Permitting Inside a Biden EPA," *Environmental Law Reporter*, Vol. 51, No. 5 (May 2021).

[37] As just one example, see Christopher Vondracek, "South Dakota's Top Regulator Far Below EPA Requirements for Inspecting Stormwater Runoff Sites," *Mitchell Republic*, April 6, 2021 (South Dakota's top water-quality regulator tells a state legislative committee that the state is doing less than 5% of the inspections for industrial stormwater sites that EPA requires).

[38] See citations earlier in this chapter in the section titled "The 1970s Federalism Model for Controlling Information Is Holding Us Back Now."

[39] Eric Schaeffer, "Co-opting Federalism," *Environmental Forum* (May/June 2020): 44.

pick up implementation of permit-writing and enforcement for one program in one state, never mind deploy that as a routine part of the regulatory arsenal. Left with a nuclear bomb as the only deterrent, EPA is forced to rely primarily on real-time review of individual permits and enforcement because that's the tool EPA has.[40] That can lead to stronger protection in those particular instances—the permit will be tighter, the company forced to comply through enforcement—but it's not a strategy that fosters systemic change. And it is extremely resource intensive.[41] That means it has to be very strategically deployed, and that EPA will never scrutinize most state permits and compliance choices to see if in fact they meet the federal requirements.

States hate EPA's real-time review of state permits and enforcement because they see it as time-consuming micromanagement and interference that diverts attention from more important work. Meanwhile, GAO and the EPA Inspector General are asking why EPA isn't making states toe the line, and Congress wants to know why EPA is trampling on state prerogatives. In other words, business as usual for environmental regulators.

It is easy to miss amidst the tumult, but a common thread does emerge from these different challenges. States want shared metrics and services (for reporting, analytics, and many other things), one of the few topics on which there is widespread agreement.[42] Electronic data available in real time to both feds and states eliminates much of the reporting shortfalls and saves states money to boot. Meanwhile, clear and unavoidable obligations have by far the strongest compliance power to achieve greater national consistency.

Next Gen aligns with these perspectives. It steps back from the "do it," "try to make me" dynamic and says maybe there is a more reliable and automatic method to get there: required real-time monitoring, common format permitting and reporting, pick lists with limited options electronically reported into a shared system, shared services that speed things up and make them more predictable, and transparency that holds sources and government accountable. The NPDES e-reporting rule, which adopted many of these strategies and was one of the biggest advances for clean water protection in recent years, was accomplished with state support. And no one sued—a rarity for major EPA regulations.

[40] EPA also does programmatic state compliance reviews—which is what states say they would prefer—but progress on serious issues identified that way can be frustratingly slow because there are so few levers to force necessary changes. And programmatic reviews done by the IG, GAO, and others continue to reveal widespread problems with state implementation, as is described throughout this chapter.

[41] As just one example, Texas issued a proposed Title V air permit to BP Amoco Chemical Company Texas City chemical plant in 2016. Petitioners filed a 49-page challenge to that permit with EPA on April 4, 2017, primarily on the grounds that the permit wasn't enforceable. EPA issued a 43-page decision largely agreeing with the challenge, on July 20, 2021. EPA, "Order re BP Amoco."

[42] ECOS, "Cooperative Federalism 2.0" (states support sharing information transparently with EPA, and EPA and states working together on shared services and implementation toolkits).

My experience talking with state regulators during my time at EPA confirmed the potential for state and federal agreement around these ideas. We had vigorous disputes about the need for and way to do enforcement. But often my biggest state sparring partners were intrigued by the possibility of Next Gen solutions that didn't sound like the same old same old. Ideas that defied classification as regulatory or anti-regulatory. Possibilities that might even make the states' job easier or faster. Although most Next Gen ideas require states to let go of the chokehold on information some use to block federal action, those days are numbered in the electronic age anyway, and maybe it's better to be at the table designing what's next than to be dragged along. Whatever the fashion may be among state political leadership, most state environmental agency staff want a cleaner environment and care about protecting the people who live in their state. If there's a way to do that without igniting the ideological wars, they are interested.

What might this look like? Here's an illustration using a problem that happens all the time: state permits that don't include emissions limits or monitoring requirements. State permits are supposed to have those things, to allow the state, EPA, and the public to know if the source is complying, but many don't. EPA occasionally objects to individual permits for this reason, which drags out the permitting timeline and supports the narrative promoted by companies and some states that environmental permitting is absurdly complicated and time consuming. Instead of continuing this trench warfare, here's a possible alternative: a common format electronic form that includes for every source a drop-down menu for emission limits and the monitoring method. There is a pick list—nothing offered that goes below the federal floor!—and the permit isn't done until all the fields are completed. Some choice and discretion are possible, but skipping something essential, or failing to put a number where a number is required, is not. The permittee trades finely tailored outcomes for speed and clarity. The state gets to move more quickly with less federal interference. The feds have more assurance that it will be done right, and also the ability to see the big picture rather than trying to steer exclusively through a few let's-make-an-example-of-you individual cases.

I am not suggesting it is some kind of panacea or that it is easy. Far from it. I am saying that if we want to see progress within our increasingly polarized national politics, everyone doubling down on entrenched positions from a 50-year-old federalism construct is pretty unimaginative. As it happens, Next Gen ideas that make sense for better compliance might also help shift our federalism discussion to a more productive level. No one gets everything they want, and wrestling over different perspectives will continue. But protection would be stronger, compliance would be much more robust, and the feds and states would both be better off than they were before. That looks a lot like what Congress envisioned when it told EPA and states to work it out.

11
Environmental Enforcement in the Next Gen Era

Violators will always be with us. No matter how robust a rule design is, there will be companies that find a way around. There will be unexpected events and problems regulators didn't think of. There will be states that don't want to insist on compliance and don't have the will to confront industries with political clout. Pollution will cross state boundaries. Environmental injustices will demand immediate attention. Environmental criminals will devise ever-changing ways to flout the law and threaten people's health. For these and other reasons, vigorous enforcement is an absolute necessity for any environmental program. It will ever be thus.

Next Gen doesn't change that. Enforcement isn't less necessary with Next Gen in rules. The need is just as profound. But with Next Gen compliance drivers built into rules, enforcers won't be expected to handle the impossible task of assuring baseline compliance, so they can be far more strategic. They can focus the powerful and creative tools of enforcement on the tough problems that Next Gen can't fix. Rules with compliance built in allow enforcement to do what it does best: tackle the daunting and unexpected and find ways to turn violations into remedies that make communities and the environment whole.

This chapter starts with an overview of the problems that will always need enforcement attention. Some problems—like emergencies, the unexpected, and the creative criminal—will forever be threats, and enforcement is the way regulators can respond with the urgency and force that is required. Federal enforcement in particular is the only way to tackle some critical problems, as this chapter explains. Next Gen can assist, but never replace, these must-do priorities. I then turn to what revitalized enforcement could look like in the Next Gen era and the many ways Next Gen innovations can increase the impact of compliance work. This includes a newly prominent role for enforcement in finding creative solutions and promoting innovation. The chapter concludes with a look at what enforcement can do now to improve protection near and middle term, while Next Gen ramps up. Throughout, I focus on the tight link between compliance, enforcement, and the mission-central challenge of environmental justice.

Next Generation Compliance. Cynthia Giles, Oxford University Press. © Cynthia Giles 2022.
DOI: 10.1093/oso/9780197656747.003.0012

Vigorous Enforcement Will Always Be Essential

The goal of Next Gen is rules that do a much better job ensuring strong implementation. Regulations that have high compliance rates because they make compliance the path of least resistance. Rules with solid compliance design don't depend on enforcement as the first line of defense for making protection real on the ground. Choosing enforcement as the only compliance strategy doesn't work, as overwhelming evidence in this book has shown. But enforcement is an indispensable part of an effective compliance program, as this section explains.

Key Situations in Which Enforcement Necessarily Plays a Leading Role

Emergencies

No matter how compliance resilient an environmental regulation is, stuff happens. Natural disasters, sudden failures, and deliberate choices to take unreasonable risks can lead to calamity. There are, regrettably, countless examples: companies that release chemicals into drinking water sources, leaving hundreds of thousands without potable water; businesses that fail to prevent toxic air emissions that sicken and kill employees, first responders, and neighbors; holes opening up at the bottom of giant hazardous waste lagoons, sending toxic waste into the region's drinking water supply; companies that operate oil wells and pipelines without proper safety precautions, spilling oil at sometimes catastrophic levels. When the worst happens, government needs a way to require quick action to stop the harm. There will be time to figure out what went wrong, and how to fix that, later. The immediate issue is how to stop the bleeding. Enforcement orders, including through a court, are often the best, and sometimes the only, way to accomplish that.

Environmental crimes

Sometimes violations are the result of deliberate or reckless behavior. Someone lies, they ignore obvious danger signs, they cheat. Yes, Next Gen tries to prevent that by making those choices tough to do or get away with. But no matter how strong a rule is, it can't prevent every criminal act. Some people will decide to take a shortcut and place others in danger. They put an unauthorized pesticide inside or too close to a home and kill or permanently injure children. They lie about lab results, so pollution continues unabated, or people are exposed to dangerous chemicals. They turn off pollution controls or ignore safety standards and ship noxious air or water contaminants into neighboring communities. They draw drinking water from a banned water source or fail to do required treatment, risking the health of the entire community. The list goes on and on. These aren't like other violations. As a society we say in these cases: no way. As companies, as individuals, you are not getting away with that. We can do a lot more than we are doing now to write rules that cut bad guys off at the pass and prevent more of these disastrous situations. But they won't ever go away. It is essential to have a strong criminal enforcement program, both to deter anyone even considering going down that road and to send an unmistakable message: that's not tolerated.

New (or newly discovered) problems

It is not unusual to find out that something regulators weren't paying a lot of attention to is actually a huge problem. The widely used pollution-control system that was designed to be 98 percent effective turns out to be nowhere near that good in real life. Hazardous waste thought to be well contained in tanks is actually releasing a lot of dangerous air pollution. Enforcement plays two important roles in these situations. The first is finding out: inspectors in the field are often the first to realize that something is far worse than assumed. The second is as a stopgap measure. It may be that some newly discovered problems require a regulatory solution. But regulatory bandwidth is limited, and writing rules usually takes years. Meanwhile the problem is urgent right now. Where these problems are violations or otherwise appropriate for enforcement orders—which they often are—enforcement can help bring a solution into focus. Through enforcement cases, the problem can be better understood and defined. Enforcement cases are public and visible, so they elevate the profile of the issue so at least other companies are aware. Regulators can amplify that message through enforcement alerts. And the resolution of an enforcement case can define a solution that others can use. Enforcement usually can't be the entirety of the answer, for all the reasons discussed throughout this book. But it has often served in this kind of problem identification, amplification, and clarification role. And it is a vital function, because the one constant in regulatory work is that there will be new problems.

Environmental justice

A mountain of evidence documents that communities of color and low-income areas face worse environmental insults than other places. Just a few examples: Black, Latino, and Asian Americans face higher levels of exposure to fine particulate matter from traffic, construction, and other sources,[1] and

[1] Juliet Eilperin and Darryl Fears, "Deadly Air Pollutant 'Disproportionately and Systematically' Harms Americans of Color, Study Finds," *Washington Post*, April 28, 2021. See also Ihab Mikati et al., "Disparities in Distribution of Particulate Matter Emission Sources by Race and Poverty Status," *American Journal of Public Health*, Vol. 108, No. 4 (April 2018): 481; Lala Ma, "Mapping the Clean Air Haves and Have-nots," *Science*, Vol. 369, No. 6503 (July 2020).

African Americans are 75 percent more likely than the average American to live in communities near polluting facilities.[2] Drinking water systems in Latino areas violate federal drinking water rules twice as much as those serving the rest of the United States.[3] Climate change will increase existing inequality in the United States,[4] and communities with large minority populations are among the most vulnerable to the effects of climate change, like extreme heat and flooding.[5] Piled on top of disproportionate exposures is the reality that these overburdened communities also face higher prevalence of conditions like cardiovascular disease and asthma, reduced access to healthcare, and other inequitable social and physical determinants of health.[6] This disparate vulnerability to health effects is the result of a host of structural problems, many of which are rooted in racism.[7] The link between air pollution and poor COVID outcomes has underscored the injustice of these long-standing disparities.[8] Enforcement cannot solve these structural deficiencies. But it will always need to be on the forefront of dealing with the consequences, because violations fall most heavily on these beleaguered communities.

Critical Enforcement Priorities That Only the Feds Can Address

All of the just mentioned roles for enforcement, which will be essential no matter what happens with Next Gen, can be performed by state or federal regulators. If the relevant states can address the problem, great. That's what we hope for from the state front-line enforcers. But if the states can't or won't, EPA has to step in. Some of the national dialog of late would have you believe that the feds aren't needed and should just defer to the states. It portrays the feds as stomping around

[2] American National Association for the Advancement of Colored People and Clean Air Task Force, "Fumes Across the Fence-Line: The Health Impacts of Air Pollution from Oil & Gas Facilities on African American Communities," November 2017, https://www.catf.us/wp-content/uploads/2017/11/CATF_Pub_FumesAcrossTheFenceLine.pdf.

[3] Emily Holden et al., "More Than 25M Drink from the Worst US Water Systems, with Latinos Most Exposed," *The Guardian*, February 26, 2021.

[4] Solomon Hsiang et al., "Estimating Economic Damage from Climate Change in the United States," *Science*, Vol. 356 (June 2017): 1363.

[5] Thomas Frank, "Population of Top 10 Counties for Disasters: 81% Minority," *Climatewire*, June 8, 2020; Tik Root, "Heat and Smog Hit Low-income Communities and People of Color Hardest, Scientists Say," *Washington Post*, May 25, 2021.

[6] Mikati, "Disparities in Distribution," 484–85. Kenneth Gillingham and Pei Huang, "Racial Disparities in the Health Effects from Air Pollution: Evidence from Ports," *National Bureau of Economic Research*, Working Paper 29108 (July 2021), https://doi.org/10.3386/w29108.

[7] Linda Villarosa, "The Refinery Next Door," *New York Times Magazine*, August 2, 2020, at 31–32; Charles Lee, "Confronting Disproportionate Impacts and Systemic Racism in Environmental Policy," *Environmental Law Reporter*, Vol. 51, No. 3 (March 2021): 10127.

[8] Hiroko Tabuchi, "In the Shadows of America's Smokestacks, Virus Is One More Deadly Risk," *New York Times*, May 17, 2020.

filing enforcement cases that the states would otherwise have handled just fine, thank you. That's nowhere close to reality. This section gives an overview of some key enforcement work that only the feds can do. These combined categories are what consume the vast majority of federal enforcement effort. No matter how quickly we shift toward Next Gen strategies, these topics will continue to require substantial federal enforcement attention.

Federal-only programs

Some programs can't be delegated to states. Congress set them up to be run by EPA. Like the Toxics Substances Control Act regulating dangerous chemicals. Or emission standards for new vehicles. Sometimes there is a role for states alongside that of EPA, like California's unique responsibility in vehicle emission standards, for example. But in general, for federal only programs EPA is the only enforcement game in town.

Interstate impacts

States will not be at their most vigorous when clamping down on companies in their state that primarily affect out-of-state communities. Air polluters that hurt downwind neighborhoods in other states are the most obvious example. In fact, in the early days of air pollution control, many states' response was to approve taller stacks so the pollution would be someone else's problem.[9] When the impacts are felt primarily outside the state where the source is located, the feds are in a position to insist on compliance, and sometimes they are the only ones who will. While pollution is the poster child for interstate impacts, it isn't confined to that. When markets are created to address environmental problems, for example, states that are lax in insisting on standards can result in noncompliant products flooding the market, to everyone's disadvantage. EPA brings a national perspective to these issues, and in many programs that's needed to ensure national program integrity, which benefits all.

National and international company violators

Many companies operate in multiple states. Individual states can't effectively take on nationwide operations; it would be both ineffective and inefficient to do it that way. The only way to get a consistent outcome across multistate companies and bring the resources to bear that match what a multinational company can muster is to handle the case at the federal level. States can join as co-plaintiffs, and that's often a big assist.

[9] "Air Quality: Information on Tall Smokestacks and Their Contribution to Interstate Transport of Air Pollution," *Government Accountability Office*, GAO-11-473 (May 2011): 1–5.

Backstopping states where political will is lacking
Some companies and industries hold significant state political power. They are a huge employer, or donor, or both. Holding them to account can be beyond the capacity of state regulators. They may wish to, but sometimes they won't be able. In those cases, EPA can step in to do what's necessary. Sometimes that will be at the state's request—whether formal or off the record—but it's something EPA needs to be prepared to do, request or not. A national program to protect everyone everywhere won't mean much if states can protect particular companies or sectors from accountability.

Protecting companies that play by the rules
States that set a low bar don't affect only their own residents. Companies in other states looking to avoid accountability often point to what others get away with in a bid for the same treatment. Meanwhile, companies doing what's required understandably feel mistreated if they have to compete with facilities not being held to the same standard. That unequal treatment isn't fair, but it is important for a more practical reason as well: if it persists, more companies will decide that it's time to be a little more relaxed about compliance. EPA needs to keep an eye on national consistency lest the anchor of low-performing states starts dragging the entire program down.

Taking on the sectors causing disproportionate harm
Most people think about enforcement as cases against individual companies or facilities. It certainly includes that, but often the problem is much broader. It isn't just one facility that is seriously violating but most of the companies in that business. That pattern has been repeated time and again: for coal-fired power plants, refineries, sewage treatment plants, mineral processing, and many other examples described in chapter 2. When a sizable percentage of an industry has similar violations, which are individually and collectively causing harm, that is often something only the feds can observe or effectively address. All the factors discussed here come into play. That is why the huge bulk of EPA's civil enforcement work focuses on these regional and national problems. A multistate approach in these cases is far more effective and vastly more efficient at deploying experts and dispersing useful solutions than a state-by-state strategy could ever be.

Elevating the urgency of environmental justice
Some states are leaders on environmental justice. New Jersey, for example, recently passed legislation showing how cumulative impacts can be included in permitting decisions.[10] California has a well-deserved reputation for being out

[10] Senate 232, 219th Legislative Session (N.J. 2020). New Jersey's leadership in addressing cumulative impacts in permits is particularly notable from a compliance perspective because it is often the

in front on environmental justice. We all can learn from innovative states.[11] EPA needs to be among those preventing the not-my-problem crowd from ignoring the profound effect of society's structural inequities on the impacts of violations. Only sustained pressure can overcome long-standing inertia that disregards disproportionate environmental harms. One of EPA's essential roles is to consistently push the collective national enforcement community to do more, starting with EPA's own enforcement agenda.[12]

For far too long we have been counting on enforcement to achieve the impossible, which—surprise!—it hasn't been able to do. When rule writers make it their job to build compliance drivers into regulations, compliance will improve, and that will create the necessary space for enforcers to do the job only enforcement can do. Enforcers often fall far short on accomplishing the mission-central roles previously described, because they are so overwhelmed by the unachievable baseline compliance task that is usually assigned to them. Lifting that burden through Next Gen in rules allows enforcement to focus on doing what it does best, aided by the additional problem-spotting data that Next Gen will create. Same resources, far better result.

Revitalized Enforcement in the Next Gen Era

Is there some nirvana where all facilities comply so enforcement isn't needed? No. Never in a million years will that happen. What Next Gen in rules can do is significantly narrow the existing huge compliance gap, so that enforcement stands a fighting chance of addressing the inevitable—but fewer—serious violators. When Next Gen becomes a part of all environmental rules, how should enforcement be deployed? Some of the ideas outlined here are already being used, and with a better baseline of compliance through Next Gen could happen more. Others that occur only at the margins now could move to center stage.

case that multiple facilities overburden a community, but enforcement can do little to help because the individual facilities are complying with their permits.

[11] Charles Lee, "A Game Changer in the Making? Lessons from States Advancing Environmental Justice Through Mapping and Cumulative Impact Strategies," *Environmental Law Reporter*, Vol. 50 (2020): 10203–215.

[12] EPA's enforcement office has been a leader in environmental justice within the agency. Jill Lindsey Harrison, *From the Inside Out: The Fight for Environmental Justice within Government Agencies* (MIT Press, 2019), 140. Through Next Gen and a strategic focus on key environmental justice priorities, it can continue in the vanguard of environmental justice innovation.

What Would Enforcement Do Differently in the Next Gen Era?

Bring more strategic focus to enforcement cases

Too much enforcement effort around the country today is spent on routine patrols and follow-up cases when violations are found. These are necessary—even if wildly insufficient—when the rule doesn't include compliance drivers, but in the Next Gen era, rule design will pick up much of this slack. That frees enforcement staff to be more strategic. Analysis of the increased amount of compliance data regulators will receive with Next Gen can reveal patterns and issues that demand compliance attention.[13] That's what EPA does today with National Compliance Initiatives (aka National Enforcement Initiatives during the Obama administration).[14] Sectors or rules with nationally widespread violations causing serious harm, in which enforcement can make a significant difference, are a good place to focus the expertise and problem-solving ability of federal enforcers. That problem-based approach lets EPA leverage business networks, the power of public visibility, and the possibility of more conclusive and systemic national solutions. EPA already spends a sizable chunk of its time on this, as it should, and Next Gen will allow EPA to do more. The same thing can happen at the state level for state-specific priorities and will be more possible when regulators aren't so dependent on boots-on-the-ground inspections to know what's going on.

Strategically deploy criminal enforcement

Criminal actors have already proved that they are willing to flout the rules and deliberately evade regulatory controls. Regulations can do much better hemming them in, but the unscrupulous will seek new paths when they find prior ones blocked. Criminals learn and adapt, and that's what regulators need to do too. EPA has a sophisticated criminal enforcement program, which will play an even larger role in the Next Gen era, particularly when deployed jointly with civil enforcers to tackle the same problems. Strategic criminal programs can be much more proactive in selecting what to investigate. The new data and analytic capability of Next Gen can help screen for evidence of criminal activity and also support an initial look into problems where crime is suspected, and the power of a criminal prosecution can motivate a change in behavior. Accuracy of self-reporting, for example, is essential to the success of nearly every environmental program; criminal strategies to ferret out reporting fraud using big

[13] For a creative and inspiring account of ways to mobilize enforcement to tackle tough compliance problems, see Malcolm K. Sparrow, *The Regulatory Craft* (Washington DC: Brookings Institution Press, 2000). Professor Sparrow specifically discusses how regulatory innovation and enforcement work together. Sparrow, 184.

[14] EPA, "National Compliance Initiatives," https://www.epa.gov/enforcement/national-compliance-initiatives.

data techniques would go a long way toward supporting nationwide program integrity.

Make environmental justice a centerpiece of strategic choices

All of the work EPA does to promote Next Gen in rules and in enforcement advances environmental justice. Overburdened communities suffer the most from widespread violations. And they will benefit the most when Next Gen strategies improve compliance across the board. Next Gen is absolutely necessary as a foundational part of addressing compliance inequities. But it isn't sufficient. In the Next Gen era, enforcement can and should do more. EPA enforcement should continue to consider environmental justice concerns at every stage of enforcement and compliance, from setting priorities and planning investigations to resolving enforcement actions.[15] That includes taking on some of the difficult compliance problems that plague communities with environmental justice concerns. For example, recent research has shown that emissions at major ports have three times the effect on health in Black communities as they do in White;[16] enforcement can help by focusing on compliance with rules requiring ships to use low-sulfur fuels, including the potential remedy of requiring continuous emission monitoring systems to track emissions and compliance far more reliably.[17] Existing emphasis on sectors that are of particular concern to fenceline neighborhoods, like refineries, chemical plants, and oil and gas wells, can be ramped up with the better information that Next Gen in rules will provide. It isn't just the supersized facilities that matter; smaller but more numerous plants located right in neighborhoods can create significant risk.[18] Rural areas that host violating industrial animal agriculture can also be an important health and justice priority.[19] And EPA should not shy away from investigating some particularly challenging problems that affect overburdened communities, like unlawful pesticide application that affects farmworkers[20] and large-scale tampering with emissions controls for diesel trucks.[21]

[15] Lee, "Confronting Disproportionate Impacts," 10125.

[16] Gillingham, "Racial Disparities in the Health Effects from Air Pollution."

[17] Seema Kakade and Matt Haber, "Detecting Corporate Environmental Cheating," *Ecology Law Quarterly*, Vol. 47, No. 3 (March 2021): 805–20.

[18] See, e.g., EPA Enforcement Alert, "Violations at Metal Recycling Facilities Cause Excess Emission in Nearby Communities," July 2021, https://www.epa.gov/system/files/documents/2021-07/metalshredder-enfalert.pdf.

[19] Wendee Nicole, "CAFOs and Environmental Justice: The Case of North Carolina," *Environmental Health Perspectives*, Vol. 121, No. 6 (June 2013): 182–89.

[20] Rafter Ferguson, Kristina Dahl, and Marcia DeLonge, "Farmworkers at Risk: The Growing Dangers of Pesticides and Heat," *Union of Concerned Scientists* (December 2019), https://www.ucsusa.org/resources/farmworkers-at-risk.

[21] "Tampered Diesel Pickup Trucks: A Review of Aggregated Evidence from EPA Civil Enforcement Investigations," EPA (November 2020), https://www.epa.gov/enforcement/tampered-diesel-pickup-trucks-review-aggregated-evidence-epa-civil-enforcement. The EPA diesel tampering report is an example of a vital role for enforcement, today and in the Next Gen era: investigating

Make the most of advanced monitoring

EPA already emphasizes using the most advanced tools to detect serious problems. That should be expanded, with an aggressive effort to discover cutting-edge monitoring technologies and to develop new ones where necessary. Monitoring expertise is growing by leaps and bounds, and so should government's ability to tap into that expertise to protect the public. Remote-monitoring strategies in particular deliver a dual benefit: they help find serious problems without alerting companies that might try to obscure what's going on, and they increase deterrent punch as companies know that they are not as invisible as they thought.[22] Funding for state equipment and communities of expertise among feds and states will help spread the most useful ideas.

Use enforcement cases to promote innovative solutions

Enforcement cases start by dragging someone to the table when they are caught in significant violation. No company likes to be there. But it is surprisingly common for businesses forced into that uncomfortable position to realize that there may be opportunity as well. Yes, they are going to have to move quickly to comply and pay a penalty. Of course. But instead of dragging their feet and making it as painful as possible, they can shift from laggard to leader. I've seen it happen. There may be a new control technology that makes both environmental and economic sense. Perhaps a more robust monitoring system could be installed that both ensures compliance and reassures neighbors. By stepping forward, companies might set a new bar for their industry and try to regain their good reputation. EPA should actively seek out these opportunities as part of its mitigation authority, as EPA has recently announced it intends to resume doing.[23] The shared government/industry experimentation that is possible under a settlement agreement can lay the foundation for a potential industry-wide improvement,

problems, drawing attention to them through enforcement cases, and scoping out how widespread the problems are. Enforcement alone can't solve many of these problems, but it can help define the issue and set it up for regulatory attention.

[22] See, e.g., Ben Chugg et al., "Enhancing Environmental Enforcement with Near Real-Time Monitoring: Likelihood-Based Detection of Structural Expansion of Intensive Livestock Farms," *International Journal of Applied Earth Observations and Geoinformation*, Vol. 103, No. 4 (2021): 012463.

[23] EPA, "Using All Appropriate Injunctive Relief Tools in Civil Enforcement Settlements," April 26, 2021, https://www.epa.gov/sites/default/files/2021-04/documents/usingallappropriateinjunctivereliefltoolsincivilenforcementsettlement0426.pdf. The Trump EPA and Department of Justice abandoned the bedrock principle that companies should have to make up for the harms they caused by violating. The Biden EPA is restoring that idea to its appropriate central position. For examples of Next Gen ideas included in settlements, see EPA, "Next Generation Compliance: Enforcement Settlement Highlights (last edited December 20, 2016)," https://www.epa.gov/sites/default/files/2016-05/documents/nextgen-enfsettlementhighlights.pdf.

as happened, for example, after enforcement cases proved that fenceline monitoring was feasible and effective for refineries.[24]

Deploy data analytics to get the most bang for the buck

Next Gen depends heavily on better company monitoring and more robust and reliable self-reporting. All that new data—plus the information EPA already has access to—can support far more sophisticated data analytics to find the worst problems. Predictive analytics can send inspectors to the locations most likely to have serious issues,[25] and analytics with satellite imagery can be used to flag sites that were not noticed with more traditional methods.[26] Potentially criminal activity can be spotted amid the noise of huge data sets, like the two recent studies finding statistical evidence that air monitors are strategically turned off[27] and water-discharge reports are deliberately not filed[28] to avoid reporting violations. EPA taps into some analytic power now but can do a lot more with the additional data that will be available under Next Gen. This is an area where limited investment will pay off big time in public health protection and extending the reach of existing enforcement staff. Analytics can be deployed at any level—federal, state, local—to help address priority issues.

Be mindful and intentional about enforcement's role in identifying issues that need programmatic attention

Inspectors in the field are often the first to suspect a previously overlooked problem. They discover hazardous waste facilities with serious air violations, including constant venting of organic vapors into the atmosphere.[29] They document the widespread practice of illegal tampering with truck emission controls—through so-called aftermarket defeat devices—which result in emissions of oxides of nitrogen (the pollutant at issue in the Volkswagen case) up to

[24] See the fenceline monitoring provisions in EPA's refinery rule, codified at 40 CFR § 63.658.

[25] Miyuki Hino, Elinor Benami, and Nina R. Brooks, "Machine Learning for Environmental Monitoring," *Nature Sustainability*, Vol. 1 (October 2018): 583–88, https://doi.org/10.1038/s41893-018-0142-9 (machine learning to better predict facilities at high risk of serious water pollution violations could double the effectiveness of environmental inspectors).

[26] Cassandra Handan-Nader, Daniel E. Ho, and Larry Y. Liu, "Deep Learning with Satellite Imagery to Enhance Environmental Enforcement," in Jennifer B. Dunn and Prasanna Balaprakash eds., *Data-Driven Insights and Decisions: A Sustainability Perspective*, (Elsevier, 2021) 205–28; James K. Lein, "Implementing Remote Sensing Strategies to Support Environmental Compliance Assessment: A Neural Network Application," *Environmental Science and Policy*, Vol. 12 (2009): 948–58.

[27] Yingfei Mu, Edward A. Rubin, and Eric Zou, "What's Missing in Environmental (Self-) Monitoring: Evidence from Strategic Shutdowns of Pollution Monitors," *National Bureau of Economic Research* (April 2021, rev. October 2021), https://doi.org/10.3386/w28735.

[28] Daniel Nicholas Stuart, "Strategic Non-Reporting Under the Clean Water Act," chapter in "Essays in Energy and Environmental Economics," PhD diss., Harvard University 2021, https://nrs.harvard.edu/URN-3:HUL.INSTREPOS:37368502.

[29] EPA Enforcement Alert, "National Compliance Initiative Focus on RCRA Air Emissions," June 2020, https://www.epa.gov/sites/default/files/2020-06/documents/ncircraairenfalert060320.pdf.

300 times the allowable limits.[30] EPA can elevate awareness of these problems through enforcement cases and can work with program offices to design Next Gen regulatory fixes where that's necessary. In the Next Gen era, enforcement can more consciously look for these opportunities, so programs have the information they need to tackle problems at the necessary scale.

Enforcement's Essential Next Gen Role in Buttressing the Strength of the Compliance Foundation

Another category of critical enforcement work in the Next Gen era, which requires far more investment than it gets today, is making sure the compliance foundation remains sound. Next Gen in rules can ensure far more effective implementation and much better compliance, but only if we build that foundation and keep it strong. What does that mean for the makeup of enforcement work?

Actively participate in writing rules

Enforcers know that a lot of companies violate. They live that every day. People who write rules don't necessarily understand that. They haven't observed close-up the many inventive and creative ways people find to avoid doing what's required. They haven't experienced the stunning willingness of some companies and their managers to put other people at risk to make a buck. Nor do they see firsthand the good old-fashioned incompetence that is so frequently an element of serious violations. That's one reason that Next Gen doesn't have a stronger foothold at EPA: lots of rule writers don't have a visceral understanding that violations abound in all sizes of companies. Enforcers' experience-supported skepticism about the likelihood of compliance, joined with education about rule design features that make violations far less likely, can add a helpful note of realism to rule design. Enforcement needs to invest a substantial chunk of effort in advocating for Next Gen ideas in the sometimes tedious and detail-oriented work of writing regulations, as I advocated in chapter 4. The enforcement mission isn't just to bring cases, it's to drive better compliance outcomes. Every time Next Gen ideas are included in a rule, that's people protected and a bunch of enforcement cases that will never be needed because the company complies on its own. All rowing in the same direction.

[30] EPA, "Tampered Diesel Pickup Trucks," 12.

Support oversight of permits to ensure they follow through on Next Gen provisions

The document that controls compliance obligations often isn't the regulation directly, it's the permit. Permits are usually written by states. If a permit isn't enforceable and doesn't include the regulations' Next Gen provisions—doesn't have a limit, a required monitoring method, or effective reporting, for example—that great-in-theory standard isn't going to cut it. It isn't enforcement's job to make sure the permits are done right. But when enforcement discovers that they aren't, that's something that programs need to pay attention to. Enforcers can collaborate with program offices to ensure there is a system for tracking permit quality and avoiding rule design holes that allow poor quality permits to slip through, undermining compliance.

Build a better system for state reporting

Many of the Next Gen features discussed in this chapter—monitoring, improved data analytics, attention to the national picture—depend on states informing EPA about violations. For example, the state and EPA analytic tool to prioritize drinking water violations for enforcement attention is for naught if states don't tell EPA about most of the violations.[31] In many programs today, state reporting is unreliable or worse. When that happens, even terrific analytic tools are useless. Garbage in, garbage out. This issue persists across many programs, despite the fact that we live in the electronic age when access to the information should be both easy and cheaper than handling it the arcane way we often do now. Enforcers should be at the front of the line insisting that this be changed so the public and regulators can know what the real compliance story is.[32]

Advocate for innovation in compliance monitoring

Enforcement is built on compliance monitoring. Some is company self-monitoring, some is government monitoring, and some comes in other ways, through nongovernmental organizations or academic research, for example. The traditional presumption that inspectors would be checking every regulated facility was always a pipe dream; inspectors can't possibly inspect even a tiny fraction of the facilities covered by environmental rules. And for many types of facilities that's not the best way to know what's going on anyway; many important problems are not observable during a one-time visit. Enforcers need to insist on accurate and reliable self-monitoring (see above: active participation

[31] GAO, "Unreliable State Data Limit EPA's Ability to Target Enforcement Priorities and Communicate Water Systems' Performance," GAO-11-381 (June 2011): 23–24.

[32] See chapter 2, section titled "For Some Important Programs, EPA's Understanding of Noncompliance Is Wrong"; chapter 10, section titled "The 1970s Federalism Model for Controlling Information Is Holding Us Back Now."

in writing rules). And enforcers should make the most of technologies that can screen for the most serious problems, deploying remote sensing, data analytics, and satellite imagery, so inspectors go where it will do the most good. Satellites can, for example, help detect industrial animal feeding operations near sensitive water bodies[33] or landscape changes in drinking water protected watersheds[34] and large releases of methane (and associated VOCs) at individual wellheads.[35] Even the comparatively low-tech strategy of remote video inspections can help extend the reach of existing inspectors. Sophisticated analytics can identify when sources may be failing to report as a strategy to avoid disclosing more serious violations.[36] Enforcement should be seeking out these new opportunities to expand the impact of government resources and, not coincidentally, increase deterrent punch.

Embrace true transparency

Sharing EPA's data with the public is a central feature of Next Gen, because it inspires better performance by industry and encourages innovative use of the data to drive better results. But true transparency means being candid about the problems with the data too. It isn't enough to drop a footnote or have an explanatory box three clicks in that informs the user that the data are wildly incomplete. Data accuracy won't improve without pressure, and there won't be pressure as long as the public gets an unduly rosy picture of how it's going.[37] Plus, over time government drinks the Kool-Aid and starts believing its own inaccurate picture, much to the detriment of protective policy. If regulators know violations aren't being reported, or that what is reported is misleading, government should say so, front and center. Yes, that's uncomfortable and invites complaints and scrutiny. That's why transparency works.

Patrol the boundaries

Well-intentioned rules can inadvertently create havens for evasion. A common place for that to happen is in regulatory exclusions, exemptions, and classifications. Many regulations attempt to encourage less impact by making requirements for the theoretically lower-risk activities less onerous. Whenever a regulation draws a line and says "on this side of the line you have tough standards

[33] Cassandra Handan-Nader and Daniel E. Ho, "Deep Learning to Map Concentrated Animal Feeding Operations," *Nature Sustainability*, Vol. 2 (April 2019): 298–306.

[34] Lein, "Implementing Remote Sensing Strategies," 953.

[35] Judy Stoeven Davies, "EDF Ready to Go into Space," Environmental Defense Fund Special Report, Spring 2019, https://www.edf.org/sites/default/files/documents/SpringSpecialReport-MarkBrownstein.pdf.

[36] Mu, Rubin, and Zou, "What's Missing in Environmental (Self-)Monitoring"; Stuart, "Strategic Non-Reporting."

[37] "Clean Water Act: EPA Needs to Better Assess and Disclose Quality of Compliance and Enforcement Data," Government Accountability Office, GAO 21-290, July 2021.

but on the other side you don't," the rule creates a motive to at least appear to be on the more relaxed side. It is well known that regulators focus the vast majority of their attention on the highest-impact facilities with the tightest standards, so the lesser regulated also have less chance of government oversight. Fewer rules, less scrutiny. Who wouldn't prefer to be there? Such line-drawing can motivate actual changes that reduce impact. But here's a sure thing: some, perhaps many, companies will falsely claim to be on the less-regulated side, figuring that the reduced obligations and oversight make it unlikely their inaccurate claims will be detected. Large generators of hazardous waste will say they only create small quantities and thereby duck more stringent obligations to protect the public from releases.[38] Companies will falsely claim to emit less air pollution and thereby avoid more stringent air pollution controls, exposing people to unlawful emissions.[39] New Source Review is a giant boundary catastrophe.[40] When this kind of rule-breaking proliferates, it can have significant impacts on communities and go largely unnoticed by regulators. Rules have to be clear-eyed about this reality and include barriers and detection systems to prevent this from routinely happening. And enforcers need to patrol those boundaries—through analytics and random sampling when appropriate—to flag the programs where boundary cheating is causing real harm and requires a regulatory fix.

Enhance oversight of state enforcement

EPA's State Review Framework sets common metrics for evaluating state enforcement performance.[41] It is extremely useful to have this established structure, but in the Next Gen era the emphasis may shift. Reporting systems shared by EPA and states will allow a more detailed examination of compliance status. Examination of state activities to ensure accuracy of self-reported data can move up the priority list. Robust analytics to examine compliance gaps and serious problems should be a more prominent feature. Environmental justice needs to be front and center in reviewing the priorities for enforcement attention. And EPA needs to reexamine the current third-rail ideas of comparing states and using transparency pressure to force attention to serious deficiencies. There aren't many levers to drive better outcomes as it is, not to mention that state residents deserve to know how their government is performing.

[38] See chapter 2, note 57.

[39] For example, see Environmental Integrity Project, "Dirty Deception: How the Wood Biomass Industry Skirts the Clean Air Act," April 26, 2018, https://www.environmentalintegrity.org/wp-content/uploads/2017/02/Biomass-Report.pdf.

[40] Chapter 1, section titled "Programs with Pervasive Violations: Four Examples," subsection titled "Air Pollution: New Source Review for Coal-fired Power Plants."

[41] "State Review Framework: Compliance and Enforcement Program Oversight," EPA, https://www.epa.gov/compliance/state-review-framework.

Enforcement Can Lead the Charge for Innovation

Enforcement has a big advantage in driving environmental innovation: it can experiment with new ideas in the context of specific cases, often with willing partners. Innovation is a central theme in many of the ideas previously discussed, like Next Gen in rules, creative mitigation, strategic focus of civil and criminal enforcement, compliance monitoring, and data analytics. Described in the following are some additional seldom-used innovations that should play a much bigger role in the Next Gen era.

Conduct field research and experimentation

It has often been lamented how thin the data are to support regulatory strategy choices. Common wisdom and assumptions—alas, too frequently wrong—are often the basis for regulatory decisions, which can lead to the unacceptably poor compliance outcomes recited throughout this book. In addition to the other Next Gen ideas discussed in this chapter, compliance should seek out opportunities for rigorous experiments to test the effectiveness of intervention options. Inertia to keep doing things the way they have always been done is powerful; sometimes a randomized controlled trial or other robust field evaluation is the blast that is needed to shake regulators out of their comfort zone. A well-designed study also provides actual proof about what works, and what doesn't, so the discussion can proceed on evidence rather than ideology. Like the inspired study that was done in India showing that random assignment of auditors resulted in a huge increase in the accuracy of the environmental data reported.[42] Government can't afford very many of these efforts, either in dollars or time expended, so they have to be chosen for maximum impact. For that reason, they should focus on the truly game-changing ideas, not just tiny increments in business as usual.

Encourage state innovation

For the reasons discussed in the previous chapter, the lack of reliable information on outcomes has discouraged EPA from supporting state compliance innovations. Once performance data are more routinely available as a result of Next Gen, that should change. Everyone will learn from state experiments, whether they succeed or fail. And an emphasis on innovation will direct regulators' combined attention to solving problems instead of focusing on the inherently fraught standoffs of cooperative federalism. A state with a theory of change—why a different approach seems likely to be more effective, based on

[42] Esther Duflo, Michael Greenstone, and Nicholas Ryan, "Truth-telling by Third-party Auditors and the Response of Polluting Firms: Experimental Evidence from India," *Quarterly Journal of Economics*, Vol. 128 No. 4 (2013): 1499–545.

more than just a hope—and the willingness to credibly measure that, should find a willing and flexible partner in EPA. We don't have to make a federal case of everything, and in the Next Gen era we won't have to.

Invest a lot more in data systems and analytics

Even though this topic has already been mentioned in this chapter, I include it again here because it is so important. The landscape for spotting and solving big problems is opening up before us at an astonishing rate, but government has only taken a few tentative steps into the new possibilities.[43] Some analytics allow regulators to do what they are already doing, just much more effectively and efficiently. Like targeting for inspectors or spotting the criminal actors. Some analytics tools have the possibility of driving an entirely new approach to regulation. If it were possible to reliably monitor compliance in real time 24/7, how might our regulatory strategy change? Sophisticated data analytics can be a form of real-time monitoring for problems that are tough to measure directly. Investing in capacity to use the newest techniques will pay off immediately in increased bang for the buck and have the potential to transform regulatory approaches.

What Should Enforcement Do Now, in the Transition to the Next Gen Era?

Even if EPA and states embrace Next Gen enthusiastically, the process of building compliance drivers into rules and permits will take many years. The rest of this chapter is about what can happen while that's underway.

Added to the mix is a new enforcement responsibility: addressing compliance with recently and about-to-be finalized climate rules. Those rules should include Next Gen provisions—right? Surely that's a no-brainer—that drive much better compliance than most EPA regulations achieve, so the enforcement lift should not be quite as heavy. But it will still be essential, both to spot issues that the rules didn't anticipate and to send a clear and unmistakable message that EPA will be both vigilant and aggressive in insisting on compliance.

Here's what's not on the list: expecting enforcement to do the baseline job of assuring compliance by millions of facilities regulated in rules that have big compliance holes. That's never going to be possible. Not now, not ever. The giant gap between what the rules expected to happen and what's really going on cannot be closed by enforcement. Acknowledging that hard truth is what drives the need for another way.

[43] Robert L. Glicksman, David L. Markell, and Claire Monteleonia, "Technological Innovation, Data Analytics and Environmental Enforcement," *Ecology Law Quarterly*, Vol. 44 (2017): 41–87.

So where does that leave us? We see the already known huge compliance problems and in our peripheral vision other looming threats that are mostly ignored because the right-in-front-of-us is already more work than is possible. Enforcers face the challenge of rules that not only fail to include compliance drivers but sometimes actively make the enforcement job harder. We look at the tiny resources with which enforcers are expected to address all those issues. Given that there is no way enforcement can accomplish the impossible task assigned in the rules, but it will take Next Gen many years to fix that, what should happen right now?

It is time to revise priorities. Traditionally enforcers nationwide have attempted to do as much as they can to tackle the giant mountain of "routine" noncompliance issues using the woefully inadequate resources at hand. No, it can't be done. But enforcers are a can-do group, so they try their best. Meanwhile, innovation that could help solve the problem over the longer-term fights for attention. Because the violations enforcers know about—and the disquieting suspicion that those are only the tip of the iceberg—exceed capacity to act by many orders of magnitude, it is tough to make room for new approaches. Ideas that could make a substantial difference in a year, or two, or three don't make it on to the agenda because there is a serious violator in front of you right now and surely that's a higher priority?

Actually, no. I am reminded of the short essay, "Mop-and-Bucket Solutions Keep Us Forever Cleaning Up."[44] If you come into a room where a faucet is spilling water all over the floor, and on the opposite side there is a mop and bucket, what do you do? It may strike you as obvious that you should start by turning off the faucet. Yet in the public policy arena too often our response is to start mopping. Next Gen seeks to turn the faucet handle, significantly reducing the violators, while enforcement is primarily about mopping. Yes, given the power of enforcement to change behavior through general deterrence, there is some faucet-turning involved. But that's nowhere close to enough. Using Next Gen strategies, it takes many years to turn the faucet, and even then it will never be completely shut, so you cannot manage without a vigorous mopping operation. But we need to put turning the faucet higher on the priority list.

What does that mean for enforcement in the near term? Here is the work that should move to center stage right away, even though it means fewer cases get done.

[44] Donella Meadows, "Mop-and-Bucket Solutions Keep Us Forever Cleaning Up," The Donella Meadows Project: Academy for Systems Change, 1995, https://donellameadows.org/archives/mop-and-bucket-solutions-keep-us-forever-cleaning-up/.

Ratchet up attention to Next Gen in rules and permits

In the perfect world, EPA programs that write regulations and issue permits, or oversee state rules and permits, would be mindful of Next Gen principles and devote time and effort to getting those compliance drivers right. After a brief pause to marvel at how wonderful that would be, we come back to reality and acknowledge that Next Gen faces an uphill climb. Without advocates willing to do the demanding work of listening to the details of program challenges and adapting Next Gen ideas to the circumstances of each program and rule, making suggestions until they find some that stick, it's not going to happen. Over time this may change as reliable results make believers of program staff, but right now the compliance team are the only Next Gen advocacy candidates. If an enforcement person isn't there pointing out that the great regulatory idea isn't going to work in real life and that outcomes will be poor unless compliance designs are built in, probably no one will. Every rule or permit program that includes Next Gen means hundreds to thousands of violations that won't happen, so from a return-on-investment point of view, it's a winner. It's all faucet, no mop.

Enhance strategic use of civil and criminal enforcement

Concentrating national focus on a particular sector or regulatory problem is powerful. It commands attention. Such efforts can drive broad-scale change in performance and improve results. And they can document how bad the problem is and why. Reacting to violations will always be an essential part of enforcement, but a bigger share of the work can be proactive: regulators pick it because it needs attention, even though the outcome is uncertain. National Compliance/Enforcement Initiatives are one example of this for civil enforcement. Criminal enforcement could dramatically increase its clout by doing something comparable for problems posing the biggest health threats. Robust data that comes from extensive field investigations, and the solutions developed through the subsequent cases, can inform a change to regulations. A much bigger share of the civil and criminal enforcement workload should be devoted to such investigations. That can feel risky; enforcement metrics reward doubling down on the sure thing far more than exploring the important but uncertain. But exploration should be seen as a core function of enforcement, far more important than speeding up an assembly line of precooked cases. Such choices will be paramount for advancing environmental justice; because there are far more deserving cases to address disproportionate burdens than there are resources to pursue them, strategies to expand impact will be vital.[45]

[45] See EPA, "Strengthening Enforcement in Communities with Environmental Justice Concerns," April 30, 2021, https://www.epa.gov/sites/default/files/2021-04/documents/strengtheningenforcementincommunitieswithejconcerns.pdf; EPA, "Strengthening Environmental Justice Through Criminal Enforcement," June 21, 2021, https://www.epa.gov/system/files/docume

Invest in innovation for advanced monitoring, data analytics, field experiments, and settlements

There will be more innovation opportunities once Next Gen is firmly in place. But a lot can be done right now. Innovation shouldn't be shoehorned in at the margins; it deserves a place right in the middle. Not only can it help solve immediate problems on today's agenda, it can also set the stage for even bigger leaps forward in the middle term. More money, staff time, and senior level attention should be devoted to finding and using the newest monitoring technologies, including satellite data. Analytics are an untapped powerhouse for turning existing, or obtainable, data into actionable information immediately, as many examples throughout this book demonstrate. Lurking in the background of analytics is the importance of upgrading EPA's data systems, both so they are ready for the flood of Next Gen data and so that they can accommodate the increased transparency and analytic load that innovation requires. There are a number of academics with both the capability and the interest to conduct field experiments that demonstrate how effective innovative ideas are at driving change; EPA should be making the most of their knowledge and design savvy to jump past ideology and land on solid results. Finally, EPA can actively use cases to build Next Gen ideas into settlements, as EPA recently said it intends to resume doing.[46]

The need for these shifts of emphasis will grow ever more pronounced as the climate changes. That's partly due to the critical need to elevate enforcement of climate rules. But a changing climate is also going to shift noncompliance problems in unpredictable ways. Droughts, floods, fires, storms, and heat will all affect both compliance and the health risks from violations. Enforcement's role as a first alert and as an investigator of both problems and solutions will be increasingly important.

All of these ideas take investment of time. And they require money. The biggest responsibility of senior leadership is deciding where to put those resources. We are way behind on the mopping already, and that's only going to get worse. More money for mopping is absolutely needed because the rising water is affecting people right now. But not at the expense of turning the faucet. That's the only way to fix the problem longer term. And, as we have so recently been reminded, depending exclusively on enforcement isn't a plan anyway; a change

nts/2021-07/strengtheningejthroughcriminal062121.pdf; EPA, "Strengthening Environmental Justice Through Cleanup Enforcement Actions," July 1, 2021, https://www.epa.gov/system/files/documents/2021-07/strengtheningenvirjustice-cleanupenfaction070121.pdf.

[46] See EPA, "Using All Appropriate Injunctive Relief Tools," https://www.epa.gov/sites/default/files/2021-04/documents/usingallappropriateinjunctivereliefto olsincivilenforcementsettlement0426.pdf.

in administrations can cut the legs out from under that approach virtually overnight. It is possible to make a far bigger difference, and one that is much more durable in both the near and long term, by embracing innovation. The more innovative approaches are tried, the more helpful applications will appear. Let's decide to learn faster.

Conclusion

What's the Bottom Line?

The push for carbon offsets—an idea that is sprinting into the regulatory arena at the present minute, often under the banner of "net zero" goals—offers a window into the compliance challenges discussed in this book. Carbon offsets are credits polluters can buy instead of cutting their own carbon emissions. In other words, instead of reducing its own carbon, the polluter funds reductions by others. Credits are intended to finance projects that will reduce or sequester carbon, thereby "offsetting" the carbon that the purchaser continues to emit. The theory is that purchase of carbon offsets will fund valuable carbon-reduction projects that otherwise wouldn't happen, at the same time they cut costs for carbon polluters transitioning to a lower emission future. The mythical universe where everybody wins. Not surprisingly, they are wildly popular.

Carbon offset programs have features that will sound very familiar to readers of this book. Carbon reductions from offsets can't be "measured"; they depend on assumptions about what would have happened in the alternate universe where the project didn't get offset funding (just like energy efficiency, chapter 7). They are largely about what happens with land use, such as forest preservation or agricultural practices (just like conventional renewable fuels, chapter 8). Decisions about what qualifies as an offset require complicated, and often individualized, determinations that involve behind-the-scenes exercise of judgment (just like New Source Review, chapter 1). Legitimacy of offset credits depends on actions many decades into the future (just like abandoned oil and gas wells, chapter 9).

And also just like many other programs analyzed in this book, offset credits create incentives to resist accountability. That's because offsets put two desirable outcomes—increased funding for carbon-reduction projects and reduced costs for polluters—in tension with the carbon-cutting purpose that originated the whole enterprise. Admitting that we are not actually getting the carbon improvements we need from offsets jeopardizes the funding streams and cost savings on which many projects, companies, and countries depend. So of course, there is huge pressure not to do that.

We know what happens. We've seen this movie before. Offset projects will routinely overestimate carbon savings, because it's in everyone's interest to do so

Next Generation Compliance. Cynthia Giles, Oxford University Press. © Cynthia Giles 2022.
DOI: 10.1093/oso/9780197656747.003.0013

(see energy efficiency).[1] It will prove close to impossible to determine if projects would have happened anyway and therefore provide no additional climate benefit, and whether outside-of-program impacts (aka "leakage") cancel out claimed reductions (see energy efficiency and renewable fuels).[2] There will be lots of fraud, because the offsets market creates an ungovernable opportunity to make a lot of money with little chance of getting caught (see renewable fuels).[3] We will find out that essential commitments to take or prevent action decades hence are not realized (see abandoned oil and gas wells).[4] Investigations that reveal all these fatal flaws will result in calls for enforcement to miraculously plug the giant compliance holes (see every program discussed in the entire book). Meanwhile, more real, we-know-for-sure carbon will be released into the atmosphere because unreliable, really-we-have-no-idea carbon offsets give companies license to do that.

My point in raising carbon offsets isn't to explore the many complexities of offset programs. It's to illustrate how ongoing and immediate the compliance

[1] Mark Shapiro, "Conning the Climate: Inside the Carbon-Trading Shell Game," *Harper's Magazine*, February, 2010 (overestimating emissions reductions from offset projects is the trapdoor in the offset program; some studies show overestimation can be as much as 100%); Lisa Song, "An (Even More) Inconvenient Truth: Why Carbon Credits for Forest Preservation May Be Worse Than Nothing," *ProPublica*, May 22, 2019, https://features.propublica.org/brazil-carbon-offsets/inconvenient-truth-carbon-credits-dont-work-deforestation-redd-acre-cambodia/ ("The hunger for these offsets is blinding us to the mounting pile of evidence that they haven't—and won't—deliver the climate benefits they promise"); Grayson Badgley et al., "Systematic Over-Crediting in California's Forest Carbon Offsets Program," *Global Change Biology* (October 2021), https://doi.org/10.1111/gcb.15943 (a systematic flaw in California's forest offsets program—the largest compliance market in operation today—results in over crediting that is larger than the buffer pool that is intended to cover risks to forest carbon).

[2] Barbara Haya et al., "Managing Uncertainty in Carbon Offsets: Insights from California's Standardized Approach," *Climate Policy*, Vol. 20, No. 9 (2020): 1117, 1122, https://doi.org/10.1080/14693062.2020.1781035 (in California's climate offsets program, offset value uncertainty for mine methane capture is so large that there might be no non-additional carbon reductions from the credits given for these projects); Kenneth R. Richards, "Environmental Offset Programmes," in Kenneth R. Richards and Josephine van Zeben eds., *Policy Instruments in Environmental Law* (Edward Elgar, 2020), 346 (leakage—carbon emission increases elsewhere—are especially challenging for land-use offset projects; studies show that between 5% and 90% of carbon reductions can be lost from leakage).

[3] "Guide to Carbon Trading Crime," Interpol Environmental Crime Programme (June 2013): 11–25, https://globalinitiative.net/analysis/interpol-guide-to-carbon-trading-crime/ (carbon offset markets are particularly vulnerable to fraud because a carbon reduction credit is so difficult to measure and verify; as demand for offsets increases, fraud is expected to rise); Richards, "Environmental Offset Programmes," 347 (a review of 10 leading forestry offset standards found that all the standards were severely flawed and could not protect against gaming by offset developers). See also chapter 6, discussing the reasons markets can't succeed when credit quality is impossible to determine.

[4] GAO, "Carbon Offsets: The U.S. Voluntary Market Is Growing, but Quality Assurance Poses Challenges for Market Participants," GAO-08-1048 (August 2008): 29 (forestry offset projects may not be permanent because disturbances such as insect outbreaks and fire can return stored carbon to the atmosphere); Evan Halper, "Burned Trees and Billions in Cash: How a California Climate Program Lets Companies Keep Polluting" *Los Angeles Times*, September 8, 2021 (trees part of carbon offset program burned in forest fire); "Driving Climate Smart Agriculture with Data," Center for Climate and Energy Solutions, May 26, 2021, https://www.c2es.org/2021/05/driving-climate-smart-agriculture-with-data/ (in agriculture there is a disconnect between what science suggests has lasting carbon benefits and what voluntary carbon markets promote).

fallacies are. We want to believe, and we have long assumed, that compliance with environmental rules is good, and that enforcement can be counted on to take care of the rest. Those assumptions have allowed policy makers to ignore the huge rift between goals and real-life implementation. The dual fictions that it's not a big problem and it isn't my problem anyway allow program authors to continue promulgating standards as if widespread violations don't matter.

Meanwhile, back in the real world, lots of admirable goals fall into the compliance gap. We see that in air not meeting safety standards, water bodies failing water-quality standards, communities across the country exposed to risks from drinking water, hazardous waste, toxic air emissions, dangerous chemicals, and industrial disasters. And those are just the problems we know about. Many of the nation's troubling health risks are obscured behind monitoring and reporting violations, so the extent of the harm will never be known. Plenty of rules leave regulators with no way to know what's really happening, although evidence suggests it isn't good.

The foundational idea of Next Gen—the basic premise of this book—is that rules have to be designed to ensure that they deliver their public health goals. That's why the book starts with examples of rules that worked well, and some that failed in spectacular fashion. Regulations can achieve extremely impressive compliance rates, as occurred in the Acid Rain Program. Or regulators can throw the barn doors wide open, then wonder where the horses went, as has happened with New Source Review and the lead in drinking water rule. These examples reveal the shared characteristics of rules that deliver in the gritty world of implementation: they make compliance the path of least resistance. Rules that are notable compliance fails also have a common theme: they offer many ways to obfuscate, avoid, or ignore.

Next Gen isn't an abstract idea. It is all about the messy, complicated world in which we actually live, which, you may have noticed, is some distance from policy theory. In the real world, many companies will look for ways around regulatory obligations, and all will be subject to the competing pressures of higher priorities, tight budgets, unexpected events, and incompetence. Some will just cheat. If regulations create space for things to go awry, they will. Those are the facts of life. To underscore how significant a problem this is, and to counter the unfounded belief that compliance with environmental rules is good, I have presented the most comprehensive data anywhere about compliance with environmental rules, showing how widespread and serious our violation problems are. Most people, including many who have spent their careers in environmental protection, find this surprising. The evidence is everywhere, but we don't see it because it doesn't align with preconceived ideas about compliance. When you put all the violations data together and acknowledge that even those dismal results are only the tip of the iceberg, it is obvious how daunting the compliance gap really is. And how much it contributes to our failure to solve problems that we thought we had addressed in regulations.

The dual false assumptions that created this mess are not just a sort of collective unconscious, they are also enshrined in written policy. Reversing the tide of commonly held beliefs is hard enough, but it becomes impossible when official guidance reinforces the error. My explanation of EPA's rule-writing process and benefit-cost analysis protocols explores how seemingly neutral policy directives can end up cementing harmful assumptions.

This isn't a woe-is-me book. Just the opposite. Solutions are available today, right now. It is gratifying to identify a problem and discover that we can fix it. We have the tools already, and more will emerge once we start looking for them. That's what most of this book is about.

The best structural designs are tailored to specific problems. They are made-to-measure solutions, which consider the unique circumstances of the issue the rule is intended to address. But there are also many ideas that will help in nearly every situation. They are the workhorses of Next Gen. Real-time monitoring is one: once everyone can know what's happening in close to actual time, compliance will improve because companies will know more about their own performance, government can efficiently follow what's happening in the field, and companies will feel the pressure of regulators and the public keeping watch. Automatic consequences for failing to report is another. Today, neglecting or choosing to skip a report is usually a freebie, with compliance and transparency the loser. The simple strategy of making a failure to report more trouble for the company than complying can rapidly turn that around. Those are two of many options that regulators have at the ready, described in chapter 5.

Next Gen isn't about inserting one compliance idea in a regulation and calling it done. Every rule requires a compliance system, a suite of ideas that work together to drive toward the desired result. Monitoring is great, but you also need a strategy to inspire companies to use the required monitors, to undertake the necessary quality control, and to accurately report results. The same is true with the other parts of a rule. Every key element has to be resilient to the compliance pressures of real life, from inattention to deliberate cheating. The Next Gen elements work together to deliver strong compliance results. No one thing will do that on its own. Only a structure that fits the pieces together can build a strong implementation system. If the preferred rule strategy—like a market or a performance standard—can't be designed to deliver robust and reliable implementation, that's telling you it is time to reconsider. Ideology can't substitute for a practical bottom line: getting the job done in the complicated reality of the world where we actually live, which is the only place that matters.

Next Gen's insistence on taking a practical, real-world look at likely regulatory outcomes faces the highest hurdles when it confronts a popular idea. Carbon offsets are a great example. Regulators, politicians, and the many companies that benefit all want to embrace policy theory that promises to achieve two important goals—funding for worthwhile projects, lower costs for regulated companies—while still accomplishing the climate purpose. That's the win-win holy grail of regulatory outcomes. But that's not what will happen. Mountains of evidence from other programs, as well as current investigations specific to carbon offsets, tell us that. Next Gen is not the most popular kid at the table when it insists on looking that kind of hard reality in the eye. But that's one of Next Gen's essential roles.

There are a host of existing problems that could benefit from Next Gen attention, but none are more critical than climate change. The widespread compliance fails we have observed for many environmental rules can't be allowed to happen with regulations to reduce climate-forcing emissions. That's why I spend three chapters of the book on essential areas for climate attention: electric generation and the reasons why the necessary work on energy efficiency should not be included in an electricity carbon reduction plan; methane from oil and gas; and renewable fuels. All of these serious problems have solutions. But they are also burdened by popular ideas that are plainly compliance losers. It is far too late in the day to deploy regulatory strategies that are iffy to possibly fatal. No can do; these regulatory programs have to work the first time, not just in the theoretical world of economics but in the actual world where the impacts of a rapidly changing climate are hitting hard. Next Gen strategies can make the difference.

Nearly all environmental rules are implemented jointly by federal and state governments. The federalism structure that Congress created has been one of the great strengths and also one of the deep frustrations of environmental implementation. At its best, federalism requires EPA to learn from leadership shown by individual states and to consider the differing on-the-ground realities across the nation. Likewise, states can learn from each other, and at the same time are forced to acknowledge that what happens in one location can reverberate around the country, and the globe. The creative tension of cooperative federalism is part of the many successes the United States has achieved in environmental protection. But it can also drag the entire system down. My chapter on federalism says let's not continue to double down on longstanding fights over regulatory authority. Instead, let's concentrate on how newly available tools—monitoring, data analytics, and information management—can orient us toward a shared focus on results and away from unwinnable and unproductive ideological battles.

Enforcement will continue to play a central and essential role in achieving compliance. Once it is freed from the impossible expectation that it can force fit compliance on millions of regulated facilities, it will be even more effective than it is today at doing what enforcers do best: addressing the large, unexpected, and impossible-to-prevent violations. Civil and criminal enforcement can be at the forefront of environmental justice and finding sometimes unexpected answers to vexing problems. It is an absolutely critical cannot-live-without element of any compliance program, but regulations relying exclusively on enforcement are doomed. Enforcement can never carry that weight on its own. Once we have rules designed to close the compliance gap, enforcers can use the powerful and creative tools the law provides to address the unanticipated and explore innovative solutions.

Next Gen will help to bridge the currently huge divide between what environmental rules are intended to accomplish and what they actually achieve. Nowhere will that matter more than in communities overburdened with environmental health risks. Pollution load isn't equally shared today; it falls far more on communities of color and low-income areas of the country. For the same reason, communities with environmental justice concerns bear far more than their share of the risks from serious violations. Widespread significant violations are an environmental justice issue. No one knows better than these communities how unreliable it is to depend exclusively on enforcement to secure a healthy environment. Regulations designed using Next Gen principles—that make compliance the default—are a more dependable route to a healthier environment.

Our biggest challenge isn't what to do to close the compliance gap—there are a wide variety of options available today, many of which are detailed in this book—it's deciding that we are going to act. My experience is that people often hold fast to conventional wisdom even when it is demonstrably false. It is extremely hard to dislodge the usual way of doing things.

That's what this book is hoping to change. There are lots of ways to build regulations that will work better in real life. Scores of options. And many more will arise once we get serious about finding them. There doesn't have to be a chasm between what regulators try to do and what actually occurs. We can get much better results than we are achieving today. That's especially important for rules to tackle the overwhelming threat of climate change. Doing our best, or living to fight another day, aren't acceptable outcomes for climate. We have to deliver.

We can have regulations that achieve what they set out to accomplish by including Next Gen strategies in every rule. How do we get rule writers to do that? Profound institutional change doesn't reliably happen because you ask nicely. Or because you say it's required. We get there by applying the lessons of Next Gen to the regulatory process. We can't just tell rule writers to do it, we have to design the process so it is impossible to avoid. Like requiring a rigorous and realistic compliance evaluation as part of the mandatory benefit-cost analysis or forcing consultation with compliance experts before the rule can be finalized. Chapter 4 lays out the institutional hurdles that currently work against including strong compliance provisions in regulations and describes changes that could push rule writers to consider compliance up front, because doing so will be the path of least resistance.

This book is about environmental problems. It includes evidence of and solutions to the widespread violations of environmental rules and the damage they do to our aspirations for protecting people's health and the natural environment on which public health depends. But regulators in other programs will also recognize their own dilemmas in these pages. The same wrong assumptions are visible in grossly underpaid taxes, drugs with dangerous impurities, and nursing homes that misrepresent their safety records. The belief that somehow enforcement can compensate for regulations that aren't designed to ensure robust implementation is nearly universal. Read the newspaper on almost any day and you will see it repeated over and over. So while Next Gen is focused on environmental regulations, its lessons apply much more broadly.

Here's my message to regulators, companies, communities, and nongovernmental advocates: make implementation design part of your policy work. Don't accept the standard view that somehow it will all be OK, as if solid implementation happens by magic. Don't defer responsibility for good compliance design with vague references to enforcement, as if it can work miracles. Insist on structural features that will drive toward the necessary action even if there never is any enforcement. Implementation isn't some janitorial function that happens off camera. Implementation is center stage. It determines if the plan works or doesn't. Policy theory is nice, until theory collides with reality. Next Gen design is about the world where we actually live and where regulations deliver or don't. Rules have to build in strong implementation design, not just hope for the best.

This book offers an inspiring message: we know how to fix this. We have the knowledge and the tools. Let's decide to use them.

Index

For the benefit of digital users, indexed terms that span two pages (e.g., 52–53) may, on occasion, appear on only one of those pages.

Tables are indicated by *t* following the page number